Reviews of

114 Physiology Biochemistry and Pharmacology

Editors

M. P. Blaustein, Baltimore · O. Creutzfeldt, Göttingen
H. Grunicke, Innsbruck · E. Habermann, Gießen
H. Neurath, Seattle · S. Numa, Kyoto
D. Pette, Konstanz · B. Sakmann, Heidelberg
M. Schweiger, Innsbruck · U. Trendelenburg, Würzburg
K. J. Ullrich, Frankfurt/M · E. M. Wright, Los Angeles

With 21 Figures and 9 Tables

Springer-Verlag
Berlin Heidelberg GmbH

ISBN 978-3-662-31160-8   ISBN 978-3-540-46754-0 (eBook)
DOI 10.1007/978-3-540-46754-0

Library of Congress-Catalog-Card Number 74-3674

© Springer-Verlag Berlin Heidelberg 1990
Originally published by Springer-Verlag Berlin Heidelberg New York in 1990
Softcover reprint of the hardcover 1st edition 1990

Typesetting: K+V Fotosatz GmbH, Beerfelden
2127/3130-543210 – Printed on acid-free paper

Contents

Indexed in Current Contents

Rev. Physiol. Biochem. Pharmacol., Vol. 114
© Springer-Verlag 1990

Molecular Properties of Calcium Channels*

HARTMUT GLOSSMANN and JÖRG STRIESSNIG[1]

Contents

---

* This work is dedicated to Professor Emeritus Heribert Konzett
[1] Institut für Biochemische Pharmakologie der Leopold-Franzens-Universität Innsbruck, Peter-Mayr-Straße 1, A-6020 Innsbruck, Austria

Abbreviations

B_{max}, maximal density of binding sites; **CHAPS**, 3-[(3-cholamidopropyl)dimethylammonio]-1-propanesulphonate; **DHP**, dihydropyridine; **ECC**, excitation-contraction coupling; **G-proteins**, GTP-binding proteins; **kb**, kilobase; **kDa**, kilodalton; **PAGE**, polyacrylamide gel electrophoresis; **PTX**, pertussis toxin; K_d, dissociation constant; k_{-1}, dissociation rate constant; k_{+1}, association rate constant; **SDS**, sodium dodecyl sulphate; **T-tubule**, transverse tubule; **SR**, sarcoplasmic reticulum

Drugs: **BAY K 8644**, methyl-1,4-dihydro-2,6-dimethyl-3-nitro-4-(2-trifluoromethylphenyl)-pyridine-5-carboxylate; **DPI 201-106**, 4-3'-(4''-benzhydril-1''-piperazinyl)-2'-hydroxy-propoxy-1H-indole-2-carbonitrile – **BDF 8784** carries a methyl group instead of the CN group; **[N-methyl-³H]LU49888** ((−)-5-[(3-azidophenethyl)[N-methyl-³H]methylamino]-2-(3,4,5-trimethoxy-phenyl)-2-isopropylvalero nitrile; **PN200-110**, isopropyl-4-(2,1,3-benzoxadiazol-4-yl)-1,4-dihydro-2,6-dimethyl-5-methoxy-carbonyl-pyridine-3-carboxylate; **202-791**, isopropyl-4-(2,1,3-benzoxadiazol-4-yl)-1,4-dihydro-2,6-dimethyl-5-nitro-3-pyridine carboxylate

1 Introduction

Our review deals with the molecular properties of voltage-dependent calcium channels. Compared with the voltage-dependent sodium or ligand-activated ion channels, structural information on channels selective for calcium is limited. It is virtually nonexistent on T-type channels, scarce on N-type channels and, with respect to L-type channels, restricted to the skeletal muscle. L-type channels have distinct but allosterically coupled receptor sites for drugs (e.g. the 1,4 dihydropyridines such as nifedipine, the phenylalkylamines such as verapamil and the benzothiazepines such as (+)-*cis*-diltiazem). These drugs, especially those of the 1,4 dihydropyridine class, were essential for the purification (and cloning) of the channel proteins. In contrast to the recently characterised archetype of cation-selective channels, the (A current) K$^+$ channel (where first the gene coding for the *Shaker* mutation was mapped in *Drosophila*, cDNA clones were isolated, and the protein later was expressed functionally in a heterologous system) isolation of the L-type channel followed a conventional path — in analogy to the sodium channel. It was perhaps this analogy which misled researchers initially to a large glycoprotein in purified preparations from skeletal muscle (drug receptors are here abundant compared with other tissues) as the pore-forming calcium channel "alpha"-subunit. This glycoprotein released a 25- to 35-kDa set of glycopeptides upon reduction of disulphide bonds. The rat brain sodium channel (heavily glycosylated) alpha-subunit is linked to (glycosylated) β_2-subunits via disulphide bonds. Quite unexpectedly, the purified calcium channel turned out to be a complex of (four to) five subunits, where the pore-forming drug receptor-carrying alpha-subunit was neither disulphide linked nor (heavily) glycosylated!

Almost simultaneously with the skeletal muscle "Ca^{2+} antagonist drug receptor" the ryanodine-sensitive calcium release channel from sarcoplasmic reticulum (SR) was isolated, characterised and reconstituted. This channel forms the foot structure, bridging the gap between the transverse tubule membrane and the SR membrane. Ironically, it was once believed that the feet had a solely structural role. Quite similarly, the Ca^{2+}-antagonist receptors in skeletal muscle were long viewed with suspicion as functionally silent drug-binding sites. Both structures are now regarded as essential constituents of a novel transmembrane communication pathway. For that reason, and despite the mysteries which still surround the process of excitation-contraction coupling, the skeletal muscle is a main theme in this review. First we present the tools which are proven or suggested to be molecular probes for calcium channels; we then mention target size analysis and photoaffinity labelling, discuss the L-type channel properties (mainly but not exclusively) from skeletal muscle in great detail and provide a critical overview of N-type channels. In the final chapter we discuss the (deduced) primary structures, models and expression.

2 Drugs and Toxins as Molecular Probes for Calcium Channels

2.1 The Voltage-Dependent Calcium Channel in Comparison with Other Ion Channels

The nicotinic acetylcholine receptor was the first ligand-activated ion channel to be characterised and even localised with radioactive toxins (see Waser 1986). Small protein toxins such as the alpha-neurotoxins from various *Naja* species and from *Bungarus multinctus* were subsequently essential in the purification of the channel and led to the elucidation of the complete amino acid sequence of its four subunits (Conti-Tronconi and Raftery 1982). Toxins are also helpful in the differentiation, biochemical characterisation and purification of potassium channels by conventional schemes. Alternative approaches with methods provided by molecular biology and exemplified by the analysis of *Shaker* mutants (potassium channels from *Drosophila*) are equally successful (Papazian et al. 1987; Tempel et al. 1987; Schwarz et al. 1988; Timpe et al. 1988).

The key role of the voltage-dependent sodium channel in information transfer has also made it a prime target for potent neurotoxins. These toxins have been classified in either six (Lazdunski et al. 1986a; Barchi 1988) or four categories (Catterall 1986) on the basis of binding sites and/or physiological effects. Polypeptide toxins which can be iodinated with iodine 125, such as the *Tityus* gamma toxin (Lazdunski et al. 1986a), or small, naturally occurring nonprotein compounds such as tetrodotoxin and saxitoxin (which can be labelled with tritium) are extremely useful probes. They can be employed for the characterisation of receptor sites in membranes from electrically excitable cells and (after chemical modification to yield affinity or photoaffinity probes) to identify the receptor-carrying polypeptides by irreversible labelling, to follow solubilization and purification and, finally, to probe for alteration of the conductance behaviour in reconstituted or even in mRNA expression systems. Identification of the toxin binding domains within the primary structure and on crystallized channel proteins are pursued. Still another aspect of the toxins is the discrimination of subtypes within the sodium channel family which can complement the rapidly increasing member of channel structures deduced by molecular biology techniques.

In contrast to voltage-dependent sodium or potassium channels and the acetylcholine-activated channel, naturally occurring toxins have not played any role in the characterisation of L-type calcium channels. Instead drugs originally synthesized as therapeutics are the keys for structural research, as they still are for differentiation of different subtypes within the calcium channel family. In contrast, some omega-conotoxins, e.g. GVIA and MVIA, are useful probes for the neuronal (N-type) calcium channel. The apparent (but perhaps not complete) neglect of the (L-type) calcium channel as a target in the everlasting struggle between organisms is not well understood. It may relate to the fact that the calcium signal has different inputs, e.g. from the extracellular space, from intracellular stores, or by changing the sensitivity of intracellular calcium-binding proteins. A blockade of the channel could be more or less compensated for by other mechanisms. Transient initial Ca^{2+} signals are often from internal stores. Only prolonged Ca^{2+} signals may require influx (Putney 1987), and there is a tissue- and species-specific variation even for one neurotransmitter to utilize these different sources. Noradrenaline contracts the rat spleen by activation of alpha$_1$-adrenoceptors (alpha$_{1B}$ type) even when the L-type Ca^{2+} channels are blocked by nifedipine. In the vas deferens (alpha$_{1A}$ type) from the same species there is almost complete inhibition of contraction by the same concentration of this L-type Ca^{2+} channel-specific blocker (Han et al. 1987). Other aspects are that neuronal L-type channels (in the majority of systems investigated) have no important role in neurotransmitter release (see review by Miller 1987), and that the overwhelming majority of L-type channels in all vertebrates so far investigated reside deeply hidden in skeletal muscle transverse tubules and have a very specialised function, where calcium influx is not required to elicit contraction. However, contractions of invertebrate skeletal muscle are highly dependent on extracellular calcium, and the membrane action potential is generated by voltage-dependent calcium channels (Fatt and Ginsborg 1958) first recorded in crab leg fibres by Fatt and Katz (1953). Blockade of these channels may be an attractive method of paralysing the prey. Ca^{2+} influx induced by mechanisms similar to those seen with many of the sodium-channel toxins (e.g. by persistent activation, enhancing activation or slowing inactivation) may be another principle of poisoning. Feedback mechanisms, Ca^{2+} pump activity, the Na^+/Ca^{2+} exchanger and intracellular storage may, at least in part, protect against the disastrous metabolic consequences of intracellular Ca^{2+} excess. L-type channels are opened by depolarisation. Receptors (e.g. the alpha$_{1A}$-adrenoceptor) − most likely directly via GTP-binding proteins (Yatani et al. 1987; Brown and Birnbaumer 1988) − and second messenger systems (Reuter 1983; Hofmann et al. 1987) can modulate channel activity. Therefore, toxins with alleged activator actions at the L-type channels may act indirectly via different mechanisms, e.g. second messengers, depolarisation, selective pore formation. L-type channel selective agents (e.g. 1,4 DHPs, (+)-cis-diltiazem or verapamil) often block venom-induced smooth muscle contraction, positive inotropic effects and hormone, mediator or neurotransmitter release, to name some of the bioassay methods. Even if it can be shown that extracellular calcium is required, this is by no means proof that the toxic principle acts directly on the calcium channel, as the cation could be crucial for binding only.

2.2 Drugs − Specific Probes for L-Type Channels

Tools to characterise, isolate and purify L-type calcium channels have been found among low-molecular-weight synthetic organic compounds, termed "Ca^{2+} antagonists" by Fleckenstein (see e.g. Fleckenstein 1983; Godfraind et al. 1986; Janis et al. 1987; Triggle and Janis 1987).

They can be classified according to criteria derived from physiology, pharmacology or therapeutics. For the present purposes, a chemical classification is appropriate. We divide the compounds (for typical structures consult Fig. 1) into five classes, named according to their basic structure(s). This division also reflects the current view (Catterall et al. 1988; Glossmann and Striessnig 1988a,b; Janis et al. 1987) that each class may recognise a distinct binding do-

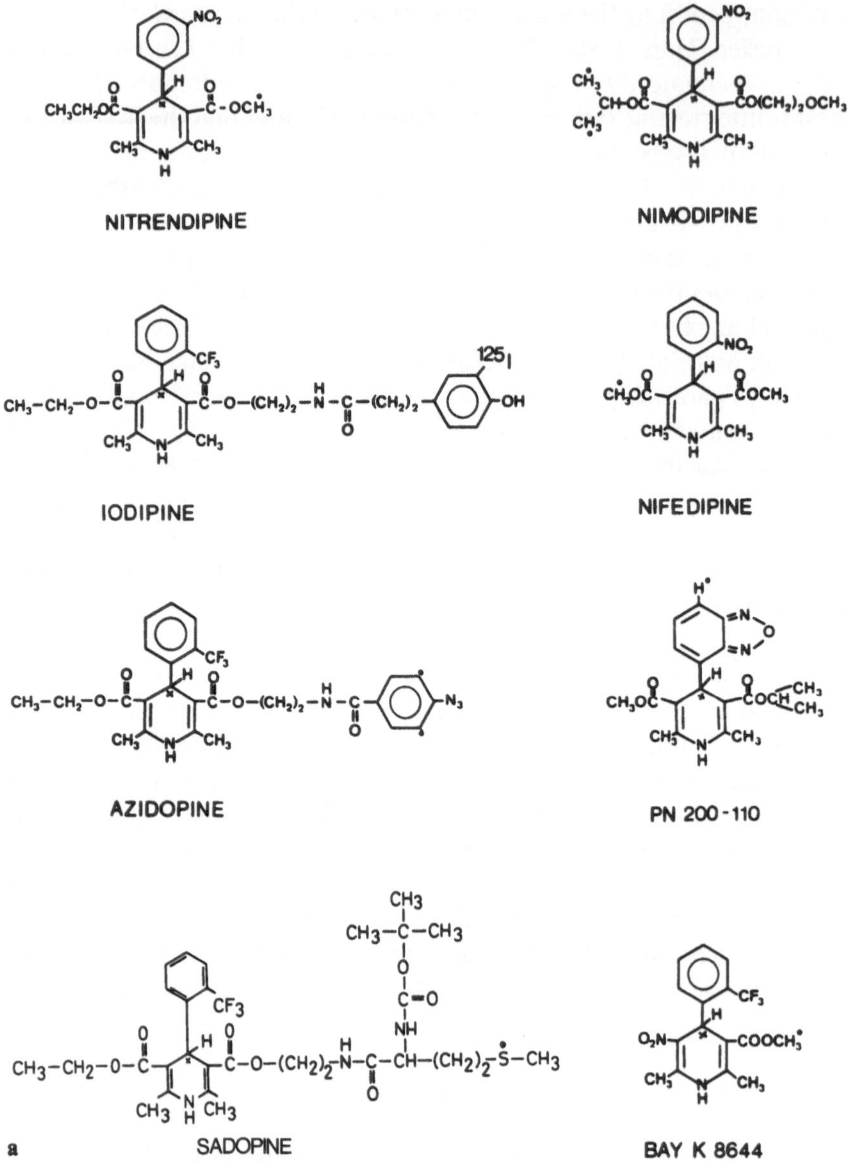

NITRENDIPINE

NIMODIPINE

IODIPINE

NIFEDIPINE

AZIDOPINE

PN 200-110

a   SADOPINE   BAY K 8644

Fig. 1a–c. L-type channel drugs.
a 1,4 DHP structures, including [^{125}I]- or [^{35}S]-labelled ligands (sadopine), the arylazide photoaffinity ligand azidopine and an agonistic 1,4 DHP (Bay K 8644). Note that in the text PN200-110 is sometimes referred to as isradipine.
b Drugs which bind to the phenylalkylamine-selective domain. (−)-Desmethoxyverapamil and LU 49888 (a reversible and photoaffinity ligand) are employed for structural research as (optically pure) tritium-labelled compounds.
c Drugs claimed to bind to the benzothiazepine-selective domain are shown (*trans*-diclofurime, MDL 12330A, Fostedil) together with the diltiazem structure. Of the four diltiazem diastereomers only (+)-*cis*-diltiazem binds with high affinity to L-type channels, and it is the tritium-labelled standard radioligand for receptor domain "3" (see Fig. 2)

(±)VERAPAMIL

BEPRIDIL

(−)DESMETHOXYVERAPAMIL

LU 49888

b

FOSTEDIL

DILTIAZEM

trans - DICLOFURIME

MDL 12,330 A

c

main on the alpha$_1$-subunit of the L-type channel (see below), although all drug-receptor domains interact with each other in a heterotropic allosteric manner in in vitro ligand binding experiments. A schematic view of the observed interactions is given in Fig. 2. Only within the 1,4 DHP class do compounds exist (e.g. S-(−)-Bay K 8644, (S)-(+)-202-791, (−)-Bay F6653) which activate L-type Ca^{2+} channels (Bechem et al. 1988). These "calcium channel agonists" have not been useful for direct structural studies (probably because

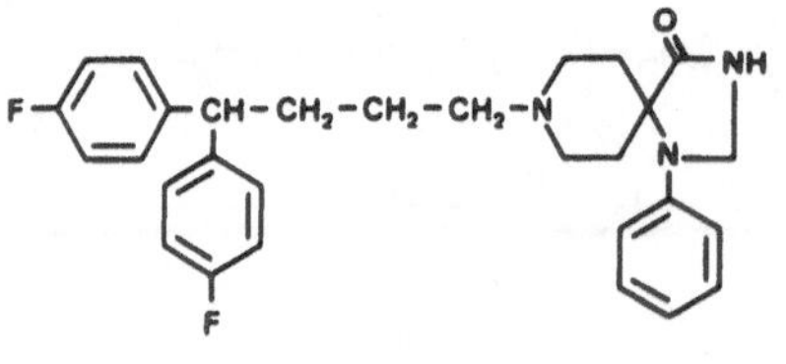

FLUSPIRILENE

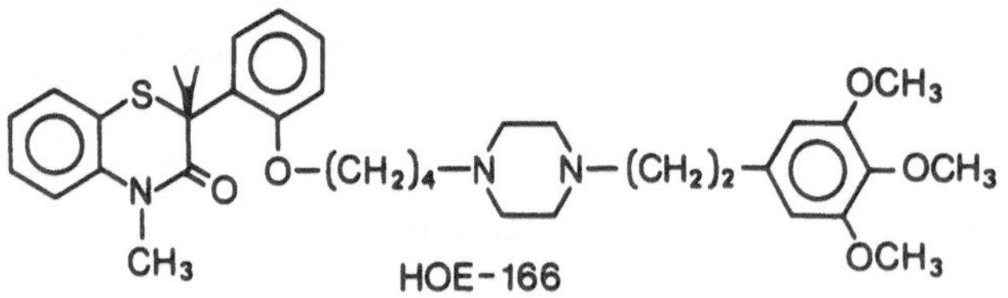

HOE-166

Fig. 1d. Compounds which define receptor domains that are distinct from those defined by drugs shown in **a**, **b** and **c**

their affinity is highest for channel states prevailing at negative membrane potentials), but they are very important for probing reconstituted channel proteins or stabilizing the channel during purification and for reconstitution (Curtis and Catterall 1986; Affolter and Coronado 1985).

Tritium-labelled derivatives (specific activities: 40–140 Ci/mmol) are available for all classes but only the 1,4 DHP group includes [^{125}I]- and [^{35}S]-labelled compounds (see Fig. 1). 1,4 DHPs with one (or more) chirality centre(s) (see Meyer et al. 1985) are useful to investigate the stereoselectivity of the receptor domains from membrane-bound, purified or reconstituted proteins. In general, the eudismic ratios (dissociation constants of the distomer divided by the dissociation constant of the eutomer) are between 10 and 300. With respect to radiolabelled 1,4 DHPs, (optically pure) eutomers (e.g. (+)-PN200-110, (−)-azidopine) are preferred, as the distomer (i.e. (−)-PN200-110 or (+)-azidopine) has much less (if any) "receptor reactivity" compared with the eutomer. The distomer complicates the analysis of equilibrium binding or kinetic data (Bürgisser et al. 1981) and decreases signal-to-noise ratios in binding experiments with racemic ligands. There is one example where the labelled calcium channel distomer (in a racemic radioligand) identified a receptor for 1,4 DHPs – unrelated to calcium channels: (+)-[^{3}H]nimodipine binds with high affinity to the nucleoside carrier in human red blood cell membranes (Striessnig et al. 1985a,b) and purified membranes from the electric organ of *Electrophorus electricus* (Glossmann and Striessnig 1988a). The L-type calcium channel, on the other hand, binds (−)-[^{3}H]nimodipine preferentially (Ferry and Glossmann 1982). Thus, the classification of eutomers and distomers in the context of this article refers to L-type calcium channel-linked receptors.

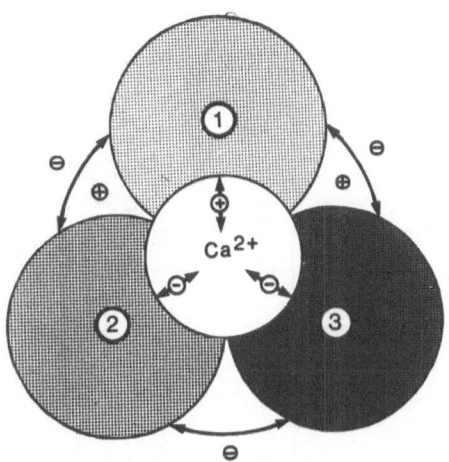

Fig. 2. The allosteric interaction model of L-type calcium channel drug receptors for 1,4 DHPs (receptor domain 1), phenylalkylamines (receptor domain 2) and benzothiazepines (receptor domain 3). The *arrows* symbolise positive (+ signs) and negative (− signs) reciprocal allosteric interactions between the respective domains which can be observed in vitro by probing with selective radioligands. All three domains are allosterically coupled to binding sites for divalent cations symbolised by "Ca^{2+}". The 1,4 DHP domain is positively coupled to high-affinity divalent cation sites. Removal of the divalent cations, i.e. by EDTA treatment, converts the domain into a very low affinity state, which is reversible by refilling with certain cations. The divalent cations present in the purified or membrane-bound channel have not been identified. Based on experiments shown in Fig. 3, most likely Ca^{2+}, perhaps also Mg^{2+}, ions occupy the Ca^{2+} sites. The divalent cation sites which are coupled to receptor domains 2 and 3 are inhibitory (note the *minus* signs). The three receptor domains shown have now been localised on the alpha$_1$-subunit of the skeletal muscle calcium channel by specific photoaffinity labelling with arylazides. The two other recently discovered receptor domains (for diphenylbutylpiperidines and benzothiazinones), which are always negatively allosterically coupled to the receptor domains 1−3, are not shown here for reasons of simplicity

In physiological experiments many (but not all) of the drugs listed in Table 1 exhibit voltage- or use-dependent channel blockade. Voltage-dependent binding of the 1,4 DHPs has been studied in intact cell systems (Kokubun et al. 1986; Porzig and Becker 1988; Kamp and Miller 1987; Schilling and Drewe 1986). The channel has the highest affinity for 1,4 DHP channel blockers when in the inactivated state. This state is favoured in depolarised cells and, of course, prevails in isolated cell membranes or solubilised preparations. High-affinity binding of the 1,4 DHPs to L-type channels in broken cell membranes or homogenates is absolutely dependent on certain divalent cations (e.g. Mg^{2+}, Mn^{2+}, Ca^{2+}; Glossmann et al. 1982; Gould et al. 1982; Luchowski et al. 1984, Glossmann and Ferry 1983a). The high-affinity conformation of the L-type channel with 1,4 DHPs is a ternary complex of divalent cations, channel proteins and the ligand. For methodological reasons only the binding isotherms of the 1,4 DHPs have yet been quantitated. It is predicted that these drugs convert the low-affinity state of the channel for Ca^{2+} into

Table 1. L-Type Ca^{2+} channel drugs used for structural characterisation

Class	Isotope	Ligand	Dissociation constants (nmol/l)			Comments	References
			Skeletal muscle	Heart	Brain		
1,4-Dihydropyridines	^{3}H	(+) PN 200-110	Guinea pig 0.29–0.7 chick 0.2	0.051 0.052	0.075[a] 0.044[b]	Most commonly employed radiolabel. Pure enantiomer. Can be also used as a photolabel	a) Striessnig et al. (1988a) Ferry et al. (1987) b) Barhanin et al. (1988)
		(−)-Azidopine (+)-Azidopine	0.35	0.030	0.096	Photoaffinity label and reversible ligand	Ferry et al. (1984a, b) Striessnig et al. (1986b) Ferry et al. (1987) Striessnig et al. (1988b)
	^{125}I	(−)-Iodipine (±)-Iodipine	0.4	n.d.	0.06	High specific activity label (2175 Ci/mmol)	Ferry and Glossmann (1984)
	^{35}S	(−)-Sadopine (+)-Sadopine	0.51 0.4	n.d n.d.	n.d. n.d.	High specific activity label (>1000 Ci/mmol)	Glossmann et al. (1988c)
Phenylalkylamines	^{3}H	(−)-Desmethoxy-verapamil (Devapamil)	1.5–2.2	1.4–2.5	1.6	Pure enantiomer, most commonly employed for the phenylalkyl-amine site	Ferry et al. (1984a) Ruth et al. (1985) Goll et al. (1984b) Goll et al. (1986) Striessnig et al. (1988a)
		[N-methyl-^{3}H]LU 49888	2.0	n.d.	1.4	Pure enantiomer, photolabel and reversible ligand	Striessnig et al. (1987, 1988b)

Benzothiazepines	^{3}H	(+)-cis-Diltiazem	39–50	40–80	37–50	Radiolabel for the benzothiazepine-selective domain	Glossmann et al. (1983b) Galizzi et al. (1986a) Balwierczak et al. (1987) Garcia et al. (1986) Schoemaker and Langer (1985) Striessnig et al. (1988a)
		(+)-cis-Azidodiltiazem	n.d.	n.d.	n.d.	Photoaffinity label	Glossmann et al. (1989)
Diphenylbutylpiperidines	^{3}H	Fluspirilene	0.100	0.070	n.d.	Reversible ligand	Qar et al. (1989) Gallizzi et al. (1986) King et al. (1989)
Benzothiazinones	^{3}H	HOE-166	0.100	n.d.	n.d.	Reversible ligand	Qar et al. (1988) Grassegger et al. (1989)

n.d., Not determined

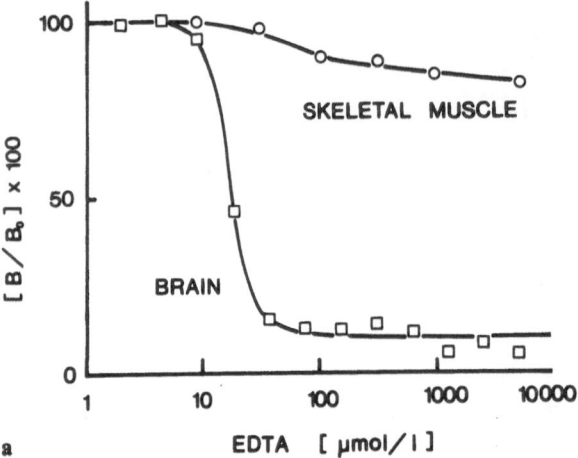

a

Fig. 3 a, b. Regulation of 1,4 DHP binding by divalent cations.
a Membranes from the guinea pig cerebral cortex (brain) and skeletal muscle T-tubule were treated at 37 °C with increasing concentrations of EDTA (as indicated), and receptor domain 1 was
subsequently probed with $(\pm)$-[^{3}H]nimodipine. Binding of the labelled ligand was almost completely inhibited in the brain, whereas the inhibition was marginal in particulate skeletal muscle
membranes. Elevated temperature and high concentrations of EDTA are needed to remove the
divalent cations coupled to receptor domain 1. The conversion to the low-affinity state is time
dependent but completely reversible. The skeletal muscle calcium channel is sensitive to EDTA
only when solubilised or purified. As seen with membrane-bound brain calcium channels, elevated temperature (>25 °C) and high concentrations of EDTA are needed for the conversion
to the low-affinity state.

a high-affinity Ca^{2+}-binding conformation (Glossmann and Striessnig
1988a). Conditions can be found under which the L-type calcium channel can
be completely depleted from divalent cations. This requires treatment of
membranes at temperatures >25 °C, high concentrations of chelators (EDTA,
CDTA) or even (in the case of skeletal muscle) additional solubilisation (see
Glossmann and Ferry 1983b; Glossmann and Striessnig 1988a). In the cation-
depleted state, high-affinity binding is completely lost for the 1,4 DHPs but
binding of ligands selective for the phenylalkylamine or benzothiazepine
domains is retained. Conversely, divalent cations in concentrations
>100 µmol/l inhibit phenylalkylamine or $(+)$-*cis*-diltiazem binding. Loss of
1,4 DHP binding (by chelation) and inhibition of the binding of the other ligands (by divalent cations) are completely reversible (see e.g. Glossmann et
al. 1988b), by adding back divalents and by chelation, respectively. Thus,
there is strong biochemical evidence that L-type channels have distinct (low-
and high-affinity) divalent cation sites coupled to the drug-receptor domains.
These features are illustrated in Figs. 3 and 4. It is appropriate to mention
here that the physiological ion selectivity for Ca^{2+} of the channel is achieved
by tight and selective binding of this divalent cation, and that high flux rates

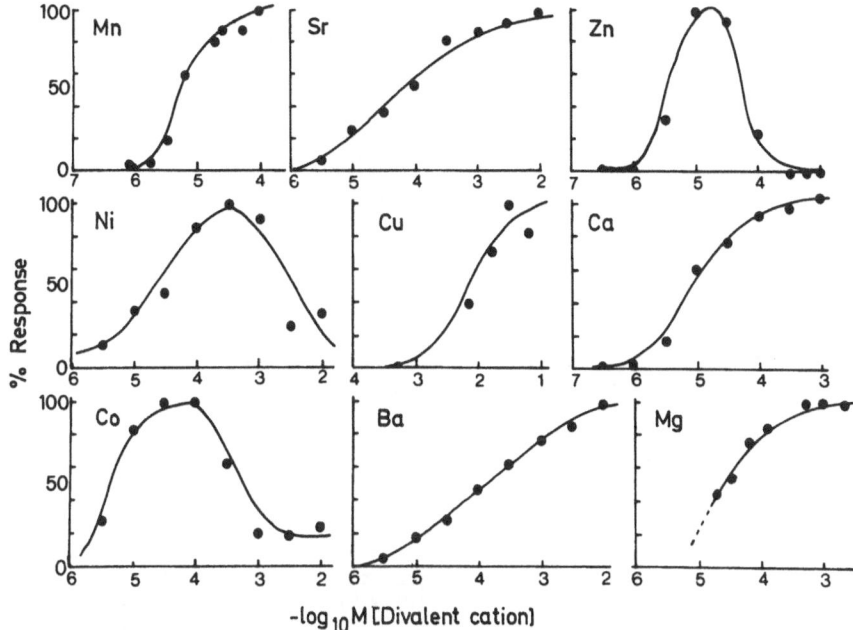

b

−log$_{10}$ M [Divalent cation]

b Divalent cation requirement of the brain 1,4 DHP receptors coupled to calcium channels. After divalent cation depletion by pretreatment with EDTA as above, no high-affinity binding is found for (±)-[^{3}H]nimodipine. High-affinity interaction with the radiolabelled 1,4 DHP was restored by addition of the divalent cations shown. The data are normalised ("response") with respect to calcium (= 100% recovery of high-affinity binding). The effect of divalent cation removal, when not complete, is a B$_{max}$ effect; i.e. the density but not the affinity of the remaining sites is reduced. Conversely, the different divalent cations, upon refilling the sites, stabilise the channel to a different extent in the high-affinity state. For instance, with Zn^{2+} at optimal concentration the channel population reaches only 35% of the maximal binding achieved with Ca^{2+}. In addition, the K$_{0.5}$ values are different. Thus, each cation is characterised by a typical shape of the refilling curve. Hill slopes are less than unity for Sr^{2+} and Ba^{2+} (which pass calcium channels easily). Co^{2+}, Ni^{2+} and Zn^{2+} (which are channel blockers) have bell-shaped curves and also differ with respect to the maximal ability to restore. The experiments illustrate that the L-type calcium channel is a divalent cation-binding protein

are achieved by repulsion between two cation binding sites (Almers et al. 1985; Lansman et al. 1986).

The data in Table 1 show that the dissociation constants for drugs from the 1,4 DHP class are the lowest compared with the other classes, namely in the low nanomolar or even in the picomolar range. This fact, together with the favourable signal-to-noise ratios in the most commonly employed dilution-filtration technique to separate receptor-bound from free ligand, has made several of them the preferred radioligands for structural research. Their broad application is perhaps the reason why the term "1,4 DHP receptor" is often used synonymously with the L-type calcium channel protein(s). This is a misnomer, as "1,4 DHP receptors" exist on many different structures including

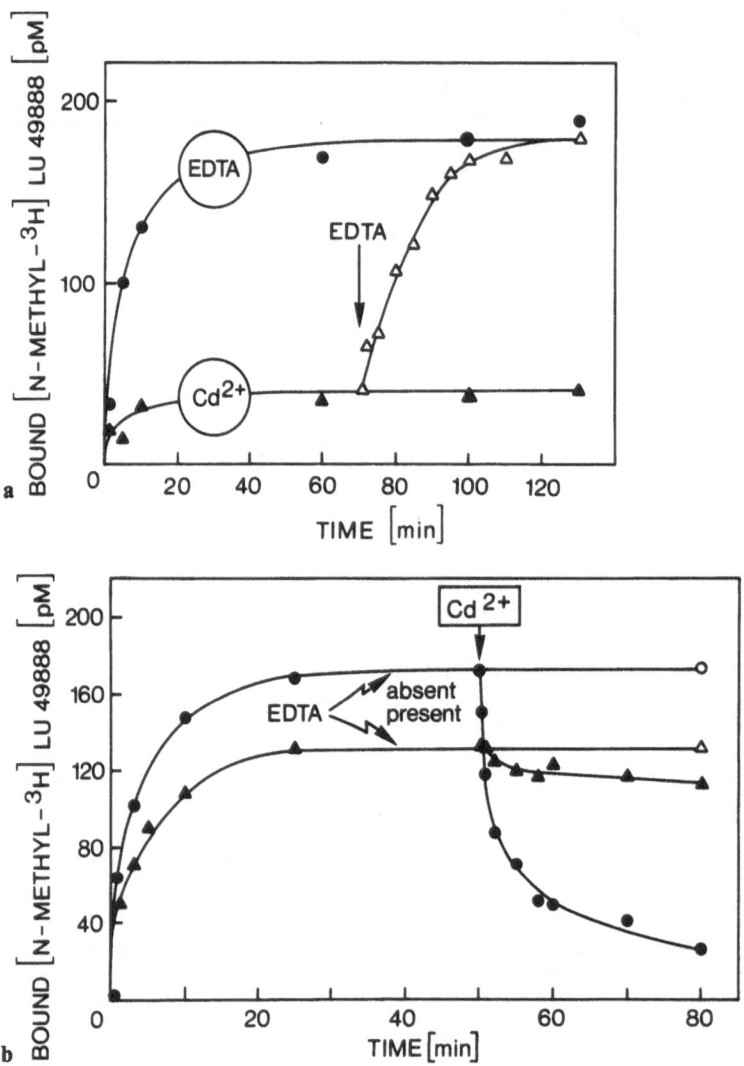

Fig. 4a,b. Reversible interaction of Cd^{2+} with the skeletal muscle calcium antagonist receptor domain for phenylalkylamines.

a Association kinetics: [N-methyl-^{3}H]LU49888, a reversible and photoaffinity ligand, was incubated (in the dark) with partially purified skeletal muscle T-tubule membrane protein (0.02 mg/ml) in the presence of 3 mM EDTA or a submaximal inhibitory concentration of $CdCl_2$ (0.1 mM). Specific binding was measured after the indicated times by rapid filtration of the incubation mixture over fibre glass filters. After equilibrium was reached, EDTA (3 mM final concentration) was added to the incubation mixture containing Cd^{2+}. The time-dependent recovery of [N-methyl-^{3}H]LU49888 binding was determined.

b Dissociation kinetics: [N-methyl-^{3}H]LU49888-calcium channel complexes were formed as described in **a** (3.2 nM [N-methyl-^{3}H]LU49888, 0.01 mg/ml of membrane protein) in the absence and presence of 3 mM EDTA. Dissociation of the complex at equilibrium was initiated by addition of Cd^{2+} (0.1 mM final concentration). For the samples with no EDTA present k_{-1} was 0.127 min^{-1} ($T1/2 = 3.9$ min). [From Glossmann et al. (1988b) with permission]

the nucleoside carrier (Ruth et al. 1985; Striessnig et al. 1985a,b), the multidrug resistance (mdr) glycoprotein gp 170 (Yang et al. 1988; Pastan and Gottesman 1987) and mitochondrial membranes (Zernig et al. 1988; Zernig and Glossmann 1988). The high affinity of the 1,4 DHPs shown in Table 1 is coupled with a very low dissociation half-life at $4\,^\circ$C (in the range of hours or days) which allows prelabelling of the L-type channel prior to solubilisation and purification. The dissociation constants of the 1,4 DHP receptors linked to L-type channels are different, depending on tissue but not on species (Glossmann and Ferry 1985; Janis et al. 1987; Gould et al. 1984). Usually, for a given 1,4 DHP radioligand, the skeletal muscle transverse tubule membrane-bound receptors have five times less affinity than those in brain or heart. In our laboratory the rank order (increasing affinity) for $(+)$-$[^3$H]PN200-110 or $(-)$-$[^3$H]azidopine is skeletal muscle > brain > heart (see, however, results from other groups in Table 1). Together with other data (e.g. chelator sensitivity in the membrane-bound state, effects of heparin, pH-dependence of 1,4 DHP binding, effects of allosteric modulators, etc.), this points to the existence of subtypes of L-type calcium channels (see Glossmann and Striessnig 1988a for review) and complements electrophysiological data.

Fluspirilene (Gould et al. 1983) and HOE-166 are members of novel classes of calcium channel blockers (Fig. 1). Although they have very high affinity, they are difficult to work with because of adsorption to glass and plastic ware, unfavourable nonspecific binding and high (buffer-lipid) partition coefficients which preclude the evaluation of binding parameters in membrane preparations where the density of L-type channels is very low (Qar et al. 1988). Within the phenylalkylamine series two optically pure ligands, including one arylazide photolabel, are available (Fig. 1). These have favourable binding characteristics and can be used to study purified and reconstituted calcium channels (Striessnig et al. 1986a,b, 1987; Flockerzi et al. 1986a,b; Barhanin et al. 1987; Sieber et al. 1987a,b). Again, as for the 1,4 DHP class (and for the benzothiazinones), compounds with a chirality center exist in which the optical antipodes (e.g. $(+)$ and $(-)$-verapamil, $(+)$ and $(-)$-gallopamil) are discriminated by the channel binding domains. Usually, the eudismic ratios are between 10 and 30. Within the benzothiazepine class, $(+)$-cis-diltiazem is the standard radioligand and $(-)$cis-diltiazem is often used to probe for the stereoselectivity of the benzothiazepine-selective domains of the L-type calcium channel. A number of compounds, chemically unrelated to diltiazem, including $(+)$-tetrandrine (a naturally occurring calcium-channel blocker; King et al. 1988) and trans-diclofurime (Mir and Spedding 1987) are claimed to bind to the same site. Figure 5 exemplifies that trans-diclofurime and an optically pure analogue of the sodium channel activator DPI 202-106 (Romey et al. 1987), termed (R)-BDF 8784 (Armah et al. 1989; see Fig. 5c) do not bind in a simple competitive manner to the benzothiazepine-selective site, as they accelerate the dissociation rate of the $(+)$-cis-$[^3$H]diltiazem-re-

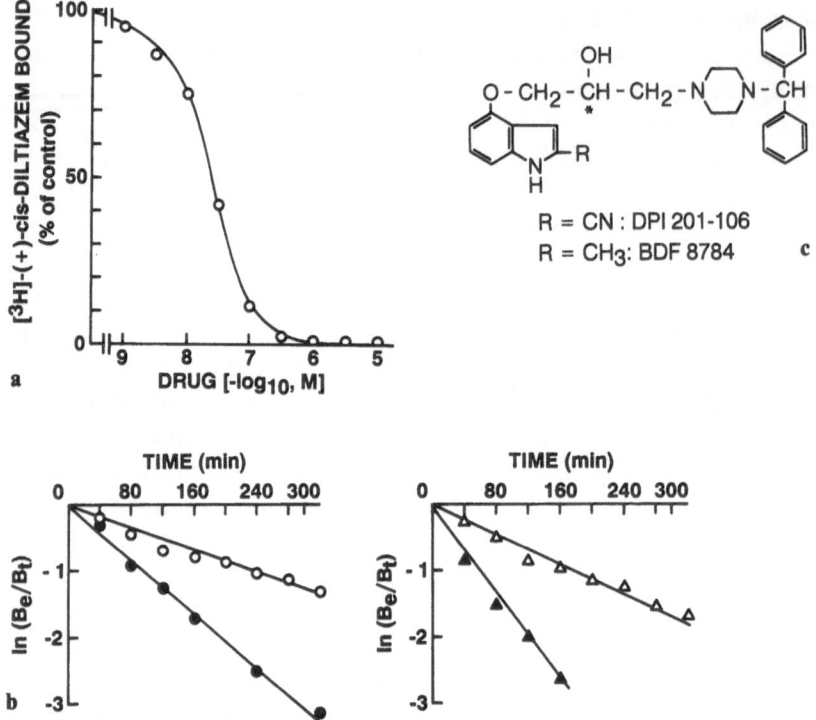

Fig. 5a–c. High-affinity interaction of the (R)-enantiomer of BDF 8784 with the skeletal muscle calcium channel benzothiazepine receptor domain.
This figure illustrates that compounds which inhibit binding of a radioligand completely and
with (pseudo) Hill slopes of unity are not necessarily simple competitive blockers, i.e. do not
necessarily bind to the same site as the radioligand. Analysis of the type of interaction is also
important for structural research, as one of the goals of molecular pharmacology is to identify
the regions within the amino acid sequence of alpha$_1$ which constitute the drug-binding domains, e.g. by photoaffinity labelling. Convenient methods for testing whether inhibition is
allosteric or competitive are dissociation experiments.
This type of analysis led to the conclusion that diphenylbutylpiperidines define a completely
novel receptor domain on L-type channels (Gallizzi et al. 1986a, b). A sodium channel ligand
[(R)-BDF 8784] and *trans*-diclofurime interact with the benzothiazepine-selective domain ("3"
in Fig. 2) with high apparent affinity. Dissociation kinetics reveal that these drugs accelerate
the decay of the [^{3}H](+)-*cis*-diltiazem-receptor complex and hence do not bind in a simple
competitive manner to receptor domain "3".
a Inhibition of [^{3}H](+)-*cis*-diltiazem binding to skeletal muscle microsomes by (R)-BDF 8784:
0.5–0.6nM of [^{3}H]-*cis*-diltiazem were incubated at 2°C with 0.07mg/ml of skeletal muscle
microsomal protein in the absence (control binding) or presence of increasing concentrations
of the drug. After 12h, specific binding was determined as described (Glossmann and Ferry
1985). Data were normalised with respect to control binding. Data of three experiments were
computer-fitted to the general dose-response equation. IC$_{50}$ = 23.8±4.8nM, apparent Hill
slope = 1.3±0.3 (means±asymptotic S.D.). This IC$_{50}$ value is lower than the K$_d$ value of (+)-
cis-diltiazem (40–50 nM).
b Effect of *trans*-diclofurime (*right*) and the (R)-enantiomer (*left*) of BDF 8784 on the dissociation kinetics of [^{3}H](+)-*cis*-diltiazem. Membranes were incubated with [^{3}H](+)-*cis*-diltiazem
as described above in a final assay volume of 0.5ml. After 12h equilibrium was reached, and
dissociation of the radioligand was started by adding 10 µM of (+)-*cis*-diltiazem (control, *open*

ceptor complex. These results are in accord only with a negative heterotropic allosteric inhibition mechanism. We show these examples as other compounds with higher affinity than (+)-*cis*-diltiazem, especially photoaffinity ligands, are needed to identify the amino acid(s) within the amino acid sequence of the channel alpha$_1$-subunit which participate in (+)-*cis*-diltiazem binding. (+)-Tetrandrine (King et al. 1988), *trans*-diclofurime and (R)-BDF 8784 (Fig. 5a) inhibited the binding of (+)-*cis*-[^{3}H]diltiazem — even with lower IC$_{50}$ values than the dissociation constant of the radioligand. Differentiation of strongly allosteric from simple competitive inhibitors in equilibrium (saturation) experiments can be difficult or impossible (Ehlert 1988). Dissociation experiments may help, as shown for skeletal muscle and in heart membranes by Garcia et al. (1984, 1986). (+)-*cis*-Azidodiltiazem (Glossmann et al. 1989; Striessnig et al. 1989), on the other hand, induces the same rate of dissociation of the benzothiazepine receptor-selective radioligand whether in the presence of unlabelled (+)-*cis*-diltiazem or not. This is strong evidence for (+)-*cis*-azidodiltiazem being a competitive ligand, selective for the benzothiazepine site. However, the only advantage of (+)-*cis*-[^{3}H]azidodiltiazem in terms of structural research is that it carries a photoreactive arylazido group (see below).

2.3 Toxins

2.3.1 Toxins as Probes – A General Comment

Peptide toxins have been suggested as tools for studying calcium channels (see Hamilton and Perez 1987). Their lipid solubility is low in contrast to some of the organic channel ligands. For the latter, high affinity is often combined with hydrophobicity (e.g. fluspirilene). Hence, nonspecific binding and signal-to-noise ratios of the peptides can be favourable. Furthermore, peptides can often be radioiodinated at histidine (e.g. apamin; Habermann 1984; Habermann and Fischer 1979) or tyrosine residues or derived with different photoreactive and/or radioactively labelled substituents without dramatic loss of receptor reactivity. Peptide toxins may have extremely low dissociation

---

symbols) or 10 μM (+)-*cis*-diltiazem together with 5 μM of *trans*-diclofurime or (R)-BDF 8784 (*closed symbols*). The concentration of specifically bound ligand was determined before dissociation was started (B$_e$) and after (B$_t$) the indicated times (see Glossmann and Ferry 1985). The (negative) slope of the line is equivalent to the dissociation rate constant k$_{-1}$. Linear regression analysis of the data gave the following k$_{-1}$ values: △, 0.004 min^{-1}; ▲, 0.015 min^{-1}; ○, 0.0038 min^{-1}; ●, 0.011 min^{-1}. The data with (R)-BDF 8784 also illustrate that voltage-dependent sodium channels and calcium channels must be structurally related (Data courtesy of Drs. C. Zech and B. Armah.)

c Structure of the cardiotonic agent DPI 201-106 (4-3'-(4''-benzhydril-1''-piperazinyl)-2'-hydroxy-propoxy-1H-indole-2-carbonitrile). BDF 8784 carries a methyl group instead of the CN group and antagonises the sodium channel-activating effects of DPI 201-106

rates and the apparent K_d can be in the picomolar range. This in turn facilitates prelabelling before solubilisation and allows even prolonged and more drastic purification schemes – conditions under which postlabelling would be difficult. Stability of the receptors (e.g. against proteolytic attack), and not dissociation of the labelled toxin, often determines the apparent half-life of such preformed ligand-channel complexes. Finally, antibodies against the toxins may be used to isolate high-affinity toxin-channel complexes (Hamilton and Perez 1987).

All these advantages have to be weighed against the disadvantages: Possible proteolytic cleavage of the ligands by proteases, difficulty of preparing the labelled toxin with sufficient purity as a homogeneous ligand, nonspecific binding of the highly charged, mostly basic toxins to glass and plastic ware, as well as to negatively charged cellular components, and, last but not least, the often restricted availability to the general scientific public are to be considered. The use of antibodies is not restricted to peptide toxins. Recent reports on high-affinity antibodies directed against 1,4 DHPs (Campbell et al. 1986; Sharp and Campbell 1987) suggest that this area needs to be explored further. Hamilton and Perez (1987) formulated the following criteria by which a toxin can be identified as calcium-channel specific:

1. The toxin should alter calcium channel function and/or the binding of channel-specific ligands. Preferably, alterations of calcium channel functions should be studied with the patch-clamp method and modulation of binding looked for in different tissue preparations as well as with drugs from different chemical classes. With respect to the L-type channel, it may not suffice to simply test 1,4 DHP binding interaction.
2. The toxin should not act via a second messenger. This can be most convincingly excluded by electrophysiological (e.g. patch-clamp and single-channel) analysis.
3. The toxin should not have any detectable enzymatic – in particular no phospholipase or protease – activity. It should be chemically homogeneous. As will be shown, the list of putative natural ligands for calcium channels is long. It also includes nonprotein structures, e.g. maitotoxin. However, the number of promising candidates is disappointingly small.

2.3.2 Toxins with Claimed but Unproven Action on Calcium Channels

2.3.2.1 Maitotoxin

Maitotoxin has been isolated from the marine dinoflagellate *Gambierdiscus toxicus*. Its structure, composition and molecular weight are not known. Maitotoxin is said to be the most potent marine toxin (see e.g. Takahashi et al. 1983; Wu and Narahashi 1988) but this statement has been contested (Kaul and Daftari 1986). Here we summarise recent evidence that the original hypothesis, namely that the toxin directly activates calcium channels, cannot be upheld. In cultured neuronal cells (NG 108-15 neuroblastoma×glioma) $^{45}Ca^{2+}$ uptake is increased with a lag period of about 2 min after application of maitotoxin. This uptake is inhibited by nitrendipine, diltiazem and verapamil but, surprisingly, was also diminished when extracellular sodium was absent (Freedman et al. 1984). Sodium (and calcium) dependence was also found for the maitotoxin-induced gamma-aminobutyric acid release from cultured striatal neurons (Shalaby et al. 1986). This release was not blocked by L-type channel-specific blockers (Pin et al. 1988).

In isolated aortic myocytes, maitotoxin ($EC_{50} = 0.3$ ng/ml) stimulates inositol phosphate production. Nifedipine, diltiazem and verapamil did not block the toxin effect (Berta et al. 1986). Phosphoinositide breakdown in neuroblastoma hybrid NCB-20 cells is also stimulated by maitotoxin. Ca^{2+} is required, but organic (nifedipine, methoxyverapamil) and inorganic (Co^{2+}, Cd^{2+}, Mn^{2+}) L-type channel blockers did not antagonise (Gusovsky et al. 1987).

In aortic myocytes large increases of free cytosolic calcium are observed after maitotoxin. Neither K^+ depolarisation nor Ca^{2+} ionophores could mimic these effects, but saponin imitated maitotoxin's action on the inositol phosphate formation (Berta et al. 1988).

Electrophysiological data on maitotoxin are scanty. In guinea pig cardiac cells the toxin produced a sustained inward current which was enhanced by adrenaline, carried by Ca^{2+} or Ba^{2+}, abolished by 1 mM Cd^{2+} and had an almost linear current-voltage relationship. When the toxin was present in the pipette solution and the cell-attached patch technique was employed, "Ca^{2+} channels" with novel properties were observed. Surprisingly, such novel channels were also been (but much less frequently) in the absence of maitotoxin. The toxin-induced (stabilised?) channels had a mean open time which was ten times longer than that of the voltage-dependent channel and were voltage independent with a unitary conductance of 12 pS with 50 mM Ba^{2+} (Kobayashi et al. 1987). Formation (or stabilisation?) of a pore was also the most likely mechanism for the toxin-induced membrane currents in neuroblastoma cells (Yoshii et al. 1987).

In summary, maitotoxin is not suitable for the characterisation of calcium channels. Its structure is unknown and its mode of action is not established. Activation of phosphoinositide breakdown, induction of tetrodotoxin-resistent sodium fluxes (Pin et al. 1988), formation of "novel" Ca^{2+} channels and stimulation of leukotriene C_4 production (Koike et al. 1986) have been observed. Perhaps, as shown for palytoxin and the sodium pump (Chhatwal et al. 1983), maitotoxin may convert a specific target in plasma membranes from excitable cells into a pore. It has even been speculated (Kobayashi et al. 1987) that the L-type channel itself may be the target.

2.3.2.2 Leptinotoxin-h (Beta leptinotarsin-h)

Leptinotoxin-h is a constituent of the hemolymph of the Colorado potato beetle, *Leptinotarsa haldemani*. The acidic 57-kDa protein (Crosland et al. 1984) stimulates acetylcholine release in the peripheral and the central nervous system in a calcium-dependent manner (McClure et al. 1980). It depolarises guinea-pig synaptosomes and neurosecretory (PC12) cells, is not antagonised by tetrodotoxin or verapamil, and induces $^{45}Ca^{2+}$ influx as well as a rise in cytosolic free calcium concentration. Depolarisation due to the toxin required the presence of external calcium, suggesting that the divalent cation was needed for binding (Madeddu et al. 1985). Leptinotoxin-h induced Ca^{2+}-dependent ATP release from resting synaptosomes from electroplax of the ray *Ommata discopyge*. The process was not blocked by omega-conotoxin GVIA (as was depolarisation-induced release), indicating different loci of action (Yeager et al. 1987). There is no direct proof that leptinotoxin-h acts directly on calcium channels (see, however, Miljanich et al. 1988); hence, its role as a tool for structural research is doubtful.

2.3.2.3 Goniopora Toxin

The polypeptide toxin *Goniopora* was isolated from the *Goniopora* coral by Qar et al. (1986). It migrated with an Mr of 19000 as a single band on SDS-gel electrophoresis.

The toxin stimulated $^{45}Ca^{2+}$ influx into chick cardiac cell cultures. The EC_{50} value was 5.3 μM and the stimulated influx was inhibited by nitrendipine. In the guinea pig ileum system contractions ($EC_{50} = 1.7$ μM) were induced. The toxin effects were blocked by the L-type channel-specific drugs nitrendipine and $(-)$-desmethoxyverapamil. The toxin also inhibited the binding of $(+)$-[^3H]PN200-110 to rabbit skeletal transverse tubule membranes with an IC_{50} value of 5.3 μM. If proven by electrophysiological methods to be an L-type channel-specific activator the toxin could be a candidate for structural research, perhaps with some specificity for organisms of marine origin (as suggested by Qar et al. 1986).

2.3.2.4 Apamin

Apamin, a potent bee venom toxin (for a review see Habermann 1984), was claimed to be a highly specific Ca^{2+} blocking agent in heart muscle. The peptide at picomolar concentration blocked naturally occurring slow action potentials in cultured chick heart cell aggregates and noncultured chick hearts (Bkaily et al. 1985). The effects depended on extracellular potassium and resisted washing. The slow action potential could be restored by the application of quinidine. Although the authors suggested that apamin might be used as a tool to study and isolate calcium channels from heart, no further data supporting this claim have been forwarded. Apamin is known to block certain types of Ca^{2+}-dependent potassium channels (Lazdunski 1983) and binds in a potassium-dependent manner to its receptor sites (Habermann and Fischer 1979).

2.3.3 Toxins Which Are Putative Candidates for Structural Research

2.3.3.1 Plectreuris tristes Toxin

The venom of the spider *Plectreuris tristes* contains excitatory and inhibitory neurotoxins which irreversibly act on *Drosophila* larval neuromuscular junctions (Branton et al. 1987). Inhibition was consistent with a specific, irreversible block of presynaptic calcium channels, and confined to fractions with M_r's of 6000–7000 upon size-exclusion HPLC. The most abundant (0.1% of venom protein) component, designated alpha-PLTX II, is a single polypeptide with an apparent M_r of 7000 and completely blocks neurotransmitter release, as evidenced by the reduction of excitatory junction potentials and also the recurrent spikes at or near nerve terminals in the double mutant *eag* Sh[rk0120]. The *eag* mutant produces reduced delayed-rectifier currents, while the *Shaker* (*Sh*) mutant has lost the transient current called "A". In these abnormally excitable, double mutants the recurrent spikes are blocked by Co^{2+} or Cd^{2+} but not by tetrodotoxin and are believed to be associated with calcium currents. Alpha-PLTX II does not block divalent cation-dependent action potentials in larval muscle and is inactive at the frog neurotransmitter junction. Its possible use as a structural probe may be restricted to arthropod presynaptic calcium channels unless other sites and mechanisms of action can be found. Crude *Plectreuris tristes* venom inhibits the binding of [^{125}I]-omega-conotoxin GVIA to rat brain membranes with a half-maximal concentration of 30 ng/ml. The inhibition was noncompetitive since both the association and dissociation rate constants are increased, and it was not observed when L-type Ca^{2+} channel ligands were employed to probe for their sites (Feigenbaum et al. 1988). Further studies are needed to clarify which venom component binds in an apparently allosteric manner to the [^{125}I] omega-conotoxin GVIA-labelled receptor sites.

2.3.3.2 Hololena curta Toxin

Whereas alpha-PLTX II appears to be a single polypeptide, a toxin isolated from the venom of the hunting spider, *Hololena curta*, may consist of two different disulphide-linked subunits with M_r's of 7000 and 9000 respectively. This toxin produces a complete and long-lasting inhibition of synaptic transmission at the *Drosophila* larval neuromuscular junction. Indirect evidence from *Drosophila* mutants (as for alpha-PLTX II) pointed to specific effects on presynaptic calcium channels in *Drosophila* motor neurons (Bowers et al. 1987). Its possible use as a structural probe must await further studies.

2.3.3.3 Taicatoxin

Taicatoxin was purified from the freshly collected venom of the Australian taipan snake (*Oxyuranus scutellatus*) as a basic, highly charged polypeptide of about 65 amino acids ($M_r = 8000$). Its name originates from *tai*pan and *ca*lcium; the abbreviation is TCX, where X stands for toxin (Brown et al. 1987).

TCX is reported to block calcium currents in the skeletal muscle BH_3H_1 line and in mammalian smooth muscle and to inhibit (in a noncompetitive manner) the binding of $(+)$ [^3H] PN 200-110 to 1,4 DHP receptors linked to L-type channels in isolated cardiac membranes.

Whole-cell patch-clamp analysis of guinea pig ventricular cells revealed a decrease of the calcium current by TCX. The effect was reversible upon washout. Potassium and sodium currents were not changed by TCX. Injection of TCX (via the intracellular pipette) produced no inhibition of calcium channels. Inclusion of the toxin in the pipette using the cell-attached, single-channel mode led to blockade, whereas in the same experimental setup application outside of the patch was ineffective. It was concluded that the extracellular mouth of the calcium channel was the TCX target.

The IC_{50} value of the toxin at a holding potential of -30 mV was 10 nM and the block was complete at saturating TCX concentrations. At -80 mV the block was incomplete and the apparent affinity of the toxin reduced, suggesting block and binding to be voltage dependent. Further evidence that the toxin did not block via second messenger systems and acted directly on the high threshold (L-type) channel came from outside-out patch-clamp experiments with ventricular cells from neonatal rat heart. TCX suppressed channel activity without changing single-channel conductance by reducing, (re)opening and increasing the frequency of records where the channel was silent.

In summary, taicatoxin appears to be a promising candidate for structural research on L-type calcium channels. Its usefulness for differentiation of subtypes, as a labelled ligand and for purification remains to be established.

2.3.3.4 Atrotoxin

Atrotoxin, a protein fraction claimed to be >15 kDa was isolated from the rattlesnake *Crotalus atrox* venom by gel filtration and ion-exchange chromatography. It increases the calcium currents in guinea pig and neonatal rat heart cells (Hamilton et al. 1985). Using the whole-cell patch-clamp method effects on potassium and sodium currents were excluded. Neither alpha- nor beta-adrenoceptors are involved in the increase of calcium current, as phentolamine and propranolol did not inhibit the toxin's action. The effects were reversible upon washout and not seen when the active fraction was injected into the cells. The activity of atrotoxin is reported to decrease during further purification (Hamilton and Perez 1987). Although atrotoxin fulfils some of the criteria of a direct calcium channel activator, it is a doubtful probe for structural characterisation.

2.3.4 Omega-Conotoxins – N-Type Channel Probes

The peptide toxins from marine snails of the genus *Conus* have been reviewed by Gray et al. (1988). Such snails, especially the fish-hunting varieties, paralyse their prey by rapid poisoning. Their venoms are a treasure box for pharmacologists, physiologists and biochemists alike. The active principles are small peptides, some with unusual amino acids (e.g. gamma-carboxy-glutamate, hydroxyproline) and with as many as three disulphide cross-bridges in a section of 13 amino acid residues (Gray et al. 1988). Their size allows synthesis by the Merrifield method and an almost unlimited variation to probe for structure-activity relationships.

Conotoxins block three types of ion channels (Olivera et al. 1985). The al-pha-conotoxins (named in analogy to the alpha-neurotoxins) block nicotinic acetylcholine receptors and may be useful for the differentiation of receptor subtypes in the brain, whereas the μ-conotoxins block voltage-dependent so-dium channels with high (1000-fold) selectivity for the skeletal muscle ("m") type versus the nerve ("n") type. Finally, the omega-conotoxins inhibit presynaptic neuronal calcium channels. Among them, the peptide GVIA (for the nomenclature the reader is referred to Gray et al. 1988), originating from *Conus geographus* venom, has been most widely used. Another toxin from *Conus majus*, MVIIA, is shown together with GVIA in Fig. 6. MVIIA can be differentiated from GVIA by toxicity tests employing different animal spe-cies and intraperitoneal (i.p.) versus intracerebral (i.c.) injection. I.p. injection of either GVIA or MVIIA leads to paralysis and death of fish but not of mice. Frogs are paralysed and killed by GVIA but not by MVIIA upon i.p. injec-tion. All of the above toxins induce, when injected into the brain of a mouse, a "shaker" syndrome. This is a persistent tremor which can last up for 5 days, depending on the dose (Olivera et al. 1985).

The lack of effect of MVIIA on the amphibian neuromuscular junction, together with experiments in which MVIIA inhibited only 50% of radioiodi-nated GVIA binding to frog brain membranes (but was able to completely block it in nammalian brain membranes), led Gray et al. (1988) to formulate a further subdivision of N-type calcium channels. This subdivision has to be briefly discussed in context with $^{45}Ca^{2+}$ uptake data, neurotransmitter re-lease experiments and mRNA expression data.

When $^{45}Ca^{2+}$ uptake, induced by K^+ depolarisation, is studied with sy-naptosomes from different species, chick and frog exhibit complete inhibition

C K S P G S S C S P T S Y N C C R + S C N P Y T K R C Y * GVIA

C K G K G A K C S R L M Y D C C T G S C R + + S G K C *   MVIIA

Fig. 6. Comparison of the amino acid sequence of the calcium channel blockers omega-conotoxin GVIA and MVIIA. Disulphide bridges are formed between the Cys residues 1 and 16, 8 and 19, and 15 and 26. *, Amidated carboxy terminal; P, hydroxyproline; +, gap for alignment

by GVIA at concentrations <1 μM. In the chick system 15% of radiocalcium uptake was blocked at 5 nM toxin, whereas the rest was (completely) inhibited with a 50 nM IC$_{50}$ value (Suszkiw et al. 1987). Complete inhibition of ^{45}Ca^{2+} uptake in chick synaptosomes with an overall IC$_{50}$ value of 10 nM was also observed by others (Rivier et al. 1987). In mammalian systems (e.g. with rat synaptosomes) the inhibition at best reaches 50% (Reynolds et al. 1986b) or 40% (Suszkiw et al. 1987) at concentrations of 0.1 or 10 μM respectively. Thus, in contrast to birds and amphibians, ^{45}Ca^{2+} uptake in mammalian systems is through calcium channels, of which only 50% or less can be blocked by this toxin. The conotoxin-resistant uptake does not occur through the L-type channel (see Miller 1987 for a review) but alternative reports have appeared (see e.g. Turner and Goldin 1985). With respect to neurotransmitters, GVIA blocks up to 80% of depolarisation-induced release of noradrenaline, serotonin and acetylcholine. At 5 nM, 50% of this toxin-sensitive component is blocked (Dooley et al. 1987, 1988). This finding agrees with the present view that N-type channels (sensitive to GVIA) are critical for delivering calcium to the neurotransmitter-release machinery (Miller 1987). However, acetylcholine release from the motor neurons is not inhibited by GVIA in mammalian systems (Kerr and Yoshikami 1984; Koyano et al. 1987; Sano et al. 1987) or by MVIIA in the frog. Interestingly, aminoglycosides such as neomycin (see below) block both motor neuron acetylcholine release (see references in Wagner et al. 1987 and Knaus et al. 1987) and neurotransmitter release from central synapses (Atchinson et al. 1988). In the various experimental systems, actions of the GVIA toxin are not reversible upon washout and are dependent (with respect to potency) on the composition of the buffer, especially on cations, as will be explained below. In cholinergic synaptosomes from electroplax of the ray *Ommata discopyge* (Yeager et al. 1987) depolarisation-induced ATP release is blocked in a reversible manner by both GVIA and MVIIA with equal potency (IC$_{50}$ value = 0.5 μM).

When rat brain mRNA is injected into *Xenopus* oocytes, a long-lasting calcium current is expressed which is blocked neither by the classical organic (L-type) calcium channel blockers (e.g. 1,4 DHPs) nor by omega-conotoxin GVIA (Leonard et al. 1987). This all leads to the hypothesis that (especially in mammalian systems) a further subdivision of neuronal (N-type) calcium channels is required to explain the species- and synapse-dependent toxin effects. The proposed N$_A$ type is insensitive to both GVIA and MVIIA in mammals (location here is on central and peripheral synapses). The channel translated from rat brain mRNA would be of this type. The N_B type, perhaps located on central synapses only, is sensitive to both toxins in mammals. Both types are blocked by GVIA and MVIIA in birds. In the frog, MVIIA is selective: it does not block the subset N_A and, as will be discussed below, inhibits only 50% of the binding of radiolabelled GVIA in frog brain membranes. Presumably, the 50% of $^{45}Ca^{2+}$ uptake that is not inhibitable by the omega-conotoxins also occurs through the N_A type of N-channels in rodent central synaptosomes.

Table 2 gives an overview of the different types of voltage-regulated calcium channels. In chick dorsal root ganglion cells the three main types (T, L, N) co-exist (see Tsien et al. 1988 for a review on neuronal calcium channels); T- and L- (but not N-type) channels are found in most muscle or endocrine cells and even in fibroblasts. The skeletal muscle calcium channels ("T"-like or "L"-like) do not fit within the T, L, N nomenclature (Tsien et al. 1988), but the main point of the table is the differentiation of the types with respect to the tools that are available for structural characterisation. The reader should realise that it is a mixed bag from electrophysiology, pharmacology and toxicology, from various species and various cell types, and obtained under often vastly different conditions. In the context of structural characterisation, the proposed high selectivity of the omega-conotoxins for "N-type" channels is put in some doubt. As exemplified in the table, there is a weak but reversible block of T-channels in heart and neuronal cells and a persistent block of neuronal L-type and N-channels. Furthermore, N-type channels are (as now postulated) again a heterogeneous group, only one of which (N_B) binds the toxin(s) in mammals with high affinity. What does one identify with a labelled toxin in membranes from the brains of different species? Is it the L_n channel, the N_B channel in rodents, or both (N_A and N_B) channels in frogs and birds or even the T-channel? For the T-channel one would, of course, expect sites of low affinity which bind the labelled toxin reversibly; L_n and N_A (plus N_B) channels should exhibit tight binding. Possibly, the L_n channel site could show some (allosteric) interactions with one or more of the hitherto identified receptor domains (e.g. for DHPs, phenylalkylamines, benzothiazepines or diphenylbutylpiperidines). In Table 3 we have summarised recent data on omega-conotoxin binding. In nonneuronal tissues (see Table 2) there should be no significant high-affinity binding, although low-affinity (T-channel) interaction, e.g. in heart, is a distinct possibility. The exclusive binding to neuronal tissue is reported uniformly as expected. Furthermore, there should be a slowly reversible or even irreversible interaction with the receptor(s). In the case of reversible interaction, kinetic equilibrium constants (derived from forward and dissociation rate constants) must be in reasonable agreement with a measured equilibrium dissociation constant obtained by saturation analysis. In the case of irreversible ("tight") interaction, formation of the complex

Table 2. Calcium channel types and subtypes

Type	Electrophysiological criteria	Location/function	Pharmacology	References
T	Low-threshold, rapidly inactivating. Single-channel conductance (100 mM Ba^{2+}): 8–9 pS	Pacemaker current. Rhythmic activity. Found in sinus node cells, heart, skeletal muscle	Gallopamil, verapamil, amiloride, flunarizine block. Cd^{2+} block: high concentrations required; Ni^{2+} is a more effective blocker. GVIA blocks weakly and reversibly (frog atria and chick dorsal root ganglion cells). 1,4-Dihydropyridines have no effect	Miller (1987) Nilius (1986) B. Nilius (personal communication) Tang et al. (1988) Tsien et al. (1988)
L	Long-lasting, high threshold. Single-channel conductance: 25 pS		Phenylalkylamines benzothiazepines, 1,4-dihydropyridines, benzothiazinones and diphenylbutylpiperidines block. 1,4-Dihydropyridine Ca^{2+} agonists activate. Cd^{2+} block: low concentrations required; Ni^{2+} is less effective as a blocker	Miller (1987) McCleskey et al. (1987) Galizzi et al. (1987) Quar et al. (1988)
L$_n$ (neuronal)		Cell soma, metabolic control?	GVIA blocks irreversibly	
L$_m$ (muscular)		Contraction in heart and smooth muscle; secretion of hormones	GVIA has no effect	
L$_{sk}$ (skeletal)	Different in gating kinetics from L$_n$ or L$_m$	Excitation-contraction coupling		

			Block by		
			Aminoglycosides	GVIA	
N	Activates like the L-type but inactivates like the T-channel. Single-channel conductance: 13 pS	In general: presynaptic location, neurotransmitter release	Cd^{2+} block: low concentrations are required; Cd^{2+} is more effective than Ni^{2+}.		Miller (1987) McCleskey et al. (1987) Gray et al. (1988) Tsien et al. (1988)
N_A		Presynaptic (i.e. in peripheral mammalian neurons)	Yes	No	
N_B		Presynaptic (i.e. in central mammalian neurons)	Yes	Yes	

Table 3. Characterisation of radiolabelled omega-conotoxins as probes for structural research

Author(s)	Radioligand	Specific activity (Ci/mmol)	Tissue conditions, temperature	Dissociation constant(s)	Kinetic constants	B_{max} [fmol/mg]	Profile (IC$_{50}$ values)	Comments
Wagner et al. (1988)	Mono[125I]-iodo omega-conotoxin GVIA	2200	Fresh rat brain (frontal cortex membranes) (0.01–2.5 µg protein/ml) 25°C	60 pM	$K_{+1} = 2.6 \cdot 10^{10}\ M^{-1}\ min^{-1}$ $K_{-1} = 0.0011\ min^{-1}$	8300	Omega-conotoxin GVIA: 0.061 nM; Omega-conotoxin MVIIA: 0.500 nM; Rat myelin basic protein: 2 nM; Polylysine: 5 nM; N, V, D have no effect; Aminoglycosides and cations inhibit	Inhibition by aminoglycosides and polylysine is non-competitive. Steady-state value of binding reached within 10 min
Barhanin et al. (1988)	[125I]iodo omega-conotoxin GVIA	1000	Chick brain membranes (1–3 µg protein/ml) 25°C	0.82 pM	$K_{+1} = 2.49 \cdot 10^{9}\ M^{-1}\ min^{-1}$ $K_{-1} = 0.056\ min^{-1}$	1030	Omega-conotoxin GVIA: 0.65 pM; N, V, D have no effect; $Ca^{2+} > TRIS^{+} > Na^{+}$ inhibit	Steady-state value of binding reached within 100 min

Reference	Ligand		Tissue/conditions	K_D	Kinetics		Effectors	Comments
Feigenbaum et al. (1988)	Mono[^{125}I]-iodo omega-conotoxin GVIA	1000	Rat brain synaptic membranes (Protein concentration not reported, but probably in the μg/ml range) 25 °C	0.78 pM	$K_{+1} = 1.3 \cdot 10^{10} \, M^{-1} \, min^{-1}$ $K_{-1} = 0.0006 \, min^{-1}$	1000	Aminoglycosides and cations (La^{3+} > Cd^{2+} > Ca^{2+} > Na$^+$) inhibit	Steady-state value is reached within 120 min; cations (e.g. Ca^{2+}) and aminoglycosides increase the K_D
							Dynorphin A [1–13] binding	Dynorphin A [1–13] increases K_{+1} and decreases the K_D
							Dynorphin A [1–13] stimulates binding	
							Plectreuris tristes venom inhibits binding	Plectreuris tristes toxin increases K_{+1} and K_{-1}
							N, V, D have no effect	
Yamaguchi et al. (1988)	Mono[^{3}H] propionyl-omega-conotoxin GVIA	105	Bovine brain membranes (20 μg protein/ml) 4 °C	site 1: 3 pM site 2: 3.5 nM	n.r.	380 2810	Cations inhibit; N, V, D do not inhibit at 3.5-nM ligand. D inhibits at 60-pM ligand. D inhibition is stereoselective	Equilibrium was not reached; 50% of ligand dissociates ($T_{1/2} = 60$ min) at 4 °C but 50% is bound irreversibly

Table 3 (continued)

Author(s)	Radioligand	Specific activity (Ci/mmol)	Tissue conditions, temperature	Dissociation constant(s)	Kinetic constants	B_{max} [fmol/mg]	Profile (IC$_{50}$ values)	Comments
Abe et al. (1986)	Mono[125I]-iodo omega-conotoxin GVIA	210 and 2100	Rat brain membranes (2–4 µg protein/ml) 4 °C	site 1: 10.3 pM site 2: 0.52 nM	n.r.	520 3400	Cations La^{2+} > Cd^{2+} > Ca^{2+} ⪢ Na$^+$, K$^+$ inhibit, N, V, D have no effect	Dissociation is very slow ($T_{1/2}$ > 10 h)
Cruz and Olivera (1986)	[125I]iodo omega-conotoxin GVIA	100 to 400	Frog and chicken embryonic brain membranes (0.5 – 1.0 mg protein/ml) 25 °C	n.r. (sub-nanomolar)	n.r.	1500 (chick)	Cations (Co^{2+}, Mg^{2+}, Ca^{2+}) prevent binding for 25 min, but do not dissociate bound ligand	
Olivera et al. (1987)	Mono[125I] iodo omega-conotoxin GVIA	100	Chick, frog and calf synaptosomes (1 mg protein/ml) Temperature not reported (25 °C?)	n.r.	n.r.	n.r.	Binding of labelled toxin is completely inhibited by the MVIIA peptide in bovine but not in frog brain membranes	
	Mono[125I]-iodo omega-conotoxin MVIIA	100	(25 °C?)	n.r.	n.r.	n.r.	This iodinated toxin dissociates with a $T_{1/2}$ < 10 min from chick brain membranes	

Abe and Saisu (1987)	N-5'-azido-2-nitrobenzoyl mono-[125I]iodo omega-cono-toxin GVIA	2100	Rat brain synaptic membranes (2–4 µg protein/ml) 4°C	n.r.	n.r.	n.r.	N, D, V do not inhibit. Divalent cations inhibit	The ligand has a $T_{1/2}$ ➤8 h at 4°C
Marqueze et al. (1988)	Mono[125I]-iodo omega-conotoxin GVIA	2100	Rat brain synaptosomal membranes (15 µg protein/ml) 37°C	—	Only forward rate constant given $K = 5.5 \cdot 10^6\,M^{-1}\,s^{-1}$	650	N, D, V do not inhibit. Divalent cations (Co^{2+} > Ba^{2+}) inhibit	The ligand binds irreversibly; loss of binding at 37°C is explained by receptor decay

n.r., Not reported; *D*, (+)-*cis*-diltiazem; *N*, nitrendipine; *V*, verapamil

should overcome any inhibition by reversible ligands, depending on time. Moreover, in the case of reversible binding the ratio of bound to free ligand (taking into account the "receptor reactivity" of the labelled ligand) at low ligand concentrations must be linearly dependent on R_T, the total receptor concentration. In the case of irreversible binding the ligand-receptor complex and not the bound-to-free ratio should increase linearly with respect to R_T. Here, bound ligand reaches asymptotically an equivalence point, $R_T = L_T$, where L_T is the total ligand concentration in the assay. Conversely, when L_T is varied at constant R_T the "apparent K_d" is a function of the total receptor concentration. Thus, in the case of irreversible binding, an almost unbeliev-able range of K_d values may be reported, simply dependent on the concen-trations of receptors present in the experimentor's test tube.

The data in Table 3 exemplify that this is indeed the case. When very low concentrations of brain membranes ($0.01 - 4$ µg protein per ml) are employed, picomolar "dissociation" constants are observed; at high concentration of re-ceptors (e.g. 1 mg of protein/ml, which is equivalent to $1 - 2$ or even 8 nM re-ceptor concentration) nanomolar "apparent K_d" values are reported (see e.g. Fig. 3 in Cruz and Olivera 1986). No authors have reported the "receptor reac-tivity" (i.e. the binding ability) of the radioligand. As a general rule, dissocia-tion experiments have to be extended for at least two or three half-lives of a complex in order to get a good estimate of K_{-1}, the dissociation rate con-stant, as well as to obtain a general idea of the type of complex decay (e.g. monoexponential, biphasic, negative cooperativity). In the two cases where this has been done the K_{-1} was either 0.001 min^{-1} (Wagner et al. 1988) or 0.006 min^{-1} (Feigenbaum et al. 1988). The corresponding kinetically derived dissociation constants were 40 fM or 46 fM, 1000 or 20 times lower than the measured "equilibrium dissociation constant". In the one other case, where the ratio of K_{-1}/K_{+1} ($= K_d$) was in agreement with the saturation-equilibri-um K_d value, dissociation was followed only for one half-life (Barhanin et al. 1988). Authors who used ligands (either tritium or ^{125}I-labelled) having a low specific activity found high-affinity sites with picomolar K_d values (similar to those laboratories reporting only one high-affinity site) and additional low-affinity sites with nanomolar dissociation constants. The sum of the densities (B_{max} values) of the two classes of sites is nearly identical to the density of a single class of high-affinity sites reported by other authors − a disturbing result.

Furthermore, in the case of mono [^{3}H]proprionyl omega-conotoxin GVIA, dissociation is observed for only 50% of the ligand (or the sites?) − the other half is irreversibly bound (Yamaguchi et al. 1988). There is general agreement, however, that cations (trivalents > divalents > monovalents) prevent the bind-ing but cannot dissociate the complex once formed. High concentrations of divalents (e.g. Ba^{2+}) in buffers can explain the nanomolar IC_{50} values often reported by electrophysiologists and pharmacologists, which contrast with the picomolar K_d values observed for receptor binding (under divalent cation-free conditions). The differences probably reflect the different incubation conditions employed. The time-dependent reversal of divalent cation inhibition of mono-$[^{125}I]$-iodo conotoxin GVIA binding is illustrated in Fig. 7. These results are in agreement with calcium being a reversible inhibitor acting on (but not necessarily directly at identical) sites where the radioligand was irreversibly fixed. The pharmacological profile of the receptor sites is interesting. Polylysine and rat myelin basic protein inhibit (in a noncompetitive manner) with nanomolar inhibition constants but are slightly less active than MVIIA (Table 3). H.G. Knaus, J. Striessnig and M. Weiler (personal communication) confirmed the polylysine inhibition and added $ACTH_{1-23}$ as well as histone type II_A to the list of the inhibitors. Together with the effects of the aminoglycosides, this all points to the critical involvement of highly charged groups, both on the receptor(s) and on the omega-conotoxins. Several explanations for the divergent results in the literature are at hand. Ligand heterogeneity is clearly one of them. For example, three distinct ^{125}I-iodinated GVIA peaks are observed on HPLC, where only the ligand in one peak (peak III in Marqueze et al. 1988) was fixed irreversibly. In contrast, radioligands separated in the two other peaks bound reversibly (M.J. Seagar, personal communication; Marqueze et al. 1988). The other possibility that needs to be investigated is receptor heterogeneity. MVIIA (when radioiodinated) is a *reversible* ligand in the chick brain membrane system. The finding that unlabelled

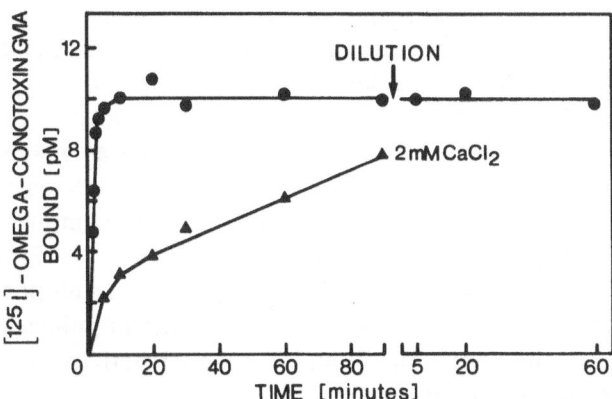

Fig. 7. Association kinetics of $[^{125}I]$omega-conotoxin GVIA and time-dependent reversal of calcium inhibition. $[^{125}I]CgTx$ (44 pM) was added to 0.024 mg/ml of guinea pig cerebral cortex membrane protein, preincubated for 30 min at 25 °C with (total binding) or without (nonspecific binding) 30 nM of unlabelled toxin. Bound toxin was separated from free toxin after the indicated times by rapid filtration over GF/C Whatman filters using the polyethylene glycol buffer method as described by Glossmann and Ferry (1985). The experiment was repeated under exactly the same conditions with 2 mM $CaCl_2$ added to the membranes during preincubation. Note that the inhibitory action of the reversible ligand Ca^{2+} is overcome by the irreversible fixation of the toxin to its receptor sites. [From Glossmann et al. (1988b) with permission]

MVIIA (most likely a reversible ligand, too) inhibits iodinated GVIA binding in frog brain by only 50% fits nicely with the subdivision (i.e. N_A and N_B) hypothesis. The inhibition kinetics of *reversible* ligands (illustrated for calcium in Fig. 7) competing with *irreversible* ("tight") radioiodinated probes suggests, however, that these results must be viewed with caution.

There is nearly uniform agreement (Table 3) that L-type calcium channel drugs do not interfere with omega-conotoxin binding, be the ligand of high or low specific radioactivity. There is one report where the tritiated derivative of GVIA bound to high-affinity sites in particulate and solubilised bovine brain membranes which appeared to be linked to benzothiazepine (i.e. (+)-*cis*-diltiazem-selective) sites. The binding of mono[³H]propionyl omega-conotoxin GVIA (at low concentrations of radioligand) is inhibited in a stereoselective manner by (+) and (−)-*cis*-diltiazem. There was no inhibition by other L-type channel drugs (see also Sect. 6).

2.4 Endogenous Ligands

2.4.1 General Remarks

Whenever drugs are found to bind with high affinity to a physiological target and exert specific effects, the hunt for the endogenous ligand(s) is on. No wonder that in view of the spectacular results in the opioid receptor field such searches seem rewarding. Triggle (1988) quoted Matthew VII, 7: "Seek and ye shall find," with respect to L-type calcium channels. The beginning of the wild-goose chase seems easy as the radioligand binding assay is a convenient and cheap test system. In the final stage one may end up with e.g. lysolecithin or an enzyme. So far, extracts from rat or bovine brain (Ebersole and Molinoff 1988), rat heart or brain (Hanbauer et al. 1988), bovine brain or lamb stomach (Janis et al. 1988) have yielded low-molecular-weight (1000–10000) active principles. Material extracted from rat brain appears to be a peptide (molecular mass 948) with the composition 55% Asp, 25% Glu, 5% Gly, 5% Thr and two unidentified peaks. 1 to 2.5 nM of the peptide inhibited [³H]-nitrendipine binding by 50% in hippocampal membranes, enhanced (with slow onset) Ca^{2+} currents in guinea pig ventricular myocytes but inhibited both T- and L-channel activity in different neuronal cells (Callewaert et al. 1989).

On the other hand, novel mediators are often claimed to act directly on calcium channels although the criteria (listed above for the toxins) have not been met. We present two (disappointing) examples below.

2.4.2 Antralin

Antralin, a protein of 16 kDa, was purified to homogeneity from acidified rat antral stomach extracts > 2000-fold. The protein inhibits 1,4 DHP binding (e.g. in heart membranes) by reducing the affinity of the radioligand but not by decreasing the number of binding sites. These effects are dependent on calcium. However, antralin also inhibits the binding of ligands to peripheral benzodiazepine receptors and was finally shown to be a phospholipase, perhaps a PLA_2 isoenzyme (Mantione et al. 1988).

2.4.3 Endothelin

Endothelin is a 21-amino acid peptide (with two disulphide bridges) isolated from the supernatant of confluent monolayer cultures of porcine aortic endothelial cells (Yanagisawa et al. 1988). It is one of the most potent vasoconstrictors known. The EC_{50} value for the vasoconstrictor effect on porcine coronary artery strips is 0.4 nM. Endothelin-induced contractions are attenuated by low concentrations of the 1,4 DHP nicardipine and completely inhibited when extracellular calcium is chelated by EGTA. In cultured rat vascular smooth muscle cells endothelin elicits an increase in both the transient phase and the sustained phase of free intracellular calcium, whereas only the latter was blocked by L-type channel blockers (Hirata et al. 1988). Endothelin contains histidine and tyrosine, amino acids which could easily by radioiodinated. In human placental membranes, mono-[^{125}I]-iodo endothelin binds with picomolar dissociation constants (10 – 20 pM) to a set of sites which have a density of 100 fmol/mg of protein. Binding is dependent on divalent cations and not inhibited by GTP. L-type calcium channel drugs (e.g. 1,4 DHPs, verapamil, nifedipine) had no effect on the binding (W. Fischli, personal communication; Hirata et al. 1988). In cultured rat vascular smooth muscle cells ^{125}I-labelled synthetic porcine endothelin bound to 13000 sites per cell (Hirata et al. 1988) with a dissociation constant of 0.4 nM. The binding was essentially irreversible; internalisation has not been excluded.

The production of endothelin appears to be regulated at the level of mRNA transcription and is induced by thrombin, Ca^{2+}-ionophores and adrenaline. Human endothelin is identical in structure with porcine endothelin. The nucleotide sequence of a clone isolated from a human placenta cDNA library indicates a high homology of the human preproendothelin with the porcine precursor (Itoh et al. 1988). It has been hypothesised that endothelin may be an endogenous agonist of the L-type calcium channel (Yanagisawa et al. 1988) but there is no evidence available which supports this speculation. As mentioned above, alpha$_{A1}$-adrenoceptor-mediated smooth muscle contractions are also blocked by L-type channel blockers, and the involvement of mechanisms other than direct binding to the L-type channel is more probable for the novel vasoconstrictor (Silberberg et al. 1989). This notion is supported by recent reports indicating that endothelin probably releases calcium from intracellular stores (Miasiro et al. 1988; Auguet et al. 1988). Sarafotoxins, 21-residue cardiotoxic peptides isolated from the venom of the snake *Atrac-*

taspis engaddensis, show a high degree of amino acid sequence homology with endothelin (Kloog et al. 1988). The peptides induce phosphoinositide breakdown in rat atrial tissue slices which is not blocked by Na^+, K^+ or Ca^+ channel blockers. As it is very likely that endothelin and the sarafotoxins share the same receptor (Gu et al. 1989), the role of endothelin as the putative endogenous ligand for L-type channels is put even more in doubt.

3 Probing the Calcium Channel with Target Size Analysis

The radiation-inactivation technique can be employed to determine the molecular size of enzymes, receptors, transporters or ion channels in crude preparations — probed either by functional tests or by ligand binding. The technique and its limitations are not further discussed here. Suffice it to mention that the decay of activity in a given sample as a function of the dose of high-energy irradiation is measured and compared with the decay of standards (e.g. enzymatic activities). Alternatively, empirical formulas are applied to derive the molecular size of the target in question. Divergent results have been obtained depending on sample preparation, e.g. for the apamin receptor site in brain membranes (Seagar et al. 1986). In addition, radiation-inactivation data are difficult to interpret, especially if regulatory components are localised on targets which are distinct from the measured activity. For instance, (+)-*cis*-diltiazem stimulates 1,4 DHP binding at temperatures > 25 °C to brain, heart or (depending on the structure of the 1,4 DHP) skeletal muscle L-type channel-linked receptors. If this heterotropic allosteric regulation requires another "channel component," the measured 1,4 DHP binding in the presence of (+)-*cis*-diltiazem is not necessarily indicative of the molecular properties of either component (see e.g. Goll et al. 1983a,b).

Leaving these complications aside, target sizes as low as 90 kDa (Gredal et al. 1987) and as high as 278 kDa (Venter et al. 1983) have been reported for the 1,4 DHP receptors linked to L-type calcium channels in different tissues (Table 4). It is still unexplained why (+)-*cis*-diltiazem preincubation decreases the size of the radiation-sensitive target (measured by 1,4 DHP binding) in three different tissues by approximately 70 kDa. The effect is stereoselective, as (−)-*cis*-diltiazem is ineffective. Furthermore, it is also unclear why both phenylalkylamine radioligands [(±)-verapamil and (−)-desmethoxyverapamil] yielded significantly smaller radiation-sensitive targets in skeletal muscle than obtained for 1,4 DHP binding in the absence of (+)-*cis*-diltiazem. Perhaps this relates to other components or subunits found in purified channel preparations. These subunits do not carry the binding domains but may be necessary for high-affinity interaction of the alpha$_1$-subunit with the respective radioligands (see Sect. 5.2).

Table 4. Radiation inactivation data on calcium channels

Tissue (species)	Ligand/method	Results (kDa)	Comments	Reference
Skeletal muscle (rabbit)	$(\pm)$-[^{3}H]nitrendipine	210		Norman et al. (1983)
Skeletal muscle (guinea pig)	$(\pm)$-[^{3}H]nimodipine	178 with $(+)$-*cis*-diltiazem: 115	The effect of $(+)$-*cis*-diltiazem is stereoselective	Ferry et al. (1983a)
Skeletal muscle (guinea pig)	$(\pm)$-[^{3}H]PN 200-110	136 with $(+)$-*cis*-diltiazem: 75	The effect of $(+)$-*cis*-diltiazem is stereoselective	Goll et al. (1983a)
Brain (guinea pig)	$(\pm)$-[^{3}H]nimodipine	185 with $(+)$-*cis*-diltiazem: 111	The effect of $(+)$-*cis*-diltiazem is stereoselective	Ferry et al. (1983b)
Brain (rat)	$(\pm)$-[^{3}H]nitrendipine	94		Gredal et al. (1987)
Smooth muscle (guinea pig)	$(\pm)$-[^{3}H]nitrendipine	278		Venter et al. (1983)
Heart (guinea pig)	$(\pm)$-[^{3}H]nimodipine	184 with $(+)$-*cis*-diltiazem: 106	The effect of $(+)$-*cis*-diltiazem is stereoselective	Glossmann et al. (1985)
Heart (rat)	$(+)$-[^{3}H]PN 200-110	185		Doble et al. (1985)
Skeletal muscle (guinea pig)	$(\pm)$-[^{3}H]verapamil	110	The target size of this receptor site is significantly smaller than that of the 1,4-dihydropyridine site	Goll et al. (1984a)
Skeletal muscle (guinea pig)	$(-)$-[^{3}H]des-methoxyverapamil	107	This high-affinity, optically pure enantiomer gives the same mol. wt. as $(\pm)$-[^{3}H]verapamil	Goll et al. (1984b)
Skeletal muscle (guinea pig)	$(+)$-[^{3}H]*cis*-diltiazem	131		Goll et al. (1984b)
Brain (rat)	^{45}Ca^{2+}-uptake	340	^{45}Ca^{2+}-uptake was measured with K$^+$-depolarisation	Gredal et al. (1987)

$^{45}Ca^{2+}$ uptake due to K^+ depolarisation can be measured in rat cortex synaptosomes. This activity decayed upon irradiation with a target size of 340 kDa (Gredal et al. 1987). No data on the target size of omega-conotoxin receptors are available at the present time.

According to the radiation-inactivation data, the L-type channel-linked receptors were considerably larger than reported initially with photoaffinity labelling employing nonarylazides or with affinity labelling (see Glossmann et al. 1987b). The $^{45}Ca^{2+}$-uptake experiments suggest that a different type of voltage-dependent calcium channel (N_A plus N_B type?) significantly larger than the L-type channel exists in neuronal tissue. This is indeed supported by photoaffinity labelling studies, as will be shown below.

4 Identification of Calcium Channel-Associated Drug Receptors in Membranes of Excitable Tissues by Photoaffinity Labelling

In addition to target size analysis, photoaffinity labelling with arylazides has provided good (and in hindsight correct) estimates on the molecular size of the channel component(s) that carry the calcium antagonist receptors. Ferry et al. (1984b, 1985) have photolabelled the calcium channel-associated 1,4 DHP receptor in skeletal muscle T-tubules from different species with the arylazide [^{3}H]azidopine. The label specifically incorporated into a 155-kDa polypeptide which did not change its apparent molecular weight upon reduction in SDS-PAGE, although the recovery of incorporated counts was lower than under alkylating conditions. Photolabelling was protected by several organic and inorganic calcium antagonists known to interact with L-channels (Ferry et al. 1985). This polypeptide was subsequently stereoselectively photolabelled with the optically pure tritiated enantiomers of azidopine, (−)-[^{3}H]azidopine and (+)-[^{3}H]azidopine (Striessnig et al. 1986b) in purified skeletal muscle calcium channel preparations and is now termed the "alpha$_1$"-subunit (see below). The alpha$_1$-polypeptide carries the drug receptors in skeletal muscle. Polypeptides from other tissues which have these drug-receptor domains are referred to in the text as "alpha$_1$-like." In the future the subscripts alpha$_{1,n}$ (for neuronal), alpha$_{1,sk}$ (skeletal muscle) and alpha$_{1,m}$ (heart muscle) may be more appropriate. As mentioned above, the recovery but not the mobility of the major [^{3}H]azidopine photolabelled band in T-tubule membranes of different species was changed upon reduction (Ferry et al. 1985). A major problem with azidopine is that a considerable fraction of the incorporated 1,4 DHP is sensitive to attack by nucleophilics, including reducing agents, which leads to a significant loss of label (Striessnig et al. 1988b; Vaghy et al. 1987). This feature is now regarded as highly characteristic for the L-type calcium channel 1,4 DHP receptors (Striessnig et al. 1988b), is not shared by some other channel arylazide probes (e.g. azidodiltiazem or [N-methyl-^{3}H]LU49888) and points to a conserved amino acid sequence in the alpha$_1$-subunit, regardless of tissue or species. The low recovery of photolabel after reduction of disulphide bonds before SDS-PAGE, however, certainly contributed to the initial confusion about which polypeptide carried the drug receptor.

In cardiac and smooth-muscle 1,4 DHP, binding sites were initially identified with the tritium-labelled isothiocyanate affinity probe 2,6-dimethyl-3,5-dicarboxymethoxy-4-(-isothiocyanatephenyl)1,4-DHP ([^{3}H]o-NCS) (Venter et al. 1983; Kirley and Schwartz 1984; Horne et al. 1984) or high-intensity ultraviolet irradiation and [^{3}H]nitrendipine (Campbell et al. 1984) or [^{125}I]BAY P 8857 (Sarmiento et al. 1986). Polypeptides of 32 to 45 kDa were claimed to be the channel-linked 1,4 DHP receptors in these tissues (see Glossmann et al. 1987b for further discussion). However, photoaffinity labelling with (−)-[^{3}H]azidopine unequivocally proved that the calcium channel-associated 1,4 DHP receptor in cardiac muscle of different species is, as in skeletal muscle, a large polypeptide (165−185 kDa; Ferry et al. 1987; Kuo et al. 1987). The smaller 1,4 DHP binding polypeptides are unrelated to calcium channels, and the majority of them are probably located on the inner mitochondrial membrane (Zernig and Glossmann 1988; Zernig et al. 1988).

[N-methyl-^{3}H]LU49888 ((−)-5-[(3-azidophenethyl) [N-methyl-^{3}H]methyl-amino]-2-(3,4,5-trimethoxyphenyl)-2-isopropylvaleronitrile is an arylazide photoaffinity ligand from the phenylalkylamine series. It is structurally closely related to verapamil and has reversible binding characteristics nearly indistinguishable from those reported for (−)-[^{3}H]desmethoxyverapamil (Striessnig et al. 1987). This phenylalkylamine photoaffinity probe specifically incorporates into polypeptides of the same molecular weight (by SDS-PAGE) as azidopine did. In membranes from skeletal muscle the photolabelled alpha$_1$-subunit is somewhat smaller than in cardiac tissue (Striessnig et al. 1987; Schneider and Hofmann 1988). In neuronal tissues alpha$_1$-like polypeptides were not demonstrated until recently. In hippocampus membranes [N-methyl-^{3}H]LU49888 irreversibly labels two large polypeptides with molecular weights of 265 and 195 kDa respectively. (−)-[^{3}H]Azidopine incorporates only into the 195-kDa polypeptide. Apparently, only this alpha$_1$-like polypeptide is a constituent of the L-type calcium channel. SDS-PAGE reveals (as for heart) that it is larger than the skeletal muscle alpha$_1$ (Striessnig et al. 1988b; Glossmann and Striessnig 1988a,b; Glossmann et al. 1988b). Whether or not the 265-kDa polypeptide is associated with another ion channel, e.g. a T- or an N-type calcium channel, is at present unclear. Evidence is presented below that a photoaffinity probe for N (i.e. N$_B$)-channels recognizes a polypeptide much larger than 195 kDa in membranes from cerebral cortex or cultured neurons.

Particulate skeletal muscle calcium channels have been photolabelled not only with arylazides but also with nonarylazide compounds such as (+)-[^{3}H]PN200-110, [^{3}H]bepridil and (+)-cis-[^{3}H]diltiazem (Galizzi et al. 1986a). (+)-[^{3}H]PN200-110 photoincorporates into alpha$_1$ in solubilised or purified preparations (Leung et al. 1987; Johnson et al. 1988). So far, no data are available about the usefulness of these radioligands for photoaffinity labelling of calcium channels in other tissues endowed with much fewer receptor sites.

A high-affinity receptor for phenylalkylamines was discovered in reversible binding experiments with *Drosophila melanogaster* head membranes. These stereoselective sites were localised by photoaffinity labelling with [*N*-methyl-^{3}H]LU49888 on a 135-kDa polypeptide (Pauron et al. 1987; Greenberg et al. 1989). The novel phenylalkylamine receptor is linked to a 13-pico Siemens (pS) calcium channel (Pelzer et al. 1989b), which is unusual: It lacks allosterically coupled high-affinity 1,4 DHP, benzothiazepine or benzothiazinone receptors as revealed by radioligand binding and functional studies (Pauron et al. 1987; Greenberg et al. 1989; Pelzer et al. 1989b). The channel, when reconstituted from head membranes into lipid bilayers, is exquisitely sensitive to phenylalkylamines which block channel activity at nanomolar concentrations, whereas there is no effect of 1,4 DHPs on the 13-pS channel. Recently, it has been found that sodium channel ligands (e.g. DPI 201-106) bind with very high affinity to the novel receptor and block photolabelling of the 135-kDa polypeptide (Zech et al. 1989). The existence of an additional low-affinity binding site on a 30-kDa polypeptide in the head membranes has been reported, but its functional significance is unknown (Greenberg et al. 1989).

As *Drosophila melanogaster* is already the focus of potassium or sodium channel research at the molecular-biology level and single gene mutations can be easily studied, the prospects for cloning and mutating the phenylalkylamine-sensitive channel are bright. The channel protein is significantly smaller than the alpha$_1$-subunit in skeletal muscle and the alpha$_1$-like subunits in mammalian neuronal or cardiac membranes. In addition, the high affinity for piperazinyl-indole compounds (as is DPI 201-106) which are active on sodium channels is interesting and points to the evolutionary conservation of drug binding domains from arthropod channels to sodium or calcium channels (Zech et al. 1989) in vertebrates.

5 Calcium Channel Structure (L-Type Channels)

5.1 Purification of Calcium Channels

5.1.1 General Remarks

The straightforward approach to the isolation of this ion channel was to purify the channel-associated drug receptors for calcium antagonists by monitoring binding activity of tritiated 1,4 DHPs. As for many other channel-linked
receptors (e.g. the nicotinic-acetylcholine receptor or the voltage-dependent
sodium channels) this effort was successful, although (in retrospect) an intact
drug receptor carrying polypeptide was not obtained in all instances. High-affinity binding sites for [^{3}H]nitrendipine and [^{3}H]nimodipine were first solubilised from guinea-pig skeletal muscle T-tubule and rat brain membranes
(Glossmann and Ferry 1983b; Curtis and Catterall 1983) and later from rat
and bovine heart (Ruth et al. 1986; Horne et al. 1986) by either the zwit-
terionic detergent CHAPS or the nonionic detergent digitonin. The solubilised receptor activity was partially purified by adsorption and biospecific elution from different lectin-affinity columns (revealing its glycoprotein nature)
and by sucrose-density gradient centrifugation. The 1,4 DHP binding activity
was determined by reversible labelling after solubilisation or by following the
receptor-ligand complex formed prior to solubilisation. The solubilised 1,4
DHP binding was allosterically regulated by the other drug-receptor domains,
i.e. for phenylalkylamines and benzothiazepines. This provided strong evidence that calcium channel-associated receptors were indeed isolated. Inclusion of (+)-cis-diltiazem, which dramatically increases the half-life of the
solubilised and membrane-bound 1,4 DHP-calcium channel complex via a
positive heterotropic allosteric mechanism, was sometimes used to stabilise
the preformed ligand-receptor complexes. In contrast to the voltage-dependent sodium channel, the digitonin-solubilised 1,4 DHP binding activity from
skeletal muscle is quite stable at low temperatures ($T_{1/2} > 10$ h), and exogenous phospholipids are not required for stabilisation. This greatly facilitated
the purification procedures. Isolation and complete biochemical characterisation of the subunits has so far been achieved only for the receptors from rabbit and guinea-pig skeletal muscle T-tubule membranes. These membranes are
easily prepared and are the richest source of Ca^{2+}-antagonist receptor sites
known (Ferry and Glossmann 1982; Glossmann et al. 1983a; Fosset et al.
1983). The density of Ca^{2+}-antagonist receptor sites in mammalian brain,
cardiac or smooth muscle membranes is 20 to 500 times lower than in skeletal
muscle (Glossmann and Striessnig 1988a; Janis et al. 1987), which makes purification extremely difficult. In addition, the stability against protease attack
is apparently a more serious problem in the nonskeletal muscle tissues. Great
efforts have been made to isolate the calcium channel from heart (Schneider
and Hofmann 1988; Cooper et al. 1987), but these preparations cannot be re-

Table 5. Some protocols for the purification of receptors associated with the skeletal muscle calcium channel

Author	Species	Methods	Radioligands	SA	PUR	SDS-PAGE (reducing conditions, kDa)	Comments
Curtis and Catterall (1984, 1985, 1986); Takahashi et al. (1987)	Rabbit	Digitonin, WGA-DEAE-WGA-SUC	NTD, PN	1950	330	175 (alpha) 143 (alpha$_2$) 54 (beta) 30 (gamma) 24–27 (delta)	Functional CA-sensitive calcium channels observed after reconstitution of purified protein into lipid vesicles. Alpha$_1$ photolabelled with [³H]azidopine and [¹²⁵I]TID. Alpha$_1$ and beta phosphorylated with cAMP-PK
Borsotto et al. (1985) Barhanin et al. (1987) Lazdunski et al. (1986b) Schmid et al. (1986a, b) Vandaele et al. (1987)	Rabbit	CHAPS, DEAE-WGA-SEC	PN	800	80	142, 33, 32	High-affinity binding after reconstitution of purified protein into lipid vesicles (Barhanin et al. 1987). No alpha$_1$-subunit described in CHAPS but in digitonin solubilised preparations. Antibodies against the 142-kDa band (alpha$_2$) identify a similar polypeptide in brain, heart and smooth muscle

References	Species	Method	Ligands		Subunit MW	Comments
Striessnig et al. (1986, 1987) Vaghy et al. (1987) Hymel et al. (1988c) Glossmann and Striessnig (1988 a, b) Glossmann et al. (1989)	Guinea pig	Digitonin, WGA-SUC	PN, – AZ + AZ DMV LU	1500 135	155 – 170 (alpha₁) 135 – 150 (alpha₂) 50 – 65 (beta) 30 – 35 (gamma)	Alpha₁ stereoselectivity photolabelled with – AZ and + AZ and LU. Differentiation of DHP agonists and antagonists in binding studies. Reconstitution of purified protein into lipid bilayers yields functional CA- and cAMP-PK-sensitive calcium channels. Alpha₁ phosphorylated with cAMP-PK and PKC
Flockerzi et al. (1986a, b) Sieber et al. (1987) Nastainczyk et al. (1987)	Rabbit	Digitonin, WGA-DEAE-(HPLC)-WGA-SUC	PN	367 122	142 – 165 (alpha₁) 122 – 130 (alpha₂) 56 (beta) 28 (gamma)	Functional CA- and cAMP-PK phosphorylation-sensitive calcium channels after reconstitution into lipid bilayers. Alpha₁ and beta identified as the physiological substrates for the cAMP-PK and PKC respectively. Alpha₁ photolabelled with – AZ and LU
Nakayama et al. (1987) Vaghy et al. (1987)	Rabbit	Digitonin, WGA-DEAE-WGA-SUC	– AZ, LU	672 30	155 – 170 (alpha₁) 135 – 150 (alpha₂) 50 – 65 (beta) 30 – 35 (gamma)	Alpha₂ N-terminal amino acid sequence determined. Proteolytic fragments of alpha₁ characterised and photolabelled with – AZ and LU. Proteolytic preparation forms functional calcium channels in lipid bilayers

Table 5 (continued)

Author	Species	Methods	Radioligands	SA	PUR	SDS-PAGE (reducing conditions, kDa)	Comments
Leung et al. (1987) Imagawa et al. (1987 a, b) Sharp et al. (1987)	Rabbit	Digitonin, WGA-DEAE	PN	n.d.	n.d.	170 (alpha$_1$) 150 (alpha$_2$) 52 (beta) 32 (gamma)	Alpha$_1$ and beta phosphorylated by cAMP-PK, Ca^{2+}/CAM-PK. Alpha$_1$ photolabelled with PN. Antibodies against all subunits reveal their co-localisation in one complex (stoichiometry 1:1:1:1). Ultrastructure shown by electron microscopy. Effects of antibodies on channel function
Morton and Froehner (1987) Morton et al. (1988)	Rabbit	Digitonin, WGA-SUC, IA	PN	n.d.	n.d.	200 (alpha$_1$) 143 (alpha$_2$) 61 (beta) 33 (gamma)	Antibody against alpha$_1$ inhibits calcium and sodium currents in BC3H1 myocytes
Hosey et al. (1987, 1988)	Rabbit	CHAPS, WGA	LU	n.d.	n.d.	165 (alpha$_1$) 140 (alpha$_2$) 26–32 ("delta")	Alpha$_1$ photolabelled with LU and phosphorylated with cAMP-PK and Ca^{2+}-/calmodulin-dependent kinase

$-AZ$, $+AZ$, $(-)$-, $(+)$-[^{3}H]azidopine; *CA*, calcium antagonist; *CAM*, calmodulin; *cAMP-PK*, cAMP-dependent protein kinase; *DEAE*, DEAE ion exchange chromatography; *DMV*, $(-)$-[^{3}H]desmethoxyverapamil; *IA*, immunoaffinity chromatography; *LU*, [*N*-methyl-^{3}H]LU49888; *PKC*, protein kinase C; *PN*, $(+)$-[^{3}H]PN 200-110; *PUR*, purification factor; *SA*, maximal specific binding activity (pmol/mg of protein); *SEC*, size exclusion chromatography; *SUC*, sucrose-density gradient centrifugation; *TID*, see text; *WGA*, wheat-germ agglutinin affinity chromatography

garded as homogeneous or yielding sufficient material − if one regards the purified skeletal muscle as a gold standard.

As shown in Table 5, many purification protocols rely more or less on the methods of the initial investigators. Lectin-affinity chromatography, then sucrose-density gradient centrifugation or ion-exchange chromatography were highly effective steps in most cases to get a preparation as pure as possible − the binding activity and polypeptide pattern after SDS-PAGE being the criteria. As for other receptors and ion channels (e.g. the adrenergic or the nicotinic-acetylcholine receptor), attempts have also been made to purify the calcium channel by affinity chromatography on matrices derived with receptor-specific ligands, i.e. 1,4 DHP structures. Specific adsorption to 1,4 DHP affinity columns was demonstrated but elution of binding activity either was impossible (Biswas and Rogers 1986) or did not result in considerable purification (Soldatov 1988).

Purified skeletal muscle calcium channel preparations have measured (or "calculated") specific binding activities for tritiated 1,4 DHPs (e.g. (+)-[^{3}H]PN200-110, (±)-[^{3}H]nitrendipine, (−)-[^{3}H]azidopine) of between 1500 and 2000 pmol/mg of protein. Assuming one single 1,4 DHP binding site per 200-kDa receptor polypeptide (taking the molecular weight of the alpha$_1$-subunit as 195000; determined by Ferguson analysis on SDS-PAGE, Glossmann et al. 1988b; see Sect. 5.1.3), a theoretical specific activity of around 5000 pmol/mg of the pure protein is expected. However, purified calcium-antagonist receptor preparations do not consist of one single (receptor-carrying) polypeptide. As will be shown below, several additional, non-covalently associated polypeptides constitute approximately 60% of the protein mass of the purified channel. Therefore, the specific activity reported for the purified preparations (see Table 5) is in good agreement with the maximal possible value of 2000 pmol/mg of protein.

5.1.2 Hydrodynamic Properties of Solubilised Calcium Channels

The hydrodynamic properties of solubilised calcium channels from cardiac tissue were reported by Horne et al. (1986). The Stokes radius (8.6−8.7 nm) for the calcium channel-detergent complex obtained by gel filtration on Sephadex 6B-Cl was independent of the detergent used. The $S_{20,w}$ determined by sucrose-density gradient centrifugation varied with the detergent. Values of 12.5, 15.4 and 21.0 S were measured in Tween 80, CHAPS and digitonin respectively. From the Stokes radius, the $S_{20,w}$ value and the partial specific volume (for details see Horne et al. 1986), a molecular weight of 595000 for the calcium channel-detergent complex was calculated. After correction for the fractional contribution of the detergent, the molecular weight of the channel protein(s) was estimated to be 370000. This value is somewhat lower than the total sum of the subunit molecular weights from skeletal muscle (see below).

$S_{20,w}$ values of $20-21\,S$ have also been found for digitonin-solubilised skeletal muscle and brain calcium channels (Curtis and Catterall 1984; Striessnig et al. 1986b; Flockerzi et al. 1986b; Morton and Froehner 1987; Takahashi and Catterall 1987a). Receptor-binding activity of the two tissues co-sediments with cardiac 1,4 DHP binding upon sucrose-density gradient centrifugation (Takahashi and Catterall 1987b). Smaller values $(12.9-14.4\,S)$ were obtained in CHAPS (Borsotto et al. 1984b; Glossmann and Ferry 1985) for the skeletal muscle system. Glossmann et al. (1987a) have reported a Stokes radius of 7.6 nm for the purified calcium channel/digitonin complex, which is close to the value in cardiac tissue. The similar hydrodynamic properties indicated a similar size and (perhaps) subunit composition of 1,4 DHP-sensitive calcium channels in brain, heart and skeletal muscle.

5.1.3 Subunit Composition of the Calcium Channel in Skeletal Muscle

Initial purification data suggested that the skeletal muscle calcium channel consists of one large polypeptide with disulphide-linked smaller subunits (Borsotto et al. 1985). Other researchers, however, consistently purified smaller polypeptides in addition to a large one. These polypeptides were not disulphide linked (Curtis and Catterall 1984; Striessnig et al. 1986b,1987; Flockerzi et al. 1986a,b). The large polypeptide was termed "alpha" polypeptide (or "subunit") by Curtis and Catterall (1984) in analogy to the voltage-dependent sodium channel. This alpha polypeptide displayed a particular and highly characteristic electrophoretic behaviour upon SDS-PAGE (Curtis and Catterall 1984). It appeared as a single band under nonreducing conditions in SDS-PAGE (~ 170 kDa). Upon reduction, the alpha region broadened; some alpha retained the same mobility as under nonreducing conditions but the majority shifted its molecular weight by $30-35$ kDa to about 135 kDa. This apparent heterogeneity in the alpha region was first explained by the incomplete reduction of intrachain disulphide bonds (Curtis and Catterall 1984) or the presence of contaminating proteins (Flockerzi et al. 1986b). The controversy was resolved by photoaffinity labelling with [N-methyl-^{3}H]LU49888 and/or $(-)$-[^{3}H]azidopine (Striessnig et al. 1987; Takahashi et al. 1987; Sieber et al. 1987b; Vaghy et al. 1987, 1988; Leung et al. 1987), demonstrating that two distinct polypeptides exist in the alpha region (now termed "alpha$_1$" and "alpha$_2$"; see Fig. 8). These polypeptides are difficult to resolve in the usual gradient gels. The newly discovered polypeptide ("alpha$_1$") did not change its apparent molecular weight upon reduction (reported ranges varied from 155 to 200 kDa). The main reason why alpha$_1$ was overlooked by many investigators was its sensitivity towards proteolytic degradation (see Sect. 5.1.8). The second large polypeptide ("alpha$_2$") displayed a higher elec-

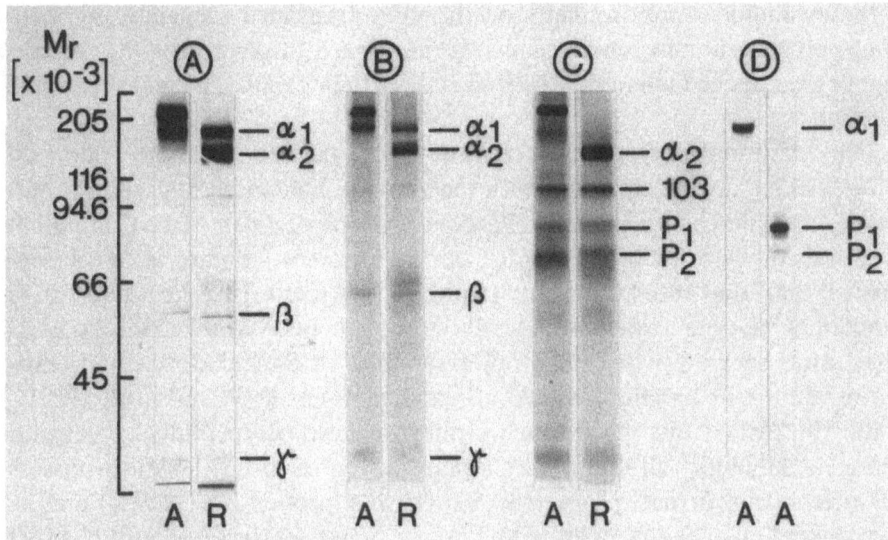

Fig. 8. Polypeptides co-migrating with calcium antagonist binding activity in purified skeletal muscle calcium channel preparations. 1,4 DHP binding activity was purified by wheat-germ lectin sepharose-affinity chromatography followed by sucrose-density gradient centrifugation from guinea pig (*panel A*) and rabbit (*panels B, C*) skeletal muscle. Separation of proteins on 8% SDS-PAGE was carried out under alkylating (*A*) or reducing conditions (*R*) and followed by silver staining (*panels A–C*). The individual "subunits" are termed alpha$_1$, alpha$_2$, beta, and gamma with decreasing apparent molecular weight. The delta subunit is not stained. The migration of standard proteins is indicated on the left hand side. The polypeptides in panels *A* and *B* were obtained from fresh tissue, whereas the polypeptide pattern in panel *C* resulted when previously frozen skeletal muscle was used as the starting material for purification. Note that in such preparations an intact alpha$_1$-subunit is almost completely absent. Instead, smaller polypeptides (*P$_1$ and P$_2$*) are visible. *Panel D*: (−)-[³H]azidopine photoaffinity labelling of the membranes used as the starting material (*left lane*: prepared from fresh muscle; *right lane*: prepared from previously frozen muscle) reveals that breakdown has already occurred during membrane preparation before solubilisation. [From Vaghy et al. (1987) with permission]

trophoretic mobility after reduction of disulphide bonds, indicating a decrease of M_r from 170000 to 135000. The loss of mass is caused by the release of several smaller disulphide-linked polypeptides (24–30kDa, termed "delta"). These "delta" subunits are only weakly stained with silver and not at all with Coomassie blue (Campbell et al. 1988a) but are easily detected with antibodies (Barhanin et al. 1987; Schmid et al. 1986a,b; Takahashi et al. 1987; Glossmann et al. 1988b). We prefer the term "alpha$_2$" for the higher-molecular-weight moiety of the reduced protein, "alpha$_2$-delta" for the disulphide-linked complex, and "delta" for the small, disulphide-linked polypeptides. Barhanin et al. (1987) presented evidence (based on limited proteolysis) that the three "delta" subunits are structurally related to each other. Lack of staining was found for the rabbit as well as the guinea pig peptides. It is not known if these subunits were originally on the same translated sequence as the alpha$_2$-polypeptide and remained disulphide linked after proteolytic processing, or are attached later and/or differ only in their glycosylation or fatty acid content.

Two additional polypeptides, termed "beta" (52−65 kDa) and "gamma" (30−33 kDa), always co-purify with the above polypeptides in digitonin buffers (Sharp and Campbell 1989). Reconstitution experiments (see Catterall et al. 1989; Vaghy et al. 1988) carried out in different laboratories have confirmed that the entire complex (alpha$_1$/alpha$_2$-delta/beta/gamma) forms functional "L-type" calcium channels, although one report claims that alpha$_1$, after separation from the other subunits by SDS-size-exclusion chromatography, can function as a 20-pS calcium channel (Pelzer et al. 1989).

For the rest of this discussion we refer to these polypeptides as calcium channel "subunits," although only alpha$_1$ has so far been shown to possess characteristic structural properties of a channel protein (see Sect. 8) and is, as it carries the drug-receptor domains, an integral L-type calcium channel component. It should be noted that none of the other subunits is labelled by azidopine. Azidopine's photoreactive group oscillates at a distance of 8.43−14.54 Å from the center of the 1,4 DHP ring − believed to be fixed at the heart of the binding domain of the alpha$_1$-subunit (Glossmann et al. 1987b).

5.1.4 Subunit Properties of the Isolated Skeletal Muscle Calcium Channel

Table 6 summarises the subunit properties. Limited proteolysis (Sieber et al. 1987b; Leung et al. 1987; Hosey et al. 1987) reveals that alpha$_1$ and alpha$_2$ are two distinct, structurally unrelated polypeptides. Other authors initially believed that the alpha$_1$-alpha$_2$ doublet is the result of incomplete proteolytic cleavage of one single polypeptide (Vandaele et al. 1987). Alpha$_1$ and alpha$_2$ are encoded by different genes, and their complete amino acid sequence in rabbit skeletal muscle has been deduced from the cloned cDNA (Tanabe et al. 1987; Ellis et al. 1988). See Sect. 8 for further details on alpha$_1$ and alpha$_2$ structures.

5.1.5 Regulatory Domains

In general, the characteristic pharmacological profile of the reversible binding of various calcium channel ligands is maintained throughout purification, although the allosteric coupling mechanisms are changed, as is the affinity, which decreases for all ligands. Fortunately, despite this affinity decrease, photoaffinity labelling with the arylazides [^{3}H]azidopine, [N-methyl-^{3}H]LU49888 and the novel (+)-cis-[^{3}H]azidodiltiazem (Glossmann et al. 1989) or even with (+)-[^{3}H]PN 200-110 is successful (Leung et al. 1987;

Table 6. Properties of calcium channel-associated polypeptides in skeletal muscle

Polypeptide	SDS-PAGE ($M_r \times 10^{-3}$)	Glycosylation	Phosphorylation	Drug[a] receptors	Hydrophobicity	Primary structure determined by	Occurrence in other tissues[b]
Alpha$_1$	155–200	(+)	cAMP-PK PKC Ca/CAM INT PK	DHP, PA, BT	+++	Tanabe et al. (1987) Ellis et al. (1988)	Heart, brain
Alpha$_2$[c]	122–145	+++	–	–	+	Ellis et al. (1988)	Heart, brain, smooth muscle
Beta	50–65	–	cAMP-PK PKC Ca/CAM INT PK	–	–	Soldatov (1988)[d] (amino acid analysis)	Heart?
Gamma	30–35	+++	–	–	+++	n.d.	?
Delta	20–30[e]	+++	–	–	+	n.d.	Heart, brain, smooth muscle

[a] By photoaffinity labelling
[b] Evidence is available that the subunit or closely related polypeptides or the respective mRNA exists in these tissues
[c] Linked to delta subunits
[d] Amino acid composition of beta indicates abundance of hydrophilic amino acids
[e] Delta-Subunits are heterogeneous with respect to size on SDS-PAGE in purified calcium channel preparations from rabbit and guinea pig
INT PK, Intrinsic protein kinase; DHP, 1,4-dihydropyridines; PA, phenylalkylamines; BT, benzothiazepines; cAMP-PK, cyclic AMP-dependent protein kinase; Ca/CAM, Ca^{2+}-calmodulin-dependent protein kinase; PKC, protein kinase C

Johnson et al. 1988). In addition, covalent modification (e.g. with $(-)$-$[^3H]$ azidopine) can be employed to monitor the structural integrity of alpha$_1$ from the first to the last step of a purification scheme (Striessnig et al. 1986b). Photoaffinity labelling also proves the co-existence of the three allosterically coupled drug receptors on one single polypeptide. Although demonstrated only for skeletal muscle, we believe that this will be found for the alpha$_1$-like polypeptides in other vertebrate tissues as well. In contrast, the *Drosophila melanogaster* head membrane high-affinity phenylalkylamine site is not coupled to high-affinity 1,4 DHP or benzothiazepine binding domains. The protection profile of the irreversible labelling of alpha$_1$ in skeletal muscle was identical with the pharmacological profile of reversible binding (Striessnig et al. 1986a, b, 1987). This rules out the possibility that non-calcium channel-associated binding sites (as often present in membranes) were isolated. Whether or not the other subunits (e.g. alpha$_2$-delta, beta or gamma) contribute to the (relatively) high-affinity state of the isolated receptors that allows photolabelling of the alpha$_1$-subunit is not yet clear.

The voltage-dependent sodium channel can be phosphorylated (Catterall 1986) and the L-type calcium channel is also a substrate for the cAMP-dependent protein kinase (A-kinase) (Curtis and Catterall 1985). Furthermore, the functional regulation of L-type channels via the cAMP second-messenger system is well established – at least in heart (Reuter 1983; Brum et al. 1983; Hescheler et al. 1987; Hofmann et al. 1987). Purified skeletal muscle Ca^{2+}-antagonist receptor complexes, when reconstituted into planar lipid bilayers (Flockerzi et al. 1986a,b; Hymel et al. 1988c, 1989) or lipid vesicles (Curtis and Catterall 1986; Catterall et al. 1988; see Fig. 9), are activated by cAMP-dependent phosphorylation. In vitro phosphorylation experiments indicate that the alpha$_1$ subunit is the preferred substrate for the A-kinase. The beta subunit is also phosphorylated, but the initial reaction rate and the extent of phosphorylation are $2-3$ times lower than for alpha$_1$. At physiological concentrations of the catalytic subunit ($0.1-1$ μM; Nastainczyk et al. 1987) 1 mol of phosphate incorporates per mol of alpha$_1$ within 10 min. Sequence analysis of the alpha$_1$ phosphopeptides identified the serine residue at position 687 in the deduced primary sequence as the preferred phosphorylated cAMP kinase substrate (Röhrkasten et al. 1988). This amino acid is localised in a putative cytoplasmic segment between the two hydrophobic repeats II and III and was predicted to be a putative A-kinase substrate (Tanabe et al. 1987; see Sect. 8). Prolonged phosphorylation leads to additional phosphate incorporation into two other amino acids. One was identified as Ser-1617, located near the C-terminal, which was not predicted to be a potential phosphorylation site (Tanabe et al. 1987). Consequently, 2 mol of phosphate are incorporated per mol of alpha$_1$ after prolonged incubation in vitro. Electrophysiological studies on protease-perfused myocytes, on the other hand, suggest that phosphorylation near the C-terminal is responsible for the modulation of

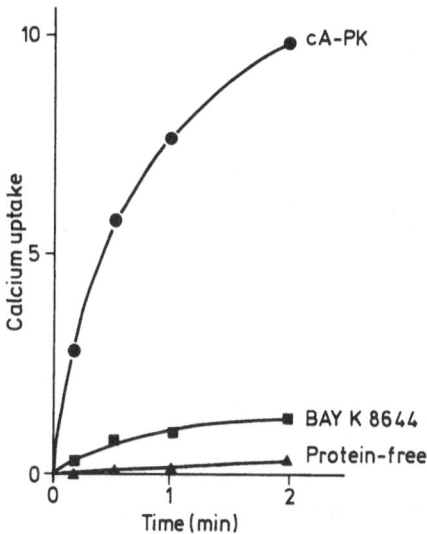

Fig. 9. Activation of purified calcium channels by cAMP-dependent protein kinase. Skeletal muscle calcium channels were solubilised and purified (Curtis and Catterall 1984) in the presence of saturating concentrations of the calcium channel-activating 1,4 DHP BAY K8644. The purified protein was reconstituted into phosphatidylcholine vesicles as described by Curtis and Catterall (1986). The reconstituted vesicles were incubated under control conditions or with 1 mM ATP and 2 µM cAMP-dependent protein kinase (cA-PK) for 1 h at 4 °C and the initial rate of ^{45}Ca influx was measured (Curtis and Catterall 1986). Control experiments with protein-free vesicles were carried out in parallel. Calcium influx into protein-containing vesicles was two to three times greater than influx into protein-free vesicles. This increase is blocked by verapamil, D600 or PN200-110 (Curtis and Catterall 1986). If the channel activator BAY K8644 is removed from the vesicle preparation by molecular sieve chromatography, ^{45}Ca^{2+} influx is markedly reduced. cAMP-dependent phosphorylation increases the initial rate and the final extent of ^{45}Ca^{2+} influx eight- to tenfold. Since the extent of influx at equilibrium is increased, phosphorylation must allow activation of eight to ten times as many purified calcium channels. [From Catterall et al. (1989) with permission]

channel function (Hescheler and Trautwein 1988). The kinetics measured by the in vitro studies in context with reconstitution experiments imply a "physiological" significance only for the phosphorylation of Serine-687. These studies also strongly suggest that alpha$_1$ is isolated mainly in an unphosphorylated state, at least with respect to the two Ser residues in positions 687 and 1617.

Alpha$_1$ and beta are also substrates for a calcium/calmodulin-dependent kinase which is an intrinsic protein kinase in skeletal muscle triads, and protein kinase C (Imagawa et al. 1987a; O'Callahan and Hosey 1988; Nastainczyk et al. 1987; Hofmann et al. 1987). In contrast to the A-kinase the preferred substrate for the protein kinase C in digitonin-purified calcium channels from rabbit skeletal muscle is the beta-subunit (Nastainczyk et al. 1987; O'Callahan and Hosey 1988). However, the membrane-bound alpha$_1$ subunit is rapidly phosphorylated by protein kinase C (O'Callahan and Hosey 1988). Protein kinase C is a known modulator of calcium currents in neurons and smooth muscle cells (for review see Glossmann and Striessnig 1988a; Tsien et al. 1988). Navarro (1987) reported that activation of the enzyme in skeletal muscle with phorbol esters causes an increase in the apparent density of 1,4 DHP receptor sites (determined by [^{3}H]PN200-110 binding) and is associated with an increase of the 1,4 DHP-sensitive ^{45}Ca uptake. Thus, protein kinase C could be involved in the promotion of the appearance of calcium channel drug receptors on the skeletal muscle T-tubule plasmalemma. The role of the beta subunit in this phenomenon is not yet clear.

5.1.6 Hydrophobic Labelling

As predicted from its amino acid sequence, the alpha$_1$-subunit contains hydrophobic (transmembrane) segments characteristic for ion channel-forming polypeptides (Sect. 8). The ratio of hydrophobic to hydrophilic amino acids is smaller for alpha$_2$ and beta than for alpha$_1$, suggesting that major portions of their structure cannot reside in the lipophilic membrane compartment. Catterall and co-workers used the hydrophobicity probe [^{125}I]TID (3-trifluoromethyl)-3-(m-iodophenyl) diazirine; Brunner et al. 1983) to irreversibly label putative transmembrane domains in purified sodium and calcium channels (Reber and Catterall 1987; Takahashi et al. 1987). [^{125}I]TID labelling revealed that alpha$_1$ and gamma are the most hydrophobic components of the purified calcium channel complex, containing significant membrane-spanning regions. As gamma never incorporates the (hydrophobic) drug-receptor photoaffinity probes which easily photolabel albumin (Glossmann and Striessnig 1988a), this provides additional good evidence that specific receptor but not simple hydrophobic interactions govern the photolabelling with these drugs. In agreement with its hydrophilic amino acid composition, the beta-subunit was not labelled by [^{125}I]TID. The alpha$_2$ polypeptide and the delta-subunits displayed only weak incorporation, indicating intermediate or low hydrophobicity.

5.1.7 Glycosylation

The most heavily glycosylated polypeptides in the isolated channel are the alpha$_2$, the gamma and the delta-subunits as revealed by glycoprotein staining with concanavalin A or wheat-germ lectin (Takahashi et al. 1987; Sharp et al. 1987; Glossmann et al. 1988b). The extent of glycosylation was estimated by removal of the oligosaccharide chains with glycosidases (neuraminidase, endoglycosydase F, N-glycanase) and subsequent determination of the molecular weights of the deglycosylated polypeptides by SDS-PAGE. Such treatment reduced the molecular weight of alpha$_2$ to 105000–112000 (20% reduction) and of gamma to 20-kDa (30% reduction) (Takahashi et al. 1987; Barhanin et al. 1987). No evidence exists for glycosylation of the beta-subunit. Alpha₁ is not (Takahashi et al. 1987) or only very slightly glycosylated (Hosey et al. 1987; Glossmann et al. 1988b). The sodium channel alpha-subunit, on the other hand, is heavily glycosylated, and this unexpected lack of glycosylation of alpha₁ certainly contributed to the initial confusion about the composition of the purified skeletal muscle calcium channel.

5.1.8 The Alpha₁-Subunit Is Sensitive to Proteolysis

Lectin-affinity chromatography is an important purification step in the isolation of the channel (see Table 5). Because alpha₁ and beta are only weakly or not at all glycosylated, their enrichment in purified calcium antagonist receptor preparations must occur via tight association and co-purification with the other carbohydrate-carrying subunits. Several investigators detected alpha₂ and delta as the single prominent glycoprotein bands without any additional evidence for the receptor carrying alpha₁ polypeptide or even other subunits (Borsotto et al. 1984a, 1985; Barhanin et al. 1987; Lazdunski et al. 1986b; Nakayama et al. 1987; Soldatov 1988). It is, of course, unlikely that the alpha₁-associated drug-receptor domains were missing, as these preparations were highly enriched with respect to 1,4 DHP binding activity (Borsotto et al. 1985; Barhanin et al. 1987) and even formed functional calcium channels in planar lipid bilayers (Smith et al. 1987). Vaghy et al. (1987) resolved the discrepancy about the different "subunit" composition found by different laboratories which had puzzled the scientific community for some time. They demonstrated that the presence of an intact alpha₁ polypeptide in the purified preparation was dependent on the method used for the T-tubule membrane isolation (Fig. 8). If previously frozen skeletal muscle was used, an intact alpha₁ polypeptide was absent. Instead, two smaller polypeptides of 67 and 94 kDa were now isolated. These 67- and 94-kDa bands were clearly alpha₁ fragments, as both were photolabelled with (−)-[³H]azidopine and [N-methyl-³H]LU49888. A direct comparison of intact and proteolytic preparations is shown in Fig. 8. Immunoreactivity at 61- to 78-kDa positions was also found in cardiac tissue (Takahashi and Catterall 1987a) and was attributed to proteolytic breakdown on an alpha-subunit. In all these experiments there was no evidence for instability of the other subunits. The alpha-subunit puzzle exemplifies that neither function (in reconstitution experiments) nor binding activity prove structural integrity of an isolated "channel." On the other hand, the interaction of alpha₁ fragments with the other subunits seems to be nearly as strong as that of intact alpha₁, since otherwise the purification scheme would not work. In addition, the 1,4 DHP and the phenylalkylamine binding domain are now predicted to reside within a region of alpha₁ comprising 20% − 50% of the total amino acid sequence.

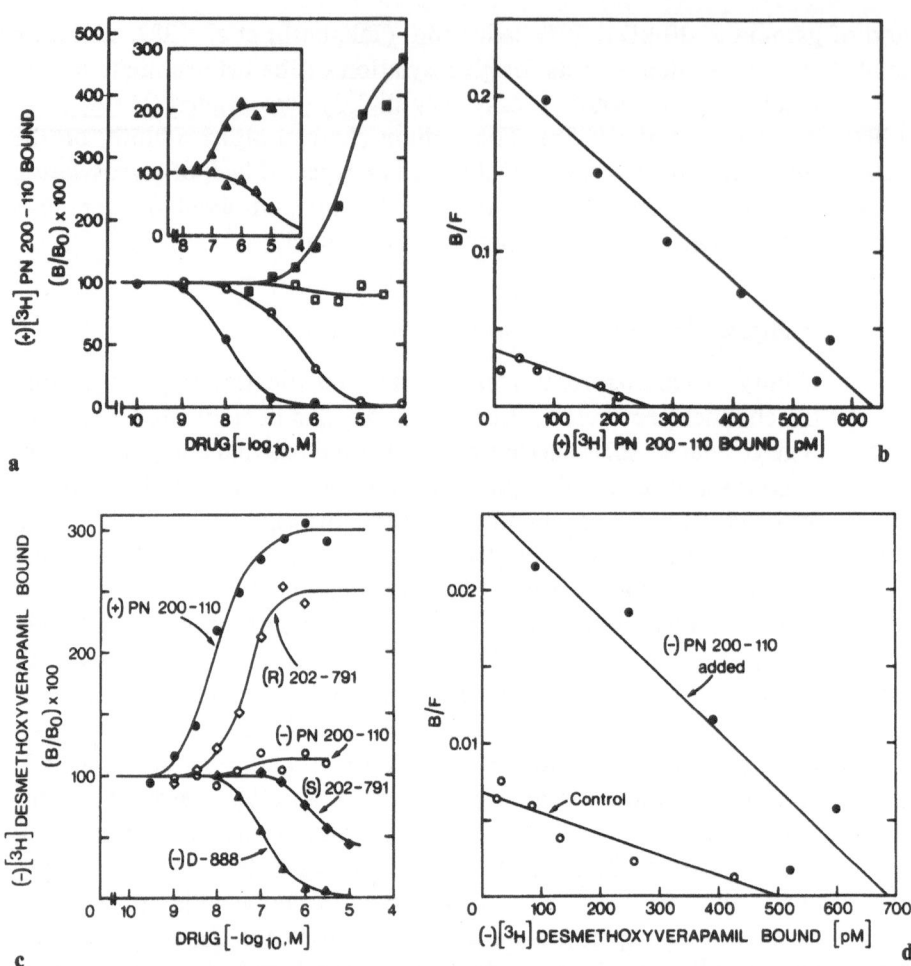

Fig. 10 a–d. Reciprocal allosteric interactions between the drug receptor sites for 1,4 DHPs, phenylalkylamines and benzothiazepines in purified skeletal muscle calcium channel preparations.

a Stereospecific regulation of (+)-[³H]PN200-110 binding by the diastereoisomers of diltiazem, and the enantiomers of PN200-110 and desmethoxyverapamil (*inset*). (+)-[³H]PN200-110 was incubated for 60 min with the channel protein at 25 °C. B_0 is the specifically bound ligand in the absence, B in the presence of unlabelled drug. The results of two to three experiments were computer-fitted to the general dose-response equation and the best fit is given ± asymptotic. S.D. (IC_{50} for binding inhibition, EC_{50} for binding stimulation): ●, (+)-PN200-110: 12.3 ± 3.5 nM, max. inhibition to 0%; ○, (−)-PN200-110: 434 ± 140 nM, max. inhibition to 0%; ■, (+)-*cis*-diltiazem: 4.5 ± 1.7, max. stimulation to 369%; □, (−)-*cis*-diltiazem: less than 10% inhibition at 10 µM; ▲ (−)-desmethoxyverapamil: 187 ± 113, max. stimulation to 221%; ▲, (+)-desmethoxyverapamil: 56% inhibition at 10 µM.

b Mechanism of binding stimulation by 100 µM (+)-*cis*-diltiazem. The Scatchard transformation of a saturation experiment is shown. The following parameters represent the best fit to a monophasic saturation isotherm: Control [absence of (+)-*cis*-diltiazem (○)]: K_d = 7.2 nM, B_{max} = 627 pmol/mg of protein. 100 µM (+)-*cis*-diltiazem (●) decreases the K_d to 2.87 nM and the apparent B_{max} increases to 1500 pmol/mg of protein.

5.2 Interaction of Purified Skeletal Muscle
Calcium Channels with Calcium Channel Drugs

Identification of an isolated complex as an L-type channel with its physiological properties requires, as a minimum, the functional integrity of the regulatory domains. It has been shown above that functional data — be it ligand binding or pore formation — and structural integrity are not necessarily the same coin viewed from different sides. Saturation analysis revealed that the affinity of the 1,4 DHP and phenylalkylamine receptor for these ligands decreases upon purification (Striessnig et al. 1986a,b; Flockerzi et al. 1986a,b; Borsotto et al. 1985) by about 10–30 times. Divergent results exist in the literature if this decrease is reversed upon detergent removal and reconstitution of the receptors into lipid vesicles (compare the data by Sieber et al. 1987a,b and the results of Barhanin et al. 1987). In purified but detergent-containing preparations the interaction of radiolabelled 1,4 DHPs or phenylalkylamines is highly dependent on the occupation of an allosterically linked receptor domain by a positive heterotropic regulator. $(+)$-[^{3}H]PN200-110 binding is stimulated two- to threefold by $(+)$-cis-diltiazem via the benzothiazepine receptor in purified preparations, mainly by increasing the number of high-affinity sites (B_{max} effect) and less so by increasing the affinity. In membranes the B_{max} effect is not observed (Striessnig et al. 1986a,b). Phenylalkylamines also stimulate 1,4 DHP binding (Fig. 10). These allosteric effects occur through reciprocal coupling mechanisms as antagonistic 1,4 DHPs stimulate the binding of $(-)$-[^{3}H]desmethoxyverapamil mainly by an increase of B_{max} and less so by a small decrease of the K_d. The effects of the 1,4 DHPs on phenylalkylamine binding are highly stereoselective. Only the antagonistic enantiomers [e.g. (R)-202-791, $(+)$-BAY K 8644] were stimulatory [as is $(+)$PN 200-110], whereas the agonistic optical antipodes [e.g. $(-)$-BAY K 8644 or (S)-202-791] were inhibitory. Agonistic 1,4 DHPs shift the channel to a very low affinity state for phenylalkylamines that is no longer detectable by the usual radioligand binding technique (apparent B_{max} decrease). Antagonists, on the other hand, recruit the channels in the high-affinity state (apparent B_{max} increase,

---

c Stereospecific regulation of $(-)$-[^{3}H]desmethoxyverapamil binding by agonistic and antagonistic 1,4 DHPs and $(-)$-desmethoxyverapamil. The results of two to three experiments were computer-fitted to the general dose-response equation and the best fit is given ± asymptotic S.D. (IC_{50} for binding inhibition, EC_{50} for binding stimulation): ●, $(+)$-PN200-110: 7.0±1.9 nM, max. stimulation to 301%; ◇, $(-)$-202-791: 44±8 nM, max. stimulation to 252%; ○, $(-)$-PN200-110: 10% stimulation at 3 μM; ◆, $(+)$-202-791: 1.46±0.48 μM, inhibition to 42% at 10 μM; ▲, $(-)$-desmethoxyverapamil: 99±18 nM, max. inhibition to 0%. Note that only antagonistic 1,4 DHPs are potent stimulators of $(-)$-[^{3}H]desmethoxyverapamil binding.

d Mechanism of binding stimulation by 1 μM $(+)$-PN200-110. The Scatchard transformation of a saturation experiment is shown. 1 μM of $(+)$-PN200-110 (●) decreases the K_d from 73.7 nM to 26.4 nM and increases the apparent B_{max}. [From Striessnig et al. 1986a with permission.]

as shown in Fig. 10). Several calcium antagonists bind to the in situ channel
in a strictly voltage-dependent manner with high affinity (nanomolar dissociation constants) for the inactivated state but with low affinity (micromolar
dissociation constants) for the resting state. Beam and Knudson (1988) performed whole-cell patch-clamp analysis of calcium channels on freshly isolated muscle fibres. They found the half-blocking concentration of PN200-110
at a holding potential of -80 mV (predominant state is "resting") to be
182 nM, whereas at -50 mV (predominant state is "inactivated") the $K_{0.5}$
was 5.5 nM. The drug-induced, long-lived stabilisation of the channel conformations in the absence of a membrane potential (as with the isolated protein
in the test tube) could therefore be viewed as analogous (but not necessarily
identical) to the transient voltage-dependent conformations in intact membranes.

No direct reversible-binding data are yet available for the benzothiazepine
receptor domain in solubilised or purified preparations. Apparently, the already low affinity (K_d values are between 50 and 100 nM) of $(+)$-cis-
diltiazem for the membrane-bound channel is further decreased upon purification. Localisation of the benzothiazepine receptor on polypeptides in the
alpha region in T-tubule membranes (Galizzi et al. 1986a) and on alpha$_1$ after purification (Glossmann et al. 1989) has nevertheless been possible by
photoaffinity labelling. Recently, a novel, distinct channel drug-receptor site
for benzothiazinones was described (Striessnig et al. 1988a; Qar et al. 1988;
Glossmann et al. 1989). This benzothiazinone-selective site also co-purifies
with the complex and retains its binding activity and pharmacological profile
throughout purification.

5.3 A Structural Model for the Skeletal Muscle
Calcium Channel Complex

The structural features of the alpha$_1$-subunit allow the hypothesis that it
alone forms the drug and phosphorylation-sensitive calcium channel pore in
skeletal muscle (see Sect. 8). A straightforward approach to proving this
would therefore be to separate alpha$_1$ from the rest of the subunits and re-
constitute it. Alternatively, the alpha$_1$ mRNA could be expressed in a system
where messages for alpha$_2$-delta, beta and gamma are missing. Unfortunately, attempts to separate the subunits have always resulted in a loss of receptor
binding activity, mainly because relatively high detergent concentrations or
even ionic detergents are necessary. However, one group (Pelzer et al. 1989a)
reported that the alpha$_1$ polypeptide alone, when separated after denatur-
ation of the complex, is sufficient to form a calcium channel after renatura-
tion. As the success rate was extremely low and the absence of other subunits
could not be excluded, this finding needs confirmation. Dissociation of the

complex can be initiated with 1% concentrations of SDS, Triton X 100 or CHAPS (Takahashi et al. 1987; Sharp et al. 1987). Removal of the alpha$_1$-subunit from the dissociated complex can be accomplished by either immunoadsorption (alpha$_1$-retained, Fig. 11a), or lectin-affinity chromatography (lectin binding to alpha$_2$-delta, Fig. 11b). Takahashi et al. (1987) estimated the affinity of the non-alpha$_1$-subunits with respect to noncovalent association with alpha$_1$. In 1% Triton X 100, alpha$_2$ completely dissociates from the complex, gamma is only partially removed and beta remains associated. The rank order of affinity for subunit binding to alpha$_1$ is therefore: alpha$_2$ < gamma < beta (Fig. 11a). The tight association of beta with alpha$_1$ is also supported by the immunoprecipitation of alpha$_1$ with monoclonal antibodies directed against beta from digitonin-solubilised membranes (Leung et al. 1988).

The subunits appear to be present in the digitonin-purified complex in a 1:1:1:1 (alpha$_1$:alpha$_2$-delta:beta:gamma) stoichiometry (Leung et al. 1988; Sieber et al. 1987b; Campbell et al. 1988a). With this stoichiometry and the best molecular mass estimates obtained (see Table 6) a molecular weight of the complex of 450 000 is calculated — in reasonable but not perfect agreement with the value of 370 000 found for cardiac tissue in hydrodynamic studies (Horne et al. 1986). In Fig. 12 a hypothetical model of the skeletal muscle calcium channel and its organisation is presented. The beta subunit is substrate for kinases, not a glycoprotein, and has no hydrophobic membrane-spanning regions. Therefore, it is positioned on the cytosolic face of the membrane but tightly associated with alpha$_1$. The gamma subunit is hydrophobic and large enough to have membrane-spanning regions with the carbohydrate residues exposed to the cell surface. The alpha$_2$ polypeptide and its disulphide-linked delta subunit(s) are glycosylated and are positioned outside the cell with some hydrophobic portions within the lipid bilayer. A similar model was recently presented by Campbell et al. (1988a).

5.4 Evidence for L-Type Channel Subunits in Other Tissues

Antibodies raised against the subunits of the purified skeletal muscle calcium channel have been used to identify immunologically related peptides in other excitable tissues. Polyclonal and monoclonal antibodies raised against skeletal muscle alpha$_2$- and delta-subunits immunoreact with polypeptides in brain, heart and smooth muscle, all of which possess the characteristic increase of the alpha$_2$ mobility on SDS-PAGE after reduction (Hosey et al. 1986; Schmid et al. 1986a,b). Together with the alpha$_2$-like polypeptides, the antibodies immunoadsorbed 1,4 DHP binding activity. Based on these data, the hypothesis was forwarded that the receptors were on the alpha$_2$ subunit. However, as in skeletal muscle, photoaffinity labelling experiments revealed

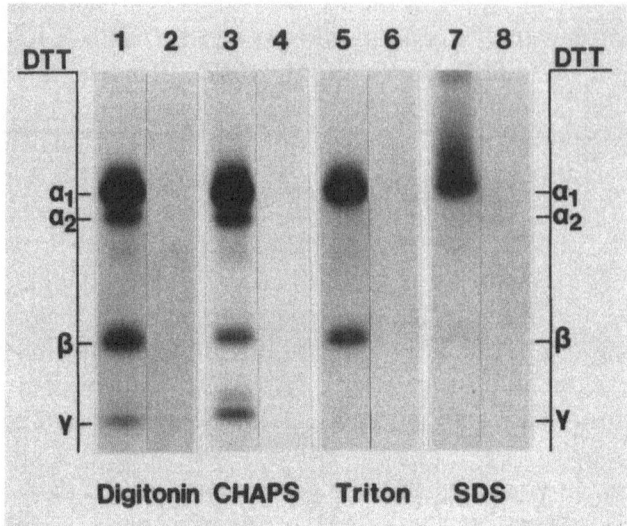

a

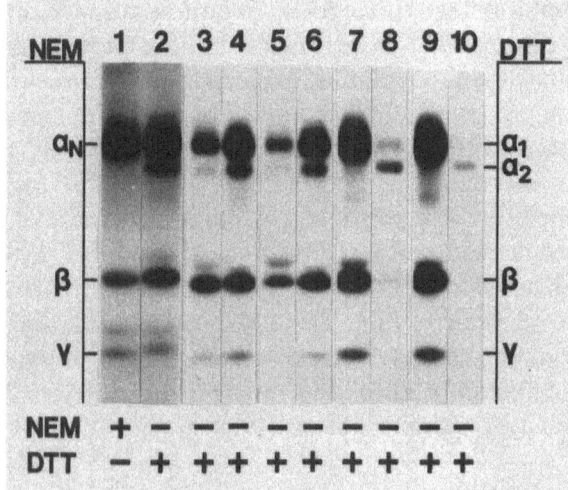

b

Fig. 11a,b. Analysis of noncovalent subunit interactions.

a Immunoprecipitation of calcium channel subunits by anti-alpha$_1$ antibodies. ^{125}I-labelled calcium channel polypeptides were immunoprecipitated with affinity-purified (Takahashi et al. 1987) anti-alpha$_1$ antibodies (*lanes 1, 3, 5, 7*) or a control preparation (*lanes 2, 4, 6, 8*) in immunoassay buffer containing the detergents indicated below. The immunoprecipitates were analysed by SDS-PAGE in the presence of dithiothreitol, followed by autoradiography. *Lanes 1* and *2*: 0.5% digitonin; *lanes 3* and *4*: 0.1% CHAPS; *lanes 5* and *6*: 1% Triton X 100; *lanes 7* and *8*: samples were incubated with 1% SDS for 2 min at 100°C and exchanged into 0.5% digitonin by gel filtration on a 2-ml Sephadex G-50 column. Note that preincubation with Triton X 100 but not with CHAPS causes partial dissociation of gamma (not easily seen in *lane 5*) and complete dissociation of alpha$_2$ from the complex. Beta is dissociated only after treatment with boiling SDS.

b Lentil lectin agarose affinity chromatography. The ^{125}I-labelled calcium channel (in 150 µl of buffer containing detergent as indicated below) was incubated for 90 min at 4°C with 50 µl of lentil lectin-agarose, equilibrated in the same buffer, under agitation. The resin was removed by

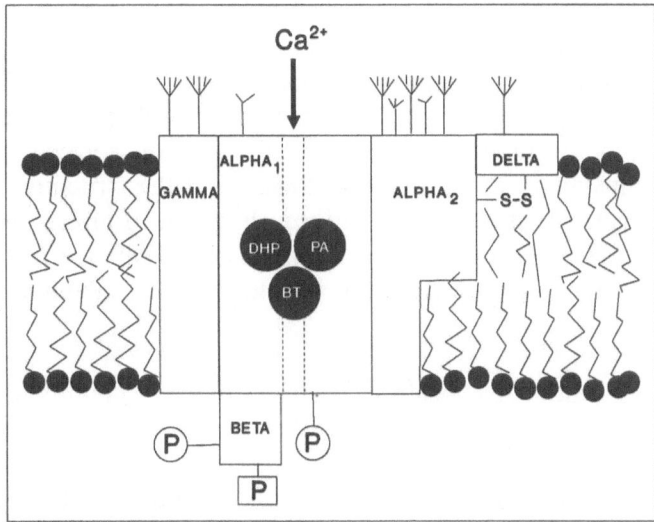

Fig. 12. Proposed model for skeletal muscle calcium channel complex. [Slightly modified from Takahashi et al. (1987) with permission]
PA, phenylalkylamine receptor; *BT*, benzothiazepine receptor; *S-S*, disulphide bond; *P*, phosphorylation sites for protein kinase C (*square*) and cAMP-dependent protein kinase (*circle*)

that the drug receptors for 1,4 DHPs and phenylalkylamines reside on alpha$_1$-like polypeptides with apparent molecular weights of 170–195 kDa in heart and brain (Ferry et al. 1987; Glossmann et al. 1987b; Kuo et al. 1987; Striessnig et al. 1988b; Chang and Hosey 1988). Schmid et al. (1986b) detected alpha$_2$-delta immunoreactivity closely distributed with (+)-[^{3}H]PN200-110 binding activity in rabbit brain slices, with the highest densities in the dentate gyrus of the hippocampus and the granular layer of the cerebellum, followed by cerebral cortex. These data suggest that, as in skeletal muscle, a tight association between a drug receptor carrying alpha$_1$ and the alpha$_2$-delta polypeptide exists in heart muscle and brain. The alpha$_1$-(Schneider and Hofmann 1988; Chang and Hosey 1988) and alpha$_2$- (Hosey et al. 1987) like polypeptides have already been partially purified from cardiac tissue. Recently functional expression studies demonstrated that a key component for L-type calcium channel activity in the heart is highly homologous to

---

centrifugation and the supernatant wash was collected. The resin was washed three times with 1 ml of buffer and resuspended in 150 μl of buffer containing 0.2 *M* methyl-alpha-D-mannoside. The eluate was collected after a 90-min batch incubation. Eluate (*lanes 4, 6, 8, 10*) and wash (*lanes 3, 5, 7, 9*) samples were analysed by SDS-PAGE and autoradiography. *Lanes 1* and *2*: ^{125}I-labelled channel without treatment; *lanes 3* and *4*: 0.1% digitonin; *lanes 5* and *6*: 0.1% CHAPS; *lanes 7* and *8*: 0.5% Triton X 100; *lanes 9* and *10*: samples treated with 1% SDS were exchanged into 0.5% Triton X 100 by filtration over a 2-ml Sephadex G-50 column and analysed in 0.5% Triton X 100. [From Takahashi et al. (1987) with permission]

the skeletal muscle alpha$_1$ subunit (Lotan et al. 1989; see Sect. 7.0 for further details).

In contrast to alpha$_2$, antibodies specific for alpha$_1$ in skeletal muscle have so far not been reported to cross-react with polypeptides in heart, brain or smooth muscle. This supports the idea of isoreceptors or isochannels (see Glossmann and Striessnig 1988a), and the structure of alpha$_1$ seems to be tissue specific and less uniform than that of alpha$_2$. Nucleic acid hybridisation studies support this view (Ellis et al. 1988; see Sect. 8). There are no data available that describe the presence of beta subunits in other tissues.

5.5 Antibodies Against Calcium Channel Subunits Modulate Channel Function

Antibodies can specifically modify the function of a membrane receptor protein by binding to defined regions of the respective molecule. Examples are monoclonal antibodies against the insulin receptor (see Forsayeth et al. 1987) or the voltage-dependent sodium channel (see Meiri et al. 1984). Effects of anti-subunit antibodies on calcium channel function have been studied in the mouse muscle cell line BC3H1, in isolated parathyroid cells, and in reconstituted skeletal muscle calcium channels.

Morton et al. (1988) developed a monoclonal antibody against the skeletal muscle alpha$_1$-subunit (Morton and Froehner 1987) and investigated whether binding of the antibody alters ion channel properties in BC3H1 cells (Morton et al. 1988). Differentiated BC3H1 cells express two types of voltage-dependent calcium channels with the pharmacological and electrophysiological characteristics of the so-called slow and fast channels found in skeletal muscle T-tubules. The antibody produced a selective, concentration-dependent attenuation (50% maximal inhibition) of only the 1,4 DHP-sensitive, high-threshold calcium current ("slow" channel). It had no direct effect on the DHP-insensitive, low-threshold calcium current ("fast" channel) or on the delayed outward K^+-currents. However, sodium current kinetics were altered at antibody concentrations similar to those required to attenuate the calcium current. This finding was explained by the sequence and structural homology between the alpha subunits of these two ion channels (Tanabe et al. 1987). Immunoblots revealed antibody binding to a polypeptide with the biochemical characteristics of the calcium channel alpha$_1$ subunit in these cells. It was suggested that the antibody binds to an extracellular epitope of the two related channels near or at their voltage sensor.

As mentioned above, cross-reaction of alpha$_1$-specific antibodies with polypeptides in heart, brain or smooth muscle has not yet been described. Fitzpatrick et al. (1988) were the first to detect alpha$_1$ immunoreactivity in cells other than those from skeletal muscle. They investigated the role of calcium channels in bovine parathyroid cells. Calcium inhibited PTH release from these cells in a concentration-dependent manner. 1,4 DHP agonists [e.g. (+)-202-791] increased calcium influx and reduced PTH release, whereas 1,4 DHP antagonists decreased calcium influx, thereby stimulating PTH release (Fitzpatrick et al. 1986). Treatment with pertussis toxin, which ADP-ribosylates a G-protein in parathyroid cells (39 kDa, Fitzpatrick et al. 1986), released the inhibitory effect of the calcium agonists. Polyclonal antisera against the skeletal muscle alpha$_1$ subunit blocked the secretion of PTH in low extracellular calcium (0.5 mM) and concomitantly increased $^{45}Ca^{2+}$ uptake into the cells. Control serum was without effect. Most of the antibody-inducible inhibition of PTH secretion was blocked by pertussis toxin pretreatment. Thus, the antibodies apparently bind to the channel and, like the 1,4 DHP Ca^{2+} agonist, induce a G-protein-dependent activation. The same antisera recognised an alpha$_1$-like 150-kDa polypeptide in immunoblots of the cell membranes which did not change its apparent molecular weight upon reduction (Fitzpatrick et al. 1988; Fitzpatrick and Chin 1988). Purified skeletal muscle calcium channels reconstituted into planar lipid bilayers were also used to study antibody effects. Monoclonal anti-beta antibodies activated the channel by more than tenfold, the activation being insensitive to inhibition by 1,4 DHPs (Campbell et al. 1988b). A polyclonal affinity purified anti-gamma antiserum was inhibitory. In contrast to the studies by Morton et al. (1988), two different monoclonals directed against the alpha$_1$ subunit were not modulatory. Stimulation was also observed with a monoclonal anti-channel antibody, with unreported subunit specificity (Malouf et al. 1987). Taken together, these findings suggest that other subunits than alpha$_1$ are necessary for channel function and/or regulation.

The fact that antibodies are able to alter calcium channel function is relevant for the understanding of human autoimmune diseases in which autoantibodies against voltage-dependent calcium channels are the underlying pathophysiological mechanism. Such antibodies directed against presynaptic voltage-dependent calcium channels have already been identified in the Lambert-Eaton myasthenia syndrome (Lang et al. 1983; Kim 1987; Kim and Neher 1988).

6 Structural Features of Omega-Conotoxin GVIA-Sensitive Calcium Channels

Although the first receptor identification was published in 1986, purification of the N-type channels has not yet been reported. Preliminary structural information about the [^{125}I]-ω-CgTx receptor comes from photoaffinity labelling and cross-liking experiments. Cruz et al. (1987) cross-linked [^{125}I]-

ω-CgTx to its binding site in chick brain membranes with the bifunctional re-
agent disuccinimidyl-suberate. They detected a specifically cross-linked band
with an apparent molecular weight of 135000 under reducing conditions of
SDS-PAGE. In the absence of reducing agents the electrophoretic mobility
decreased, as previously observed for the skeletal muscle alpha$_2$ polypeptide.
Abe and Saisu (1987), Yamaguchi et al. (1988), and Glossmann and co-workers (Knaus 1988; Glossmann and Striessnig 1988a, b; Glossmann et al. 1988b)
used a different approach to irreversible labelling. The [^{125}I]-iodinated toxin
was coupled with either N-5-azido-2-nitrobenzoyl-oxysuccinimide or N-
hydroxysuccinimidyl-azidobenzoate and the resulting photoaffinity probe
([^{125}I]azido-ω-CgTx) was incubated with the brain membranes. The azido derivative has approximately ten times lower (apparent) affinity than the
nonderived [^{125}I]-labelled toxin (Abe and Saisu 1987; Yamaguchi et al. 1988).
After irradiation and electrophoresis of the incubation mixture, two high-molecular-weight polypeptides were specifically photolabelled in rat (310000 and
230000, Abe and Saisu 1987), bovine (310000 and 230000, Yamaguchi et al.
1988) and guinea pig cerebral cortex (245000 and 195000, Glossmann and
Striessnig 1988a,b; see Fig. 13). The electrophoretic mobilities were unaffected by reduction. An additional smaller band (30–45 kDa) was also consistently labelled in the different tissues. Control membranes lacking [^{125}I]-
ω-CgTx binding activity (erythrocyte membranes, skeletal muscle microsomes) could not be specifically photolabelled. It was observed that the larger
of the two high-molecular-weight polypeptides was more heavily labelled
(Glossmann and Striessnig 1988a, b; Yamaguchi et al. 1988). (+)-cis-Diltia-
zem suppressed photolabelling for all three polypeptides in membranes and
for the two high-molecular-weight ones in digitonin-solubilised membranes.
The low-molecular-weight polypeptide was not photolabelled in the latter material (Yamaguchi et al. 1988).

Divergent results were obtained with cross-linking and photoaffinity labelling experiments in rat and chick brain membranes. In chick (but not rat)
brain membranes [^{125}I]-ω-CgTx was cross-linked to a 170-kDa polypeptide,
which shifted its apparent molecular weight upon reduction in SDS-PAGE
(Barhanin et al. 1988). A 220-kDa polypeptide was, however, photoaffinity labelled with [^{125}I]azido-ω-CgTx in synaptosomes of both species (Barhanin et
al. 1988; Marqueze et al. 1988). The apparent molecular weight was slightly
higher (245–300 kDa) in cultured rat brain embryonic neurons (Marqueze et
al. 1988). Thus, the molecular weights and the electrophoretic properties of
the putative receptor polypeptides seem to depend on the methods employed
for irreversible labelling. Whilst photoaffinity labelling in different tissues always identifies three polypeptides, it is not clear if the smaller ones arose by
proteolysis of the largest one or are a priori essential constituents of N-type
channels. The finding of (+)-cis-diltiazem blocking photoincorporation into
three polypeptides including a 310-kDa band is intriguing, but so far only a

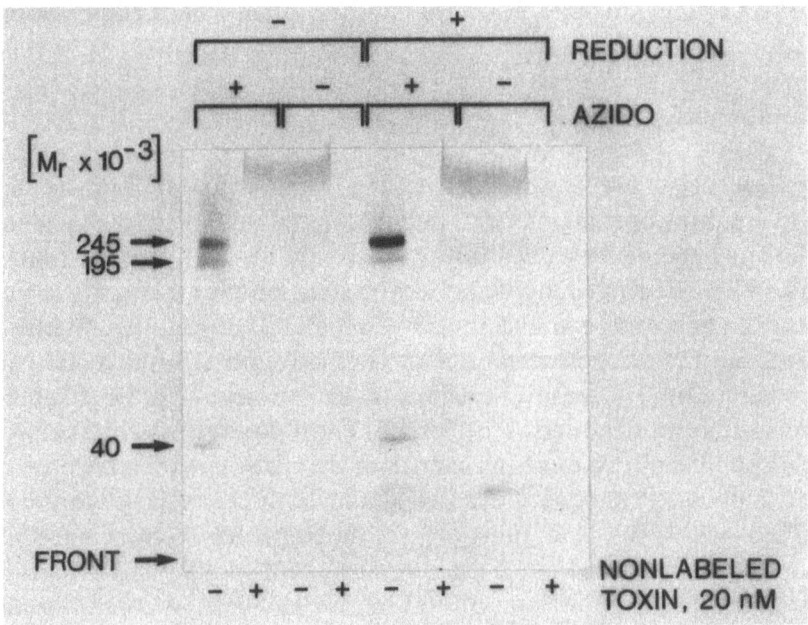

Fig. 13. Identification of ω-CgTx binding sites in guinea pig cerebral cortex membranes by photoaffinity labelling. [^{125}I]Azido-CgTx was synthesised by incubating [^{125}I]-omega-conotoxin GVIA with an excess of N-hydroxy-succinimidyl-azidobenzoate for 60 min on ice. An aliquot of the photolabel ($AZIDO$ +) was then incubated with 0.5 mg of guinea pig cerebral cortex membrane protein in the absence (indicated by $-$) and presence (indicated by +) of 20 nM unlabelled ω-CgTx at 25 °C. Underived [^{125}I]CgTx ($AZIDO$ $-$) was employed as a control. After 25 min the membranes were irradiated with UV light (Striessnig et al. 1986b, 1987) and collected by centrifugation. The pellets were solubilised in electrophoresis sample buffer in the absence (reduction $-$) or presence (reduction +) of 10 mM dithiothreitol. The samples were separated on a 5% – 15% SDS polyacrylamide gel, stained with Coomassie blue and dried, and the radioactive bands were revealed by autoradiography. Their apparent molecular weights were obtained from the relative mobilities of standard proteins on the same gel. [From Glossmann and Striessnig (1988b) with permission]

phenylalkylamine site with such a high M_r has been identified in mammalian brain (Striessnig et al. 1988b). On the other hand, the alpha$_1$-like, i.e. the L-type channel-associated, photoaffinity-labelled bands in brain are indistinguishable by size on SDS-PAGE from the second largest putative N-channel polypeptide (195000 in Glossmann and Striessnig 1988b; 230000 in Abe and Saisu 1987 or in Yamaguchi et al. 1988). The two other laboratories that were successful with cross-linking (see, however, Yamaguchi et al. 1988) identified polypeptides which, based on their behaviour on SDS-PAGE, are putative alpha$_2$-like (i.e. alpha$_2$-delta) structures. Any further speculation about the composition of N-type channels should rest on better and unambiguous structural data.

7 The Calcium Channel Structure and Excitation-Contraction Coupling (ECC)

7.1 Introductory Remarks

The sequence of events which convert the voltage-signal of T-tubule depolarisation into the opening of Ca^{2+} channels in the sarcoplasmic reticulum (SR) is largely hypothetical. The evidence is strong, however, that the same structure(s) that elsewhere function as L-type channels play a key role in this process. The now most popular theory about ECC between the T-tubule membrane and the sarcoplasmic calcium release suggests a functional coupling through a 1,4 DHP-sensitive calcium channel protein. The SR channel structure has been delineated, the morphological equivalents for the proposed tight coupling of two calcium channels in the triad junction have been discovered, and co-reconstitution experiments of both channels are forthcoming.

Thus, with respect to function, the situation has been dramatically reversed. Once regarded as an obscure and silent binding site for drugs, the Ca^{2+}-antagonist receptor in skeletal muscle has been awarded a functional role. Although this review deals mainly with structure, we cannot disregard two equally important aspects: First, ultrastructural data on purified L-type channels are available only for the skeletal muscle. Second, a genetic "channel" defect has been corrected by expression of a plasmid that carries the cDNA coding for the $alpha_1$-subunit. Therefore, we have to discuss ECC, some structural and biochemical features of the SR channel and the restoration of ECC in dysgenic muscle by the $alpha_1$-cDNA. The features of the two opposing channels (i.e. the SR and the T-tubule-L-type channel complex) which are striking, topologically as well as functionally, may or may not be unique for skeletal muscle. In any case, the structures represent a novel pathway of transmembrane communication (see Bean 1989).

7.2 ECC in Skeletal Muscle — Effects of Cations and Drugs

In cardiac muscle contraction is initiated by the opening of voltage-dependent, drug-sensitive (L-type) calcium channels in the sarcolemma. These channels provide essential calcium ions, which in turn trigger ryanodine-sensitive SR calcium channels to release calcium for contraction. This process is also termed "calcium-induced calcium release". The density of the L-type channels (as measured by 1,4 DHP binding) in highly purified heart membranes is two orders of magnitude lower than the density of Ca^{2+}-antagonist receptors in T-tubule membranes from skeletal muscle, where, paradoxically, no extracellular calcium is needed to support ECC. Numerous experiments over several decades have confirmed the fundamental difference with respect to extracellular calcium between heart and skeletal muscle, but a major, critical point was overlooked, namely the role of inorganic cations per se. If one substitutes the divalent cation Ca^{2+} with Na^+, K^+, Rb^+, Li^+, Sr^{2+}, Mg^{2+} or Ba^{2+} in the extracellular fluid, skeletal muscle contraction is unimpaired. However, if these inorganic cations are substituted by the organic cations tetraethylammonium or dimethonium, ECC is not supported (Brum et al. 1988; Rios and Pizarro 1988; Rios et al. 1989). When calcium release from the SR is measured with an optical signal derived from the interaction of Ca^{2+} and the dye antipyralazo III (as described by Brum et al. 1987; Melzer et al. 1987) the peak release induced by the different cations can be compared and an estimate of the cation affinity delineated. For divalents the affinity sequence is $Ca^{2+} > Sr^{2+} > Mg^{2+} > Ba^{2+}$ and for monovalents $(Ca^{2+}) > Li^+ > Na^+ > K^+ > Rb^+ > Cs^+$. The rank order for the divalents is identical to the rank order of their $K_{0.5}$ values on cation-depleted, calcium channel-linked 1,4 DHP receptors to restore the high-affinity binding state in the presence of $(+)$-*cis*-diltiazem (Glossmann and Striessnig 1988a). The rank order of affinities, including the monovalents, is very similar to the relative permeabilities determined on the cardiac L-type channel (Hess et al. 1986; Lansman et al. 1986). However, the L-type channel blockers Co^{2+} and Cd^{2+} and even La^{2+} support ECC (Rios et al. 1989). These cations bind very tightly to L-type channels and permeate, if at all, only at a very slow rate. Undoubtedly, the mechanism by which the voltage change in the transverse tubule membrane is converted into a signal leading to SR calcium release requires *bound* inorganic cations with a specificity profile similar to L-type channels or channel-linked 1,4 DHP receptors. The release of SR calcium is regulated by voltage and is turned on and off within milliseconds. Schneider and Chandler (1973) first described a "charge movement" (see Schneider 1981) across the T-tubule membrane, thought to be intimately connected to the potential-driven conformational changes of a "voltage sensor". The charge movement or voltage sensor is sensitive to L-type calcium channel drugs (Rios and Brum 1987) from the three main chemical classes, namely the 1,4 DHPs, the phenylalkylamines and the benzothiazepines. Pizzaro et al. (1988) have recently summarised the effects of calcium channel drugs on charge movement. In general, Ca^{2+}-channel-active drugs and Ca^{2+} agonists as well as Ca^{2+} antagonists inhibit ECC. Depending on the drug under investigation, this occurs in a voltage- or use-dependent manner. Stimulation by $(+)$-*cis*-diltiazem or 1,4-DHPs, for example, has also been reported and may be rationalised as a left-shift of the activation curve of ECC (Pizarro et al. 1988). These drugs also modulate the Ca^{2+} current (I_{slow}) through skeletal muscle T-tubule calcium channels. The calcium flowing through the channel is not responsible for ECC, and both processes have different properties. Whereas the I_{slow} is potentiated by the DHP agonist $(+)$-202-791, the SR Ca^{2+} release is inhibited in parallel. The slow inward current and Ca^{2+} release also have different

voltage dependence of activation and recover at different rates after inactivation (Pizarro et al. 1988). As will be outlined below, the evidence − despite
these different properties − is strong that the same protein mediates both
functions. The charges which move can be quantitated and divided into two
classes, namely those which are, for example, nifedipine-sensitive and those
which are not. For the former, the calculation gives $250-300$ per μm^2 of
transverse tubule membrane. The density of 1,4 DHP sites in this membrane
is 360 per μm^2 on average − a remarkably similar number (Glossmann and
Striessnig 1988a).

7.3 The Triad Junction and the Calcium Release Channels

Once released from the SR in response to an action potential, calcium has to
meet the contracting elements of the muscle fibre. To keep the distance as
short as possible the reticular membrane system of the SR surrounds the sar-
comere. However, it also keeps in close contact to the cell surface which delivers the signal. Finally, SR terminates tension by re-uptake of calcium. SR in
skeletal muscle is composed of the terminal cisternae and the longitudinal SR,
which is contiguous with the former. The terminal cisternae are in close association with the T-tubules, i.e. deep invaginations of the plasmalemma into the
sarcoplasma. The T-tubular and the SR lumina are separated by the junctional gap $(10-20\ nm)$ which is spanned by the so-called feet. Their electron-
dense structures were described by Revel (1962). The characteristic junctional
association of the T-tubule and two cisternae is termed the "triad junction".
   Calcium release and re-uptake are controlled by different biochemical enti-
ties. Calcium re-uptake is mediated by the Ca^{2+}-Mg^{2+} ATPase of the SR,
which has been characterised in detail, including the amino acid sequence
analysis (MacLennan et al. 1985). The biochemical equivalent of the calcium
release mechanism was elucidated only recently. Fleischer and co-workers
have characterised the different membrane components of the SR after their
isolation from the microsomal fraction (Saito et al. 1984). Terminal and longitudinal cisternae can be distinguished by their different morphology (revealed
by electron microscopy) as well as by some functional and biochemical characteristics. Isolated terminal cisternae vesicles consist of two distinct types of
membranes, namely the junctional face membrane and the membrane that
contains the calcium pump (Fig. 14a). In contrast, the longitudinal cisternae
possess only the calcium pump-containing membrane, and more than 90% of
the Ca^{2+}-Mg^{+} ATPase activity is present here. The junctional face membrane amounts to ~ 16% of the surface area of a terminal cisternae vesicle
and carries the feet structures spanning the junctional gap. The ultrastructural characteristics of the calcium pump protein are not detected between the

feet. They are limited to the nonjunctional membrane of the vesicles. The calcium-binding protein calsequestrin is found inside the terminal cisternae.

The alkaloid ryanodine, from the plant *Ryania speciosa* Vahl has become
an important tool for the characterisation of the calcium release process in
the SR. It is toxic in low doses to both vertebrates and invertebrates, resulting
in irreversible skeletal muscle contractures and flaccid paralysis of cardiac
muscle (see Fill and Coronado 1988). Employing the morphologically well-
characterised membrane fraction of the SR (Saito et al. 1984), the ryanodine
sensitive-calcium release channels were localised to the junctional terminal
cisternae (Fleischer et al. 1985). Ruthenium red (a blocker of calcium release)
had no effect on the already maximal calcium loading rate in the longitudinal
SR but did stimulate the low calcium loading rate of the terminal cisternae
about fivefold. This suggested that terminal cisternae were "leaky" for calcium ions under the experimental conditions. Ryanodine had no effect on the
calcium loading rate or the ATPase activity in the longitudinal system but
blocked the effect of ruthenium red in the terminal cisternae at nanomolar
concentrations. Therefore, the actions of ruthenium red and ryanodine were
clearly not on the calcium pump protein but on the calcium release system
(Fleischer et al. 1985). Ruthenium red locks calcium release channels in the
closed state, whereas ryanodine at low concentrations stabilised the open
state, thereby preventing the effect of later-added ruthenium red. The major
candidate for the calcium release channel structure was a high-molecular-
weight polypeptide (M_r 350000−450000) present in SDS gels of terminal cisternae membranes but not in longitudinal SR membranes (Saito et al. 1984).

Tritiated ryanodine was then used to characterise its receptor on the putative channel and to isolate the latter. [^{3}H]Ryanodine binds with nanomolar
dissociation constants to skeletal muscle and cardiac SR vesicles. Binding is
calcium dependent and stimulated by millimolar concentrations of adenine or
guanine nucleotides and caffeine. As expected, based on functional studies,
ruthenium red inhibited [^{3}H]ryanodine binding at micromolar concentrations (Pessah et al. 1986, 1987; Fleischer et al. 1985; Campbell et al. 1987;
Michalak et al. 1988). Pharmacological specificity is maintained after receptor solubilisation with CHAPS (Pessah et al. 1986; Inui et al. 1987a,b; Lai et
al. 1987, 1988a,b) or digitonin (Campbell et al. 1987; Imagawa et al. 1987b).
The solubilised ryanodine receptor was purified by a single sucrose-density
gradient centrifugation or immunoaffinity chromatography from skeletal
muscle (Lai et al. 1987, 1988b; Imagawa et al. 1987b) and cardiac SR (Lai et
al. 1988a). More complex chromatographic isolation procedures have also
been described (Inui et al. 1987a,b). Independent of the procedure employed,
a single, very-high-molecular-weight polypeptide (350−450 kDa) was purified
with the [^{3}H]ryanodine binding activity. Electron microscopy of the purified
ryanodine receptor showed particles with the characteristic size and shape of
the feet structures in the terminal cisternae, indicating that ryanodine binds

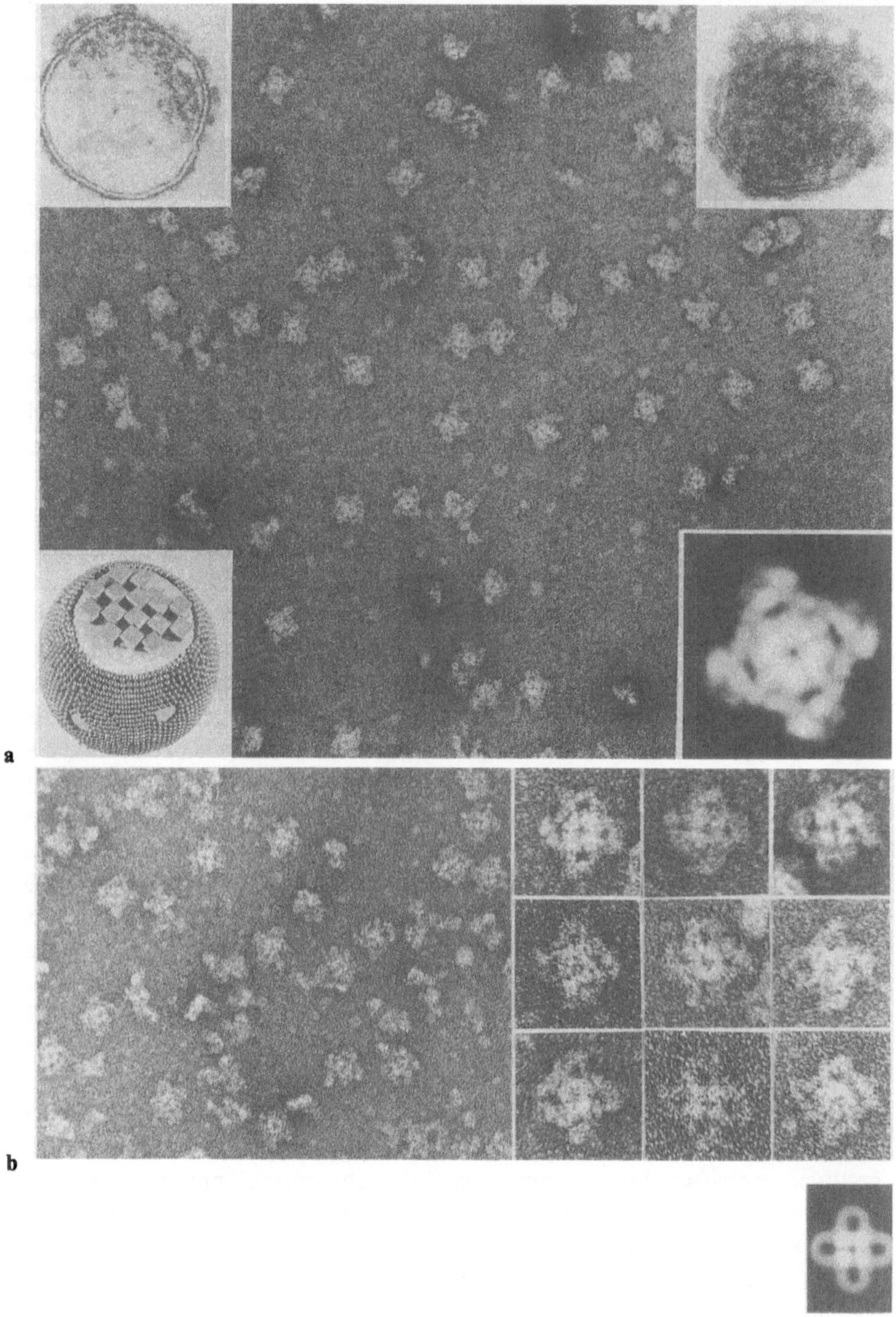

a

b

directly to the feet structures. Ultimate proof that ryanodine receptors and feet structures are identical with the calcium release channels came from functional reconstitution experiments. After reconstitution into planar bilayers the purified receptor forms calcium-specific channels with regulatory characteristics consistent with the calcium release observed in isolated terminal SR.

7.4 The Foot Structure Is an Oligomer of Calcium Release Channels

The arrangement of the feet in situ on the terminal cisternae is shown in Fig. 14a (upper right corner, lower left corner) and described in detail by Ferguson et al. (1984) and Saito et al. (1988). The figure also shows the ultrastructural details of the purified ryanodine receptor published by Fleischer and co-workers (Saito et al. 1988). They found a fourfold symmetry of an outer and a denser inner core with a central hole $1-2$ nm in diameter (see Fig. 14, legend). Meissner and co-workers (Lai et al. 1988a) independently described a similar, but not identical consensus image (Fig. 14b). Their purified channel complexes showed four-leaf clover structures (quatrefoils). The dimensions were 34 nm from the tip of one leaf to the tip of the opposite one. Each leaf was 14 nm wide and apparently consisted of a 5-nm-thick loop of protein around a central depression or hole $2-4$ nm long. At the centre of the quatrefoil there was a region of higher protein density forming a tetramer with a central hole $1-2$ nm in diameter. This central high-density tetramer of the complex may consist of transmembrane domains from each subunit that jointly ensheath the ion channel. Several questions arise. First, how many ryanodine receptors are located on a single functional release channel? Second, how many 350-

---

Fig. 14a, b. Morphology of the ryanodine receptor/calcium release channel.
a Negative-staining electron micrograph of the purified ryanodine receptor is shown in the centre field. Considerable structural detail can be obtained from the square structures of 25 nm/side (foot structure). A computer-averaged view of the receptor of 240 images (*lower right inset*) displays fourfold symmetry and impressive ultrastructural detail (Saito et al. 1988). The dense central mass, divided into four domains, is enclosed by an outer frame that has a pinwheel appearance. There is a 2-nm hole in the center. In the thin section of the terminal cisternae vesicles (*upper left insert*), the foot structure extends 12 nm from the surface of the junctional face membrane. The tangential section to the surface of the junctional face membrane of the terminal cisternae (*upper right inset*) reveals the square shape of the feet structures and their association with one another. The model of the terminal cisternae (*lower left inset*) shows the two types of membranes in the vesicle. The junctional face membrane at the upper surface contains the feet structures. The remainder is calcium-pump membrane (Saito et al. 1984). [From Saito et al. (1988) with permission]
b Electron micrograph of the negatively stained calcium release channel complex (Lai et al. 1988b; with permission). *Left*: Particles, some with the characteristic cloverleaf shape and others distorted ($\times 250000$). *Right*: Selected quatrefoils ($\times 500000$). The schematic drawing, *lower right*, is the authors' interpretation of the consensus image (stain, dark; protein, white)

to 450-kDa polypeptides form one foot. Third, how many 350- to 450-kDa polypeptides form one physiological channel?

Assuming a molecular weight of the monomer of 400 kDa (Lai et al. 1987, 1988b), purified preparations should have a maximal specific activity for [^{3}H]ryanodine binding of 2200 pmol/mg of protein. The highest values reported in the literature were 650 pmol/mg and 393 pmol/mg of protein (Lai et al. 1988b; Inui et al. 1987b). This suggests that four (or even six) monomers are required to bind one [^{3}H]ryanodine molecule. Alternatively, the experimental conditions for [^{3}H]ryanodine binding may not yet have been optimised (S. Fleischer, personal communication). Thus, the first question is still open.

Size-exclusion chromatography shows that four polypeptides form an oligomeric complex in CHAPS with an apparent molecular weight of $1.20-1.22 \times 10^6$ (Inui et al. 1987b; Pessah et al. 1987). Assuming a protein density of 1.37 cm^3/g, a molecular weight of 4.8×10^6 per foot was calculated (Inui et al. 1987b) from its dimensions (Fig. 14a). To account for the apparent fourfold symmetry a maximum of four oligomers (e.g. 16 monomers) could fit into one foot. However, if one takes into account the large clefts and holes revealed by the more recent ultrastructural images, the above mentioned molecular weight is clearly overestimated and a minimum of one oligomer (i.e. four monomers) seems more reasonable (compare Inui et al. 1987b; Lai et al. 1988b; McGrew et al. 1989). Further ultrastructural, biochemical and immunochemical studies will allow more accurate estimates.

The third and most important question cannot yet be answered and requires detailed functional studies of the purified receptor, i.e. reconstitution studies. Several groups have investigated the conductive properties of the purified channel and found multiple conductance states (Lai et al. 1988a,b; Ma et al. 1988; Imagawa et al. 1987b) rather than a single unitary channel conductance. Schindler and co-workers used the fast-dilution technique/lipid vesicle-derived planar bilayer technique (Schindler 1989) for reconstitution. By this method the reconstituted polypeptides become randomly and singly distributed in the plane of the bilayer before they form larger oligomers by lateral diffusion. Their density in the bilayer can be estimated from the protein concentration and the specific (ryanodine binding) activity of the preparation. The smallest conductances, recorded preferentially at the beginning of the experiment, are assigned to the smallest functional units (elementary conductance, "monochannels"). The presence of time-correlated subconductance states was explained by the time-dependent association of such units to larger cooperative switching aggregates ("oligochannels"). Analysis of the data thus obtained reveals a smallest well-defined conductance state of 3.8 pS, and larger conductances and gating transitions are often integer multiples of this value (Hymel et al. 1988a,b, 1989). The structural fourfold symmetry is also evident in these functional data, as the most often seen "oligochannels" have conductances of four times the elementary conductance, i.e. 15 pS and multiples thereof. A clear structural assignment of the elementary conductance (3.8 pS) level is not yet possible. It could be attributed to a 400-kDa polypeptide and the 15 pS events to the cooperative opening of the four-peptide oligomer as found in gel filtration. Alternatively, the elementary conductance might be generated by a four-peptide oligomer and the 15 pS events by the cooperative opening of a 16-peptide oligomer.

7.5 ECC in Skeletal Muscle − Functional Association of Two Oligomeric Calcium Channels Across Membranes

When elements of the T-tubule-SR junction are analysed by high-resolution electron microscopy small intramembranous particles are found which are organised as tetramers consistently facing the underlying feet. Rios and Pizarro (1988) have given this remarkable structure the picturesque name "a double zipper in register", where the tetramers are organised in such a way that every ovoidal particle tetramer is always surrounded by three unopposed (SR) tetramers. The density of feet is twice that of the particles. Campbell's group, together with Franzini-Armstrong and co-workers (Leung et al. 1988; Block et al. 1988), has investigated the ultrastructure of the purified 1,4 DHP-sensitive calcium channel. Electron microscopy of freeze-dried, rotary-shadowed samples reveals a homogeneous population of 16×22 nm ovoidal particles (Fig. 15). It was concluded that a particle is primarily composed of two components of similar size separated by a small central gap, suggesting a twofold symmetry. The chances are good that the ovoidal structures are identical with the particles which are organised in tetrads and oppose the feet. The ratio of morphologically identifiable particles to feet (1:2) is equal to the ratio of 1,4 DHP receptors (or nifedipine-suppressible charges) to the feet (see above). The question now arises, how many Ca^{2+}-antagonist receptors/channels are in one particle? Taking the density of protein as 1.37 g/cm^3, one particle could accommodate a molecular weight of 2.4×10^6 and contain − as an upper limit − about four (alpha$_1$, alpha$_2$-delta, beta, gamma) complexes. If one particle represents a single complex this would fit the ratio of 1:2, but the question of an oligomeric organisation remains, as the organisation in situ is tetradic − similar to that of the feet. Evidence for such an oligomeric nature of reconstituted 1,4 DHP-sensitive calcium channels was indeed found (Glossmann et al. 1988a,b; Hymel et al. 1988c, 1989) using the same reconstitution procedure as described for the purified calcium release channel. An elementary conductance level of 1 pS was recorded and the higher conductance levels were preferentially conductance jumps of 12 pS (Glossmann et al. 1989). This indicates that, as for the ryanodine receptor, simultaneously switching oligochannels are formed.

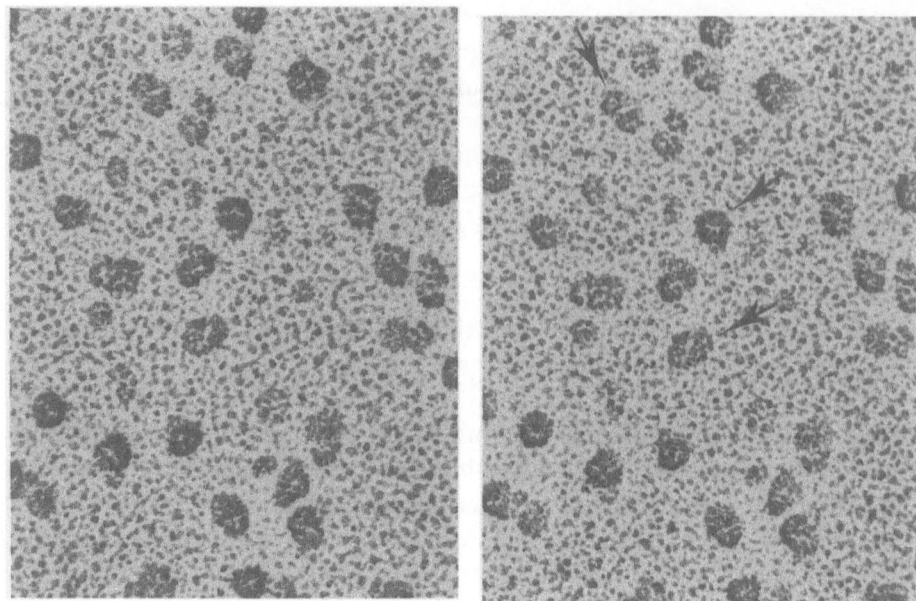

Fig. 15. Stereomicrographs of freeze-dried, rotary-shadowed purified calcium-antagonist receptor complexes from rabbit skeletal muscle. The molecules were rotary-shadowed at 15° with carbon-platinum and imaged in a Philips 410 electron microscope (for details see Leung et al. 1988). Micrographs taken under eucentric conditions were used for measurements. Note variations in shape, from round to elongated, of the globular molecule and separation into two subunits (*arrows*). [From Leung et al. (1988) with permission]

For both the isolated 1,4 DHP-sensitive channel and the ryanodine receptor the functional cooperativity evidenced by their gating properties supports the morphological findings about their fourfold structural symmetry. A voltage-driven conformational change of four structural elements of the T-tubule channel, for example, could convey the signal to the four underlying feet elements across the triad junction (as suggested in Fig. 16), resulting in the opening of the calcium release channels. The liberated calcium may then open the adjacent nonparticle (i.e. not voltage-sensor-linked) channels in the SR (Rios and Pizarro 1988). Direct evidence for a dualistic role of the T-tubule calcium channel polypeptides comes from experiments with dysgenic mice (mdg/mdg mouse, see below). The largely hypothetical model (shown schematically in Fig. 16) does not preclude a role for other possible second messengers, e.g. inositol trisphosphate (IP_3). An IP_3-sensitive calcium release mechanism exists in the SR (Vergara et al. 1985; Volpe et al. 1985, 1986). This mechanism is sensitised by depolarisation, and very low concentrations of IP_3 are required for induction of calcium release, thus reducing the need for IP_3 production (Donaldson et al. 1988).

Fig. 16a, b. Hypothetical model of the structural (**a**) and functional (**b**) organisation of the coassociated calcium channels at the triad junction. Note that there are also feet (not shown) which lack opposing T-tubule membrane calcium channel complexes that carry 1,4 DHP receptors. The *large arrows* in **b** indicate the flow of information mediated by Ca^{2+} ions, protein conformational transitions, or electrostatic redistribution. The 1,4 DHP-sensitive Ca^{2+} channel complex (*DHP REC*) is thought to serve as a sensor and the ryanodine-sensitive Ca^{2+} channel (*RYAN REC*) as the receiver of the message. The *smaller arrows* in **b** represent the exit of Ca^{2+} ions from the SR terminal cisternae through the foot structure of the calcium release channel. [From Hymel et al. (1989)]

Coassociated Calcium Channels At the Triad Junction

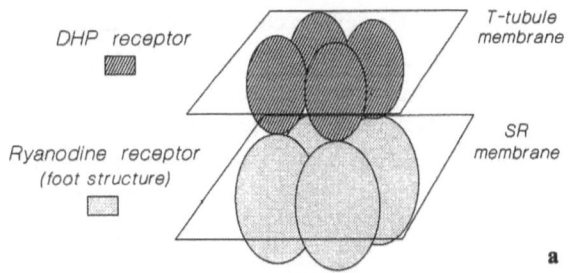

a

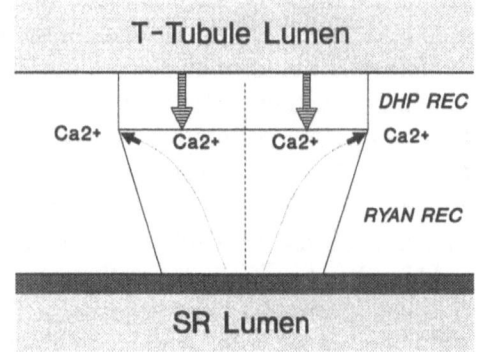

b

The role of GTP-binding (G) proteins in ECC remains to be clarified. A nonhydrolysable GTP analogue (GTP-γ-S) elicits the development of isometric force in chemically skinned fibres. The concentration dependence of this effect is shifted to the right by treatment with pertussis toxin (PTX) (Di Virgilio et al. 1986). Recently, a 40-kDa PTX substrate (immunologically related to the alpha-subunit of G_o from brain) was identified in skeletal muscle T-tubules (Toutant et al. 1988). G-proteins are involved in the regulation of calcium channel function in brain, cardiac and skeletal muscle (Rosenthal and Schultz 1987; Brown and Birnbaumer 1988; Yatani et al. 1987), but further studies are necessary to analyse their importance for ECC.

Taken together, key elements of the more classical signal transduction pathways (e.g. G-proteins, calcium ion channels) participate in a novel communication system between two membranes. The regular arrangement of the two cation channels is regarded as significant in the process of ECC. State models of the SR channels and speculations on the dualistic role of the 1,4 DHP-sensitive channel have been forwarded by Rios and Pizarro (1988). The question of cooperativity between channel molecules and across two membranes is provocative and deserves further study.

7.6 Pathology of the Triad Junction and Restoration of ECC in the mdg/mdg Mouse by Alpha$_1$-Subunit cDNA

Recently, genetic defects causing skeletal muscle dysfunction have been shown to be due to abnormalities of sarcolemmal and/or triad components. They manifest as muscular dysgenesis (mdg) in mice and Duchenne muscular dystrophy (DMD) in man. Pathological abnormalities of the L-type calcium channel (Ervasti et al. 1989) and/or the ryanodine-sensitive calcium release channel may be responsible for the malignant hyperthermia syndrome (Mickelson et al. 1988), but sequence data comparing normal and affected individuals are not yet available.

Dystrophin is a very low abundance protein of 427 kDa (0.002% of the total muscle protein) which is associated with the cytoplasmic side of the sarcolemma and the T-tubule membrane (Arahata et al. 1988; Watkins et al. 1988; Arahata et al. 1989) and co-purifies with 1,4 DHP and ryanodine receptor binding activity (Hoffmann et al. 1987; Knudson 1988). DMD is caused by quantitative and/or qualitative changes in dystrophin, which is coded (Koenig et al. 1987, 1988) by a very large gene (0.05% of the entire human genome). The function of dystrophin is not known, although it is speculated that it stabilises the T-tubule membrane (Brown and Hoffmann 1988).

Muscular dysgenesis is a lethal, spontaneous, recessive, autosomal mutation characterised by the absence of ECC in mice (Powell and Fambrough 1973; Beam et al. 1986; Romey et al. 1986; Pincon-Raymond et al. 1985). The dysgenic embryonic muscles generate normal action potentials and possess intact voltage-dependent sodium channels. There is no biochemical evidence for abnormal calcium release and storage in the SR. However, their SR fails to release calcium in response to sarcolemmal depolarisation. Triads in the homozygous mutants (mdg/mdg) are less numerous and disorganised (Rieger et al. 1987). The genetic defect is associated with the absence of the slow calcium current (Pincon-Raymond et al. 1985; Beam et al. 1986) in skeletal muscle that is carried by the 1,4 DHP-sensitive calcium channel. The apparent density of 1,4 DHP receptors is reduced by about five times in dysgenic skeletal muscle but not in heart. Although the calcium current is not required for ECC (see Sect. 7.5), these findings taken together indicate a structural defect of the skeletal muscle calcium channel structure, whereas neuronal and cardiac calcium channels are unaffected (Beam et al. 1986). Lazdunski and coworkers (Rieger et al. 1987) concluded that the defect lies in the spinal cord motoneurons innervating the skeletal muscle rather than in the calcium channel protein itself. Normal calcium channel activity, triads and contractile capacity were all restored in myotubes of mdg/mdg mice co-cultured with normal spinal cord neurons, suggesting a role for neuronal trophic factors or muscle differentiation components not produced by mdg/mdg mice. However, the conclusions drawn from these studies are now in doubt, as more recent work (Tanabe et al. 1988), detailed below, proves that the cause of this disease is a structural abnormality of $alpha_1$. Restriction endonuclease digests of genomic DNA extracted from the liver of newborn mdg/mdg, heterozygous (+/mdg) and normal (+/+) mice were blot-hybridised with nine probes derived from different regions of the $alpha_1$-cDNA. Two probes revealed characteristic differences in the restriction fragments. It was concluded that at least two regions of the $alpha_1$ gene are altered in the diseased mice. As normal L-type calcium currents are observed in heart or sensory neuronal cells, the structural genes for these channels must be quite distinct from $alpha_1$ (i.e. $alpha_{1,sk}$ is distinct from $alpha_{1,n}$ or $alpha_{1,m}$). Muscle poly (A)$^+$ RNA from normal (+/+) mice contained a 6.5-kb RNA species which hybridised with a rabbit skeletal muscle cDNA-derived probe, whereas this species was not found (or greatly reduced, depending on the probe) in poly (A)$^+$ RNA extracted from (mdg/mdg) muscle. The conclusion is allowed that the genetic defect leads to greatly diminished amounts (or even absence) of fully functional $alpha_1$-subunits. This is supported by the finding that in dysgenic muscle $alpha_1$- (but not $alpha_2$-) immunoreactivity is absent (Knudson et al. 1989), whereas other proteins involved in excitation-contraction coupling (e.g. the ryanodine-sensitive calcium release channel, SR calcium pump, calsequestrin) are present. Tanabe et al. (1988) then approached the correction of the defect. To this end an expression plasmid, termed pCAC6, was constructed which contained the entire protein-coding sequence of $alpha_1$-cDNA (see Sect. 8) linked with the simian virus 40 (SV40) early gene promoter. The circular pCAC6 DNA was microinjected into nuclei of multinucleate dysgenic myotubes in primary tissue culture. In control experiments the vector plasmid (pKCRH2), which is identical to pCAC6 (except for the $alpha_1$ cDNA) was employed.

It was found that pCAC6, which was injected into ~1000 myotubes (of which ~400 survived), restored ECC in 47 cells 2–10 days after injection. Approximately one half of these myotubes contracted spontaneously, the other half only after electrical stimulation. It is not known why the success rate of restoration was only 10%. However, none of the ~200 myotubes that were injected with the vector plasmid and survived contracted either spontaneously or after electrical stimulation. In normal skeletal muscle cells (including normal mouse myotubes) slow (I_{slow}) and fast, transient (I_{fast}) calcium currents are observed that have properties similar to L- and T-type currents in other tissues. Significantly, the slow current is completely absent from myotubes of mdg/mdg mice but I_{fast} is not affected. This strongly argues for different genes encoding for the different channel types.

If the same protein, encoded by the expression plasmid pCAC6, restored the I_{slow} current the postulated double role of $alpha_1$ would be further strengthened. Indeed, in mdg/mdg myotubes, where either electrically induced or spontaneous contractions were observable after pCAC6 injection,

I_{slow} was observed. Half-maximal block with $(+)$-PN 200-110 was obtained with 204 nM, close to the value (182 nM) reported for a normal holding potential (-80 mV) on freshly isolated muscle fibres (see Sect. 5.2).

As outlined in Sect. 7.2, divalent cations (Co^{2+}, Cd^{2+}) and even La^{3+} support ECC but "block" ion flux through the channel. Cd^{2+} block of I_{slow} (and I_{fast}) was observed in the injected myotubes, leaving ECC unimpaired. The similarity between the "corrected" muscle cells and normal cells is evident.

These studies provide the first example in which a genetic disease in a vertebrate is caused by a defect in a structural gene for an ionic channel and is corrected in cell culture when the "normal" cDNA is provided. The results obtained by Rieger et al. (1987) were explained by the contamination of the spinal cord neurons with nonneuronal cells (e.g. fibroblasts), which fuse with the myotubes expressing skeletal muscle-specific proteins (see Tanabe et al. 1988 and references cited therein). The experiments were not designed to answer questions about the role of the other channel subunits (e.g. alpha$_2$-delta, beta, gamma) which are presumably present in the *mdg* muscle cells and complemented the newly expressed alpha$_1$-subunits.

8 Molecular Cloning, Models and Expression

Molecular cloning has provided the deduced primary amino acid sequences of a few ion channels, out of more than 60 (Hille 1984). Ligand-activated channels (e.g. the nicotinic-acetylcholine receptor, the glycine receptor-chloride channel and the GABA$_A$ receptor-chloride channel) consist of several (homologous) subunits (each of some 500 amino acids). For these channels and the voltage-gated ion channels, models for secondary and even tertiary structure (see e.g., Guy and Seetharamulu 1986 for a tertiary structure model of the sodium channel) have been proposed. As a first step in building models the primary sequences are screened for alpha-helices which can span the hydrocarbon core of a phospholipid bilayer. Such a helix must consist of at least 22 amino acids and can be composed mainly (or entirely) of hydrophobic residues. However, as recently pointed out by Lodish (1988), there is no need for helices which are internal to be 22 amino acids in length. "Internal" helices do not contact the phospholipid bilayer; they interact with other helices of the protein. Laterally amphipathic helices (one face hydrophobic, the other polar) are also sought which could form the inner lining of a water-filled pore by their polar side chains.

Potential N-glycosylation and phosphorylation sites are identified which should face the extracellular fluid and the cytosol respectively.

For one of the best studied examples, the nicotinic-acetylcholine receptor, different models have been derived for the *Torpedo* delta-subunit, with three, four or even five transmembrane foldings (see McCrea et al. 1988 and Maelicke 1988 for critical comments). The five subunits, in constructing the channel, are arranged in such a way that amphipathic helices surround the pore. The accuracy of these models has been contested by Maelicke (1988), who also suggested that the hydropathy profiles and the alignments commonly employed to compare ionic channels (and other membrane proteins, e.g. receptors or transporters) should be viewed with some caution: "They could reflect properties of structural design more basic than functional specialisation". With respect to voltage-regulated ion channels, the deduced primary amino acid sequences of potassium channels from *Drosophila melanogaster* (see Sect. 2.1), of sodium channels (alpha-subunit) and of two subunits of the rabbit skeletal muscle (L-type) calcium channel/voltage sensor are now known.

The structure of the $alpha_1$-subunit was here of prime interest, as it contains the known regulatory sites (i.e. the receptors for L-type channel drugs and phosphorylation sites) of the channel. Its close structural relationship with the voltage-dependent sodium channels from rat brain is evident if one compares the respective amino acid sequences derived from the cloned cDNAs in Numa's laboratory (Noda et al. 1986a). Figure 17 shows the 6083-nucleotide sequence (excluding the poly (dA) tract) encoding the alpha$_1$-subunit from rabbit skeletal muscle. An open-reading frame encodes 1873 amino acids having a calculated molecular weight of 212018. Similarity matrix analysis of the amino acid sequence reveals the presence of four internal repeats (termed domains or motifs I – IV) that exhibit sequence homology (Fig. 18a,b). The homology with the sodium channel, most pronounced in the regions comprising the four internal repeats, is also shown in Fig. 18a. As many as 29% of the residues are identical for the two proteins, and the overall similarity is 55% if both identical residues and conservative exchanges are taken into account. The transmembrane topology predicted from the hydropathy profile (Fig. 18c) is also very similar for the two channels. Each of the four internal repeats has six presumably alpha-helical, membrane-spanning segments; five are hydrophobic (S1,S2,S3,S5,S6) and one is positively charged (S4). Segment S4 in every repeat of the alpha$_1$-subunit contains five or six Arg or Lys residues at every third (or fourth) position.

In the original sodium channel model by Noda et al. (1984) it was suggested that the S4 segments were located on the cytoplasmic side, but they are now believed to span the membrane (Kosower 1985). Interest in the "S4 regions" originates from theoretical considerations about voltage-dependent activation models of ionic channels (Armstrong 1981; Catterall 1986) and the finding that their amino acid sequence is essentially conserved in all sodium channel alpha-subunits from eel electroplax to rat brain (see Catterall 1986 and 1988

```
                                                    5'-----TTCCACCTACATGTTGGCCTGGACAGCAGGGAGCCGAGGGGAGGCTAATTTTACTGCTGGGAGCAGCTAGCATAA -151
TCCTCCCGCCCCCACCCCGCTGGCTCAGCAGGGCAGGCTTCGCCCGGCCAAGCTCAGCGGCCCAGTCCCCAAGGCGGGGAACACTGGGGACGCAGGGAAGAGAGGGCCGCGGGGTGGGGGAGCAGCAGGAAGCGCCGTGGCCAGGGAAGCC   -1
ATGGAGCCATCCTCACCCCAGGATGAGGGCCTGAGGAAGAAACAGCCCAAGAAGCCCCTGCCCGAGGTCCTGCCCAGGCCGCCGCGGGCTCTGTTCTGCCTGACCCTGCAGAACCCGCTGAGGAAGGCGTGCATCAGCATCGTGGAATGG   150
Ca  M E A P S S P Q D E G L R K K Q P K K P L P E V L P R P P R A L F C L T L Q N P L R K A C I S I V E W    50
Na  M A P P G P D S F R F F T R E S N K P K I S R F S A T S A L Y I L T P F N P I R K L A I K I V H           128
```

[The remainder of this page consists of an aligned nucleotide and deduced amino-acid sequence comparison between Ca and Na channel proteins, with boxed regions labelled IS1–IS6, IIS1–IIS6, IIIS1–IIIS6, and IVS1–IVS4, and numbered insertion markers (▲). The dense sequence alignment spans the full page with position numbers in the right margin.]

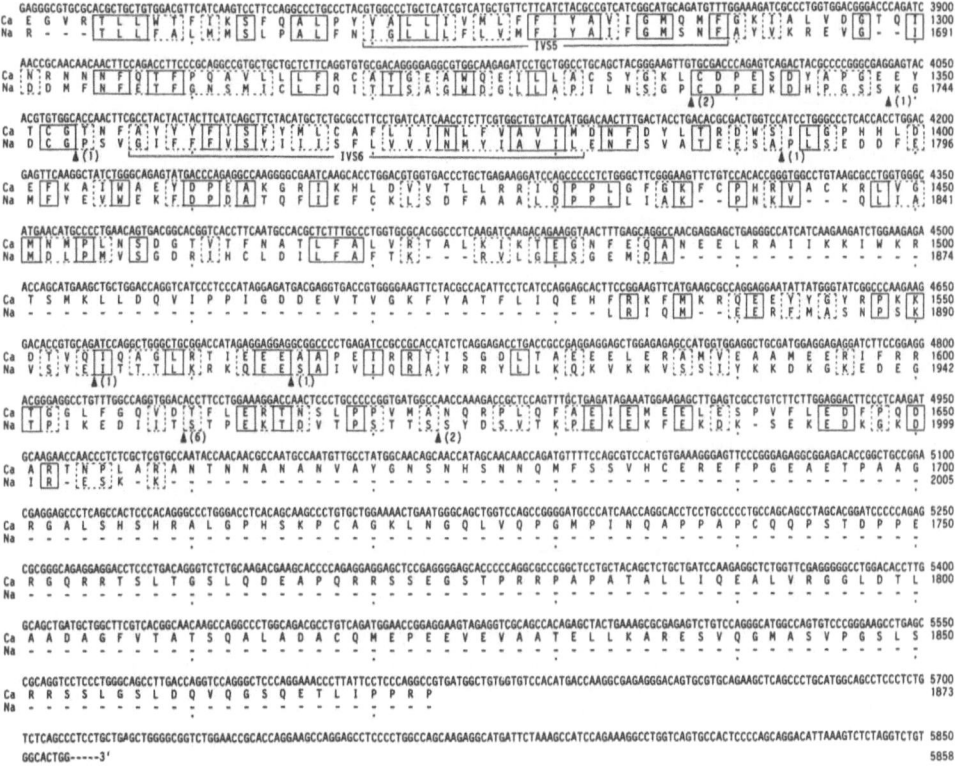

Fig. 17. Nucleotide sequence of cloned cDNA encoding the alpha₁-subunit from rabbit skeletal muscle and alignment of the deduced amino acid sequences of alpha₁ and rat sodium channel II alpha-subunit. For details see Tanabe et al. (1987). Sets of identical amino acid residues are enclosed with *solid lines* and sets of residues considered to be conservative substitutions with *broken lines*. Deletions and insertions in the amino acid sequence of the sodium channel II, as compared with that of the alpha₁-subunit, are indicated by gaps (−) and triangles (with the number of inserted residues in parenthesis) respectively. The putative transmembrane segments S1−S6 in each of the repeats I−IV of the alpha₁ sequence are indicated. Different residues were found by Ellis et al. (1988) in position 1808 (Thr to Met), 1815 (Ala to Val) and 1835 (Ala to Glu). [From Tanabe et al. (1987) with permission]

for further details). It is now generally believed that the "S4 region" is part of the voltage sensor. In the resting stage of the membrane, at negative potentials, all its positive charges are assumed to form ionic pairs with negatively charged amino acid residues on (yet unidentified) transmembrane segments.

Upon depolarisation, the forces holding the positive charges in inward position are reduced and the S4 helix is proposed to undergo a screwlike movement and outward displacement by a few angstroms, leaving one unpaired negative charge on the cytoplasmic face and revealing an unpaired positive charge on the extracellular face of the membrane. The movement of the helix is assumed to induce a conformational change in its domain as one step in

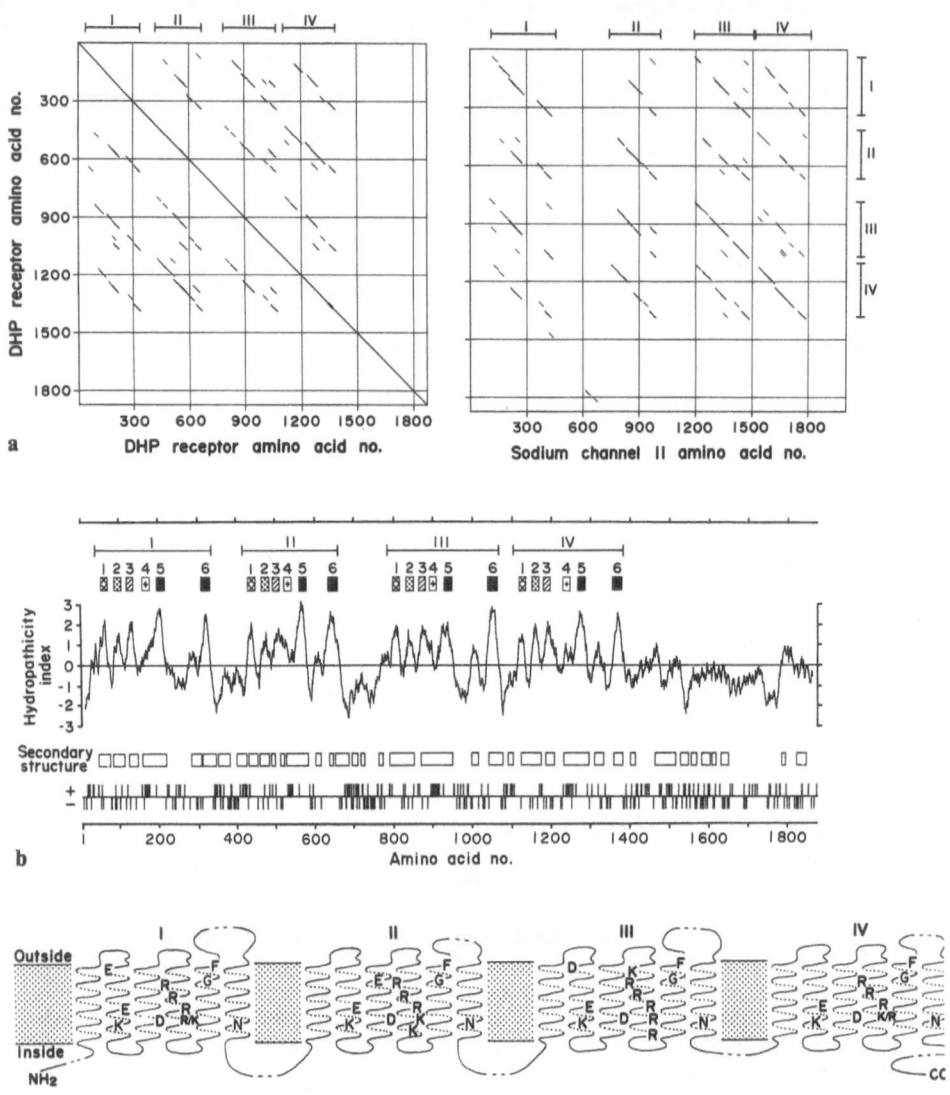

Fig. 18a. Similarity matrix analysis. The amino acid sequence of the alpha₁ polypeptide (*ordinate*) is compared with the sequences of itself (*abscissa, left*) and rat sodium channel II (*abscissa, right*). *Diagonal lines* indicate that 25 residues show sequence similarity (conservative substitutions as well as identities).

b Averaged hydropathicity index of a nonadecapeptide composed of amino acid residues i-9 to i+9 plotted against the amino acid number, i. The positions of the predicted structures of alpha-helices and/or beta-sheets that have a length of ten or more residues are shown by *open boxes* below. The positions of positively (Lys, Arg) and negatively (Asp, Glu) charged residues are indicated respectively by upward and downward vertical lines.

c 1–6, Putative transmembrane segments (S1–S6, see Fig. 17) of the repeats I to IV. Structural characteristics common to the alpha₁ subunit and the voltage-dependent sodium channel. The four units of homology spanning the membrane, which are assumed to surround the ionic channel, are displayed linearly. (For arrangement in pseudosymmetrical fashion, viewed perpendicu-

the activation of sodium channels. This model, which may also apply for other voltage-controlled structures (Armstrong 1981), is termed the "sliding helix" model (Guy and Seetharamulu 1986; Catterall 1986, 1988).

The *Drosophila melanogaster* potassium channels (*"Shaker"*) also contain the "S4 region" (see Miller 1988), which strengthens the view that this helix is elementary to voltage-regulated ion channels as opposed to ligand-activated ionic pores.

In contrast to the alpha$_1$-subunit of the skeletal muscle calcium channel and the alpha-subunit of the sodium channel, where one motif with the postulated six transmembrane segments is repeated four times, the clones of *Shaker* code for only one domain (or motif) with some 600 amino acids (see e.g. Tempel et al. 1987).

Within the alpha$_1$ sequence only weak similarity is found to parts of calcium-binding proteins, e.g. those of the EF-hand type or calcium-dependent membrane-binding proteins. Thus, with respect to ion selectivity (not to mention pharmacological specificity), there is as yet no clue from the deduced primary amino acid sequence (or the derived model) to characterise the structure as a "calcium" channel. The possible exception is the very large (cytosolic) C-terminal region of the alpha$_1$-subunit, which is extended compared with the sodium channel. However, it remains to be seen if this is characteristic for all alpha$_1$ polypeptides, be it from skeletal muscle, neurons, or smooth and heart muscle.

The alpha$_2$-subunit of the calcium channel has no homology with other known protein sequences (Ellis et al. 1988). Figure 19 shows 3802 nucleotides of the cDNA sequence. The open reading frame of 3318 nucleotides encodes a sequence of 1106 amino acids (calculated molecular weight 125018). Comparison of the previously determined N-terminal amino acid sequence of alpha$_2$ (Nakayama et al. 1987) with the cloned structure suggests that alpha$_2$ has a 26-amino-acid signal sequence. The hydropathy analysis reveals that

---

larly, consult Fig. 21.) The putative transmembrane alpha-helical segments in each repeat (I–IV) are illustrated. The Arg and/or Lys residues in segment S4 (see Figs. 17 and 18b) conserved in the individual repeats of the alpha$_1$-subunit and the three sodium channels of known sequence are shown in one-letter codes. The residues (one-letter code; D = Asp, E = Glu, F = Phe, G = Gly, K = Lys, N = Asn, R = Arg) in the other segments conserved in all repeats of alpha$_1$ and the three sodium channels are also shown and are as follows (amino acid numbers indicating those of alpha$_1$): Glu residues 100, 478, 846, 1164 and Lys residues 104, 482, 850 and 1168 in S2; Asp residues 126, 500, 872 and 1186 in S3; Gly residues 214, 577, 946, 1285 and Phe residues 218, 581, 950 and 1289 in S5; Asn residues 327, 654, 1058 and 1374 in S6. In addition, Glu 90 and Asp 836 in S2, conserved in repeats I and III respectively, and Glu 510 in S3, conserved in repeat II of the four proteins, are indicated. The proposed model is consistent with two of the five potential N-glycosylation sites (Asn 79 and Asn 257) on the extracellular side and all the potential cAMP-dependent phosphorylation sites (Ser 687, 1502, 1575, 1757, 1772, 1854 and Thr 1552) located intracytoplasmically. Phosphorylation of Ser 687 occurs in vitro with a rapid time course (Röhrkasten et al. 1988). Additional, but slower phosphorylation of Ser 1617 is also observed (see text). [From Tanabe et al. (1987) with permission]

```
5'      AGAAGGGAGGGCGAGCGTGGTGTGTGCGCGCTCGGCCGCCGGCGGCACCGCCGAGGTCTGTTGGCAAAAGT      -238
CGCCCTTGATGGCGGCGGAGGCGAGGCAGCCGCGGCGCCGAACAGCCGACGCGCGCTAGCGGGGTCCGCCCGCCCCTTT      -159
CCCAGAGCCCGAGCGCCGCCGTTCGCCGCCGCCGCCGCCGCCGCCGCCGTTCGCCGCCGCCGCCGCCGGGGTGGC      -80
AGCGCCGCTCGGTCCCCGGCCCCGGGGCCGGCTGGGGGGCGGTCGGGGCGTGTGAGGGGCTTGCTCCCAGCTCGCGAAG      -1
-26                                                                          -7
ATG GCT GCG GGC CGC CCG CTG GCC TGG ACG CTG ACA CTT TGG CAG GCG TGG CTG ATC CTG      60
Met Ala Ala Gly Arg Pro Leu Ala Trp Thr Leu Thr Leu Trp Gln Ala Trp Leu Ile Leu
                                                                              14
ATC GGG CCC TCG TCG GAG↓GAG CCG TTC CCT TCA GCC GTC ACT ATC AAG TCA TGG GTG GAT     120
Ile Gly Pro Ser Ser Glu Glu Pro Phe Pro Ser Ala Val Thr Ile Lys Ser Trp Val Asp
        -1  +1                                                                 34
AAG ATG CAA GAA GAC CTG GTC ACA CTG GCA AAA ACA GCA AGT GGA GTC CAT CAG CTT GTT     180
Lys Met Gln Glu Asp Leu Val Thr Leu Ala Lys Thr Ala Ser Gly Val His Gln Leu Val
                                                                              54
GAT ATT TAT GAG AAA TAT CAA GAT TTG TAT ACT GTG GAA CCA AAT AAT GCA CGT CAG CTG     240
Asp Ile Tyr Glu Lys Tyr Gln Asp Leu Tyr Thr Val Glu Pro Asn Asn Ala Arg Gln Leu
                                                                              74
GTG GAA ATT GCA GCC AGA GAC ATT GAG AAG CTT CTC AGC AAC AGA TCT AAA GCC CTG GTG     300
Val Glu Ile Ala Ala Arg Asp Ile Glu Lys Leu Leu Ser Asn Arg Ser Lys Ala Leu Val
                                                                              94
CGC CTG GCT TTG GAA GCA GAG AAA GTT CAA GCA GCC CAC CAA TGG AGG GAA GAT TTT GCA     360
Arg Leu Ala Leu Glu Ala Glu Lys Val Gln Ala Ala His Gln Trp Arg Glu Asp Phe Ala
                                                                              114
AGC AAT GAA GTT GTC TAC TAT AAC GCG AAG GAT GAT CTT GAT CCT GAA AAA AAT GAC AGT     420
Ser Asn Glu Val Val Tyr Tyr Asn Ala Lys Asp Asp Leu Asp Pro Glu Lys Asn Asp Ser
                                                                              134
GAA CCA GGC AGC CAG AGG ATC AAA CCT GTT TTC ATT GAC GAT GCT AAC TTT AGA AGA CAA     480
Glu Pro Gly Ser Gln Arg Ile Lys Pro Val Phe Ile Asp Asp Ala Asn Phe Arg Arg Gln
                                                                              154
GTA TCC TAT CAG CAC GCA GCT GTC CAT ATC CCC ACT GAC ATC TAT GAA GGA TCG ACA ATC     540
Val Ser Tyr Gln His Ala Ala Val His Ile Pro Thr Asp Ile Tyr Glu Gly Ser Thr Ile
                                                                              174
GTG TTA AAC GAA CTC AAC TGG ACA AGT GCC TTA GAT GAC GTT TTC AAA AAT CGA GAG     600
Val Leu Asn Glu Leu Asn Trp Thr Ser Ala Leu Asp Asp Val Phe Lys Lys Asn Arg Glu
                                                                              194
GAA GAC CCT TCA CTG TTG TGG CAG GTG TTT GGC AGT GCC ACT GGC CTG GCC CGG TAT TAC     660
Glu Asp Pro Ser Leu Leu Trp Gln Val Phe Gly Ser Ala Thr Gly Leu Ala Arg Tyr Tyr
                                                                              214
CCA GCT TCT CCA TGG GTT GAT AAT AGC CGA ACC CCA AAC AAG ATT GAT CTT TAT GAT GTA     720
Pro Ala Ser Pro Trp Val Asp Asn Ser Arg Thr Pro Asn Lys Ile Asp Leu Tyr Asp Val
                                                                              234
CGC AGA AGA CCA TGG TAC ATC CAA GGT GCT GCA TCC CCT AAA GAT ATG CTT ATT CTG GTG     780
Arg Arg Arg Pro Trp Tyr Ile Gln Gly Ala Ala Ser Pro Lys Asp Met Leu Ile Leu Val
                                                                              254
GAT GTG AGT GGA AGC GTT AGT GGA CTG ACA CTC AAA CTC ATC CGG ACA TCC GTC TCC GAA     840
Asp Val Ser Gly Ser Val Ser Gly Leu Thr Leu Lys Leu Ile Arg Thr Ser Val Ser Glu
                                                                              274
ATG TTG GAA ACC CTC TCA GAT GAT GAT TTT GTG AAC GTG GCT TCA TTT AAC AGT AAT GCT     900
Met Leu Glu Thr Leu Ser Asp Asp Asp Phe Val Asn Val Ala Ser Phe Asn Ser Asn Ala
                                                                              294
CAG GAT GTA AGC TGC TTT CAG CAC CTT GTC CAA GCA AAT GTA AGA AAT AAG AAA GTG TTG     960
Gln Asp Val Ser Cys Phe Gln His Leu Val Gln Ala Asn Val Arg Asn Lys Lys Val Leu
                                                                              314
AAA GAT GCA GTG AAT AAT ATC ACA GCA AAA GGA ATC ACA GAT TAT AAG AAG GGC TTT AGT     1020
Lys Asp Ala Val Asn Asn Ile Thr Ala Lys Gly Ile Thr Asp Tyr Lys Lys Gly Phe Ser
                                                                              334
TTT GCT TTT GAG CAG CTG CTT AAT TAT AAT GTA TCC AGA GCC AAC TGC AAT AAG ATT ATC     1080
Phe Ala Phe Glu Gln Leu Leu Asn Tyr Asn Val Ser Arg Ala Asn Cys Asn Lys Ile Ile
                                                                              354
ATG TTG TTC ACG GAC GGA GGA GAA GAG AGA GCC CAG GAG ATA TTT GCC AAA TAC AAT AAA     1140
Met Leu Phe Thr Asp Gly Gly Glu Glu Arg Ala Gln Glu Ile Phe Ala Lys Tyr Asn Lys
                                                                              374
GAC AAG AAA GTA CGT GTA TTC ACA TTC TCA GTT GGT CAA CAT AAT TAC GAC AGA GGA CCT     1200
Asp Lys Lys Val Arg Val Phe Thr Phe Ser Val Gly Gln His Asn Tyr Asp Arg Gly Pro
                                                                              394
ATT CAG TGG ATG GCT TGC GAA AAT AAA GGT TAT TAT TAT GAA ATT CCA TCC ATT GGA GCC     1260
Ile Gln Trp Met Ala Cys Glu Asn Lys Gly Tyr Tyr Tyr Glu Ile Pro Ser Ile Gly Ala
                                                                              414
ATA AGA ATT AAT ACT CAG GAA TAC CTA GAT GTT CTG GGA AGA CCG ATG GTT TTA GCA GGA     1320
Ile Arg Ile Asn Thr Gln Glu Tyr Leu Asp Val Leu Gly Arg Pro Met Val Leu Ala Gly
                                                                              434
GAC AAA GCT AAG CAA GTC CAA|TGG ACA AAT GTG TAC CTG GAT GCA CTG GAA CTG GGA CTT     1380
Asp Lys Ala Lys Gln Val Gln|Trp Thr Asn Val Tyr Leu Asp Ala Leu Glu Leu Gly Leu
                                                                              454
GTC ATT ACT GGA ACT CTT CCG GTC TTC AAC ATA|ACT GGC CAA TTT GAA AAT AAG ACA AAC     1440
Val Ile Thr Gly Thr Leu Pro Val Phe Asn Ile|Thr Gly Gln Phe Glu Asn Lys Thr Asn
                                                                              474
TTA AAG AAC CAG CTG ATT CTT GGA GTG ATG GGA GTG TCT TTG GAA GAT ATT AAG     1500
Leu Lys Asn Gln Leu Ile Leu Gly Val Met Gly Val Asp Val Ser Leu Glu Asp Ile Lys
                                                                              494
AGA CTG ACA CCA CGT TTT ACA CTC TGC CCC AAT GGC TAC TAT TTT GCA ATT GAT CCT AAT     1560
Arg Leu Thr Pro Arg Phe Thr Leu Cys Pro Asn Gly Tyr Tyr Phe Ala Ile Asp Pro Asn
                                                                              514
GGT TAT GTG TTA TTA CAT CCA AAT CTT CAG CCA AAG CCT ATT GGT GTA GGT ATA CCA ACA     1620
Gly Tyr Val Leu Leu His Pro Asn Leu Gln Pro Lys Pro Ile Gly Val Gly Ile Pro Thr
                                                                              534
ATT AAT TTG AGA AAA AGG AGA CCC AAT GTT CAG AAC CCC AAA TCT CAG GAG CCA GTG ACA     1680
Ile Asn Leu Arg Lys Arg Arg Pro Asn Val Gln Asn Pro Lys Ser Gln Glu Pro Val Thr
                                                                              554
TTG GAT TTC CTC GAT GCA GAG TTG GAG AAT GAC ATT AAA GTG GAG ATT CGA AAT AAA ATG     1740
Leu Asp Phe Leu Asp Ala Glu Leu Glu Asn Asp Ile Lys Val Glu Ile Arg Asn Lys Met
                                                                              574
ATC GAT GGA GAA AGT GGA GAA AAA ACA TTC AGA ACT CTG GTT AAA TCT CAA GAT GAG AGA     1800
Ile Asp Gly Glu Ser Gly Glu Lys Thr Phe Arg Thr Leu Val Lys Ser Gln Asp Glu Arg
                                                                              594
TAT ATT GAC AAA GGA AAC AGG ACA TAC ACG TGG ACT CCT GTC AAC GGC ACA GAT TAT AGC     1860
Tyr Ile Asp Lys Gly Asn Arg Thr Tyr Thr Trp Thr Pro Val Asn Gly Thr Asp Tyr Ser
                                                                              614
AGT TTG GCC TTG GTA TTA CCA ACC TAC AGT TTT TAC TAT ATA AAA GCC AAA ATA GAA GAG     1920
Ser Leu Ala Leu Val Leu Pro Thr Tyr Ser Phe Tyr Tyr Ile Lys Ala Lys Ile Glu Glu
```

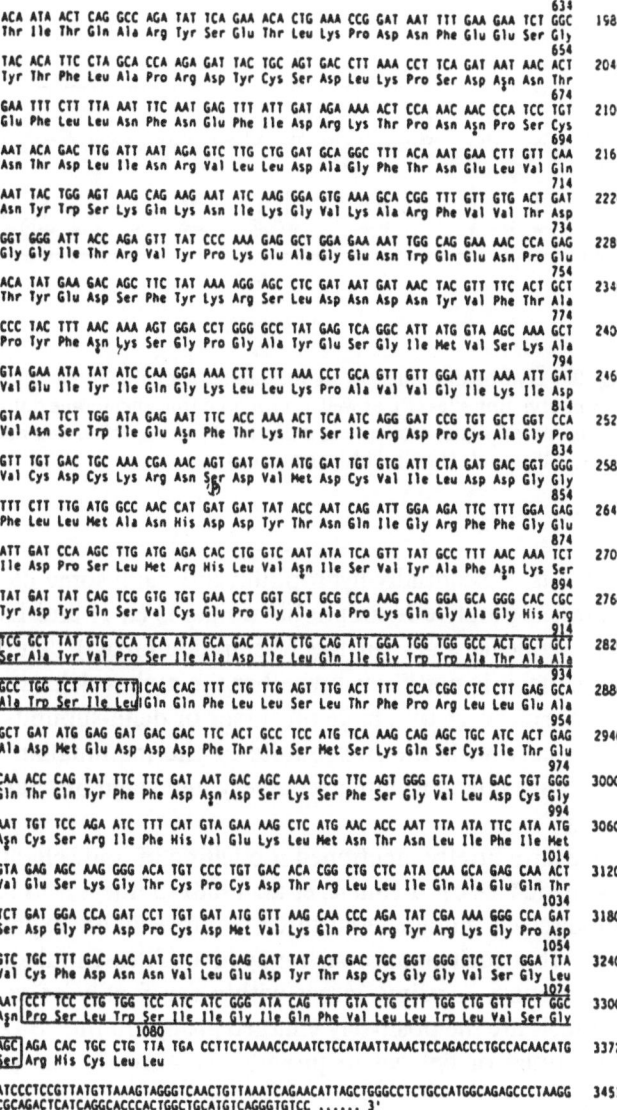

Fig. 19. cDNA sequence of alpha₂ from rabbit skeletal muscle and the predicted amino acid sequence. A signal peptide is assumed to consist of the first 26 amino acids (residues −26 to −1). An *arrow* identifies the proposed cleavage site, and positive numbering starts with the proposed N-terminal residue of the mature protein. The N-terminal amino acid sequence previously determined (Nakayama et al. 1987) is equivalent to residues 1−17. An in-frame upstream stop codon is underlined as well as the start and stop codons of an upstream short open reading frame. Three putative transmembrane regions are enclosed in boxes. Potential N-glycosylation or phosphorylation sites are identified by an *asterisk* (*) below Asn residues or a *P* symbol below the Ser and Thr residues respectively. [From Ellis et al. (1988) with permission]

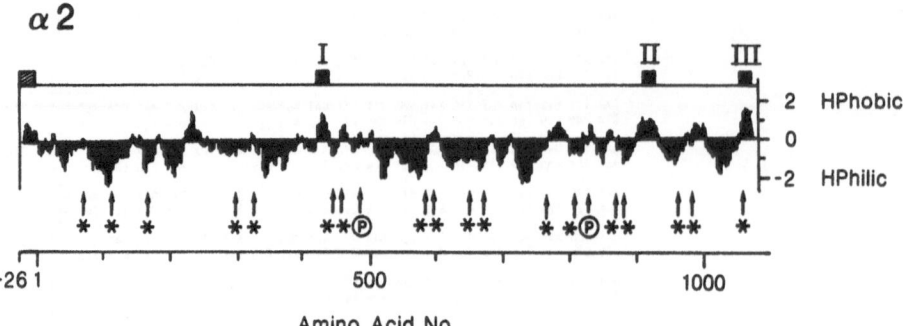

Fig. 20. Hydropathy analysis of the alpha$_2$-subunit. Potential sites of glycosylation (*) and phosphorylation sites (*P*), as well as proposed transmembrane domains (*I, II, III*), are indicated. The proposed signal sequence is indicated by the *hatched box* above the residues −1 to −26. [From Ellis et al. (1988) with permission]

alpha$_2$ is substantially hydrophilic, although some parts of the sequence may represent putative hydrophobic domains (I, II and III in Fig. 20). As amino acid sequence data on the (disulphide-linked) delta subunit(s) are not yet available, it is not known whether the alpha$_2$-delta structure arises from proteolytic cleavage of a large precursor or delta-subunits are attached later. This is the case with the β_2-subunits of the mammalian nerve sodium channel, which are linked to alpha at a very late stage of channel assembly (Catterall 1986).

We have briefly mentioned modelling of ionic channels, as attempts are being made (see e.g. Leonard et al. 1988) to inject normal and modified (e.g. by site-directed mutagenesis) mRNA into suitable expression systems. It is believed that such an approach may identify essential amino acids in the primary structure. A prime target within the family of voltage-dependent cation channels are the positively charged groups in the "S4 region", which could be substituted or deleted to provide evidence for the proposed gating mechanism.

Ultimate proof that the structure of interest has been cloned requires expression of mRNA synthesised in vitro from cloned cDNA probes in systems like *Xenopus* oocytes that allow electrophysiological or even biochemical recognition (Lester 1988). Expression of sodium channels has been achieved in this system, indicating that the alpha-subunit of the rat brain sodium channel alone is sufficient to form a voltage-dependent channel protein (Goldin et al. 1986; Noda et al. 1986b). A single *Shaker* messenger RNA suffices to direct the synthesis of functional "A"-type potassium channels (Timpe et al. 1988). By analogy with the other ionic channels it is deduced that the *Shaker* products form homomultimeric structures in oocytes (Fig. 21). It is also speculated that in the fly, potassium channels are made up of heterologous subunits. In-

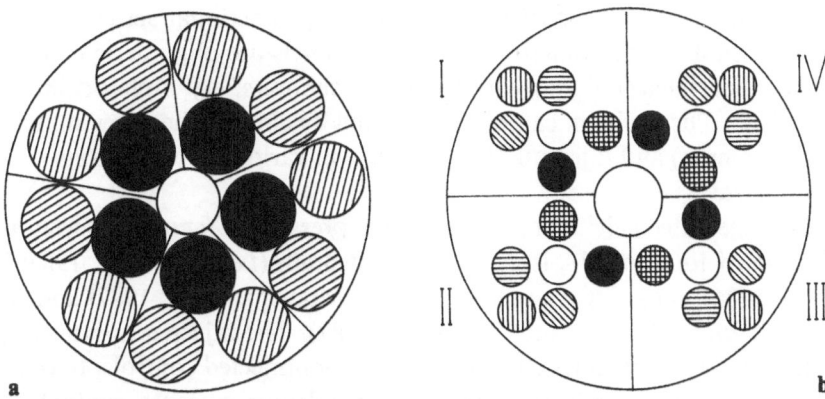

Fig. 21a,b. Models showing the arrangement of repeated domains or motifs and of homologous subunits of ion channels to form pores.

a Model for a ligand-activated ion channel. The *Torpedo* nicotinic-acetylcholine receptor (shown schematically) consists of five homologous subunits (hetero-oligomer) in pseudosymmetric pentagonal array. Putative transmembrane elements (only three for each subunit are shown) are depicted as *circles* (in different shades) in this cross-sectional view. Amphipathic helices (*shaded*) with polar residues facing the water-filled pore form the inner lining. GABA$_A$ receptors are hetero-oligomers in vivo. Surprisingly, when alpha and beta subunits are expressed individually in *Xenopus* oocytes any one of the RNAs (synthesised from the cloned DNA) expressed a GABA-sensitive chloride channel. This suggests that hetero-oligomers are not necessary for basic channel functions, including ligand activation (Blair et al. 1988). A synthetic peptide from the rat brain sodium channel S3 segment with 22 amino acids is sufficient to form a cation-selective channel and was modelled as a bundle of four amphipathic helices, surrounding a central pore (Oiki et al. 1988). This supports but does not prove the central dogma that pores are formed either by symmetrical arrangement of homologous subunits or by pseudosymmetrical arrangement of homologous domains (see **b**).

b Voltage-gated ion channels consist of one large polypeptide (alpha-subunit of the sodium channel, alpha$_1$-subunit of the calcium channel) with four homologous motifs or domains (labelled I–IV) each consisting of six putative transmembrane elements (see, however, Guy and Seetharamulu (1986) for an alternative model with eight segments) arranged in pseudosymmetrical fashion around the central pore. The common "S4 region" (part of the voltage-sensing element) is depicted as an *open circle*. The potassium channel (i.e. Sh A1 protein) from the *Shaker* locus of *Drosophila melanogaster* is proposed to have seven membrane-spanning regions including the voltage sensor (Tempel et al. 1987). It is believed (see above) that a single 70-kDa potassium channel protein is too small to form a channel and the pore may be formed by four identical subunits (tetrameric structure) instead of four repeated motifs. Thus, the potassium channel may be an archetype of cation-selective voltage-dependent channels from which sodium and calcium channels evolved by gene duplication and fusion

deed, cross-linking experiments with the dendrotoxin binding sites (putative potassium channels) indicate a heterologous subunit composition in rat and chick synaptic membranes (Dolly 1988).

Expression of functional calcium channels with mRNA derived from the cloned cDNA of rabbit skeletal muscle alpha$_1$ has not yet been reported in *Xenopus* oocytes.

One reason for the difficulty of expressing functional activity with alpha$_1$ alone could be that functional L-type channels are oligomeric complexes of (phosphorylated?) alpha$_1$ with the other subunits, e.g. alpha$_2$-delta, gamma and beta. The problem can be approached by cloning the other subunits (e.g. beta, gamma) and injecting their mRNAs in various combinations into a single oocyte.

Voltage-gated calcium channels were expressed in *Xenopus* oocytes after injection of total mRNA isolated from rat brain (Leonard et al. 1987; Dascal et al. 1986), rat heart (Dascal et al. 1986; Moorman et al. 1987) and *Torpedo* electric lobe (Umbach and Gundersen 1987). The injected oocytes displayed additional calcium currents not present in noninjected controls. In oocytes injected with heart mRNA, two distinct calcium currents could be distinguished. A fast-inactivating, transient component (similar to the T-type) and a slowly inactivating, more dominant component (similar to the L-type) could be distinguished pharmacologically, as only the slow component was sensitive to inhibition by calcium antagonists, e.g. nifedipine. Furthermore, the total current through the calcium channels was increased by isoproterenol, by the adenylate cyclase activator forskolin, and by injection of cAMP. Thus, the calcium channels incorporated after total mRNA injection display regulatory properties very similar to those of native cardiac calcium channels. Expression of a high-threshold (L-type) cardiac calcium channel was also confirmed by single channel recordings (Moorman et al. 1987). Taken together, the injection experiments with total mRNA give reason to hope that the "ultimate proof" for the cloned calcium channel cDNAs is only a question of time. Lotan et al. (1989) provided indirect evidence for the existence of a polypeptide in heart muscle which is highly homologous to the skeletal muscle alpha$_1$ subunit and of crucial importance for channel function. When two cDNA oligonucleotides complementary to mRNA coding for S4 and S3 segments of the skeletal muscle alpha$_1$ subunit were preincubated with heart mRNA prior to injection into oocytes expression of calcium channel activity is blocked. This can be explained by an RNAse H-like degradation of the mRNA moiety of the cDNA-mRNA hybrid formed upon incubation. The effect is specific as the expression of other voltage-dependent ion channels (Na- or K-channels) is not suppressed.

9 Future Prospects

The voltage-dependent calcium channel from skeletal muscle is only one isochannel from the family of L-type channels. Perhaps by the time this review appears, the alpha$_1$-subunits from smooth and heart muscle or neuronal tissue will have been cloned. It is predicted from photoaffinity labelling experiments that their size may be slightly larger in heart or brain than in skeletal muscle. It does not make much sense to predict in detail what will happen next. Molecular biology (in concert with expression methods) and immunology have already set the stage with ligand-activated or voltage-dependent sodium channels. There is, however, an important difference to all other ionic channels studied so far, namely the "double role" (i.e. to function as a pore and as a sensor for excitation-contraction coupling) of $alpha_1$ in skeletal muscle. With the unique mdg/mdg mouse system, $alpha_1$-subunits from tissues where a single (pore-only) function is assumed may be tested for restoration of contraction. This is no easy task, as the partial-length clones must be ligated first, inserted into a vector and injected. The long cytosolic carboxy terminal of $alpha_{1,sk}$ is suspected (Tanabe et al. 1987) to be functionally important in ECC, and truncation will be a logical approach. The serine residue 687 is the main substrate for in vitro phosphorylation by cAMP-dependent protein kinase and seems (at least for some reconstituted systems) to be essential to pore function. Site-directed mutagenesis could reveal whether this holds for ECC as well.

One would very much like to inspect $alpha_1$ primary structures from other species – evolutionarily as distant as possible. Although more spectacular gains are within reach this approach is scientifically sound and justified as exemplified for the sodium channel: Here a short cytosolic segment, connecting domains III and IV, is highly preserved from *Drosophila* to mammalian brain and was suspected to be involved in the inactivation process ("h"-gate). Site-directed antibodies generated against the synthetic peptide segment confirmed the role of this domain (see Catterall 1988).

Cloning (and/or expression) is at present the only way to get access to T-channels, as no selective high-affinity drug or toxin is in sight. Prospects for a conventional purification of N-channels may be better, and the comparison of the pore-forming subunits of all three types (L,T,N) will be most interesting. A cloned phenylalkylamine-sensitive calcium channel from *Drosophila* (about twice the size of the *Shaker*-coded potassium channel) will presumably present more insight into the evolution of calcium channels and their drug-binding domains.

$Alpha_1$ from heart and brain appears to be associated with $alpha_2$-delta, as first shown for skeletal muscle. While $alpha_1$ is most likely encoded by different genes in different tissues, $alpha_2$ may be encoded by one gene. The role of $alpha_2$-delta and of the other subunits found in the skeletal muscle-calcium channel complex needs to be clarified. Ankyrin, spectrin and a 33-kDa protein bind with high affinity to sodium channels (Srinivasan et al. 1988; Edelstein et al. 1988). These proteins are believed to segregate the channel to specialised regions in membranes. The L-type calcium channels, in contrast to the sodium channels, in skeletal muscle are concentrated in the T-tubule membrane, and a different localisation of N- versus L-type channels in neurons has been postulated (Miller 1987). Perhaps alpha$_2$-delta guides alpha$_1$ into a specialised domain (as suggested by Ellis et al. 1988). However, the putative sodium channel immobilisation or segregating proteins never co-purify in a 1:1 stoichiometry as is observed for alpha$_1$ and alpha$_2$-delta.

Of great interest are the effects of G-proteins on calcium channel activity. The ultimate criteria (reconstitution of purified components) have not yet been met here. Again, as emphasised above, if alpha$_1$ is the effector one would like to compare as many sequences as possible to get clues as to which cytosolic region binds activated G-proteins. This could widen the knowledge on the structural requirements for G-protein binding. For the best-studied effector, adenyl cyclase, the complex of activated $G_{s,\,alpha} \cdot Gpp(NH)p$ with the enzyme survives purification to homogeneity (see Gilman 1987). Such data on the calcium channel are still missing.

A major goal is to identify the domains on L-channels which bind the drugs. For the best-studied class, the 1,4 DHPs, low-affinity (but nevertheless stereoselective) sites have been shown to exist on sodium channels (Yatani et al. 1988). This points to related domains on both voltage-sensitive pores. The radiolabelled photoaffinity probes which are now available for the three drug classes will be useful here. Very recent work by Catterall's group on the localisation of the sodium-channel alpha-scorpion toxin-binding site with site-directed antibodies (Catterall 1988) is interesting in this context. Alternatively, a novel approach (Scheffauer 1989; Striessnig et al. 1989) may be feasible. Antibodies directed against a benzothiazepine photoaffinity ligand bind with high affinity to alpha$_1$-subunits which are photolabelled with the non-radioactive ligand. Inexpensive enzyme-linked immunosorbent assays could help to identify the binding domains in digested samples. For further prospects the reader may consult the article by Lester (1988) on heterologous expression of excitability proteins.

Acknowledgements. We thank the following colleagues for their friendly cooperation by making available their preprints and figures or supplying us with their most recent data: Drs. K.G. Beam (Fort Collins, Colorado, USA), K.P. Campbell (Iowa City, Iowa, USA), W.A. Catterall (Seattle, USA), W. Fischli (Basel, Switzerland), F. Hofmann, D. Pelzer, V. Flockerzi (all Homburg/Saar, F.R.G.), M. Seagar and colleagues (Marseille, France), S. Fleischer (Nashville, Tennessee, USA), G. Meissner (Chapel Hill, North Carolina, USA), A. Schwartz (Cincinnati, Ohio, USA), and E. Rios (Chicago, Illinois, USA). We appreciate the stimulating cooperation of L. Hymel, H.G. Schindler, W. Schreibmayer and H.A. Tritthart within the *Schwerpunktprogramm "Ionenkanäle"* of the *Fonds zur Förderung der wissenschaftlichen Forschung* (Austria). Research by the authors was funded by the *Deutsche Forschungsgemeinschaft*, the Dr. Legerlotz Foundation and by a project grant from the *Bundesministerium für Wissenschaft und Forschung*. Many colleagues in the pharmaceutical industry were helpful with unlabelled and labelled drugs. K. Hofer provided secretarial assistance and C. Trawöger did our art work.

References

Abe T, Saisu H (1987) Identification of the receptor for omega-conotoxin in brain. Probable components of the calcium channel. J Biol Chem 262:9877–9882

Abe T, Koyano K, Saisu H, Nishiuchi Y, Sakakibara S (1986) Binding of omega-conotoxin to receptor sites associated with the voltage-sensitive calcium channel. Neurosci Lett 71:203–208

Affolter H, Coronado R (1985) Agonists Bay-K8644 and CGP-28392 open calcium channels reconstituted from skeletal muscle transverse tubules. Biophys J 48:341–347

Almers W, McCleskey EW, Palade PT (1985) The mechanism of ion selectivity in calcium channels of skeletal muscle membranes. Prog Zool 33:61–73

Arahata K, Ishiura S, Ishiguro T, Tsukahara T, Suhara Y, Eguchi C, Ishihara T, Nonaka I, Ozawa E, Sugita H (1988) Immunostaining of skeletal and cardiac muscle surface membrane with antibody against Duchenne muscular dystrophy peptide. Nature 333:861–863

Arahata K, Ishiura S, Tsukahara T, Sugita H (1989) Dystrophin digest. Nature 337:606

Armah BI, Pfeiffer T, Ravens U (1989) Reversal of the cardiotonic and action potential prolonging effects of DPI 201-106 by BDF 8784, a methyl-indol derivative. Br J Pharmacol 96:807–816

Armstrong CM (1981) Sodium channels and gating currents. Physiol Rev. 61:644–683

Atchison WD, Adgate L, Beaman CM (1988) Effects of antibiotics on uptake of calcium into isolated nerve terminals. J Exp Pharmacol Toxicol 245:394–401

Auguet M, Delaflotte S, Chabrier PE, Pirotzky E, Clostre F, Braquet P (1988) Endothelin and Ca^{++} agonist Bay K 8644: different vasoconstrictive properties. Biochem Biophys Res Commun 156:186–192

Balwierczak JL, Johnson CL, Schwartz A (1987) The relationship between the binding site of [^{3}H]-d-cis-diltiazem and that of other non-dihydropyridine calcium entry blockers in cardiac sarcolemma. Mol Pharmacol 31:175–179

Barchi RL (1988) Probing the molecular structure of the voltage-dependent sodium channel. Annu Rev Neurosci 11:455–495

Barhanin J, Coppola T, Schmid A, Borsotto M, Lazdunski M (1987) The calcium channel antagonists receptor from rabbit skeletal muscle. Reconstitution after purification and subunit characterization. Eur J Biochem 164:525–531

Barhanin J, Schmid A, Lazdunski M (1988) Properties of structure and interaction of the receptor for omega-conotoxin, a polypeptide active on Ca^{2+} channels. Biochem Biophys Res Commun 150:1051–1062

Beam KG, Knudson CM (1988) Calcium currents in embryonic and neonatal mammalian skeletal muscle. J Gen Physiol 91:781–798

Beam KG, Knudson CM, Powell JA (1986) A lethal mutation in mice eliminates the slow calcium current in skeletal muscle cells. Nature 320:168–170

Bean BP (1989) More than a channel? Trends Neurosci 12:128–129

Bechem M, Hebish S, Schramm M (1988) Calcium agonists – new sensitive probes for calcium channels. Trends Pharmacol Sci 9:257–261

Berta P, Sladeczek F, Derancourt J, Durand M, Travo P, Haiech J (1986) Maitotoxin stimulates the formation of inositol phosphates in rat aortic myocytes. FEBS Lett 197:349–352

Berta P, Phaneuf S, Derancourt J, Casanova J, Durand-Clement M, Le Peuch C, Haiech J, Cavadore JC (1988) The effects of maitotoxin on phosphoinositides and calcium metabolism in a primary culture of aortic smooth muscle cells. Toxicon 26:133–141

Biswas CJ, Rogers TB (1986) Synthesis of carboxy-nifedipine and its use in the preparation of an affinity resin for the 1,4-dihydropyridine receptor. Biochem Biophys Res Commun 134:922–927

Bkaily G, Sperelakis N, Renaud JF, Payet MD (1985) Apamin, a highly specific calcium-blocking agent in heart muscle. Am J Physiol 248:H961–H965

Blair LAC, Levitan ES, Marshall J, Dionne VE. Barnard EA (1988) Single subunits of the, GABA$_A$ receptor form ion channels with properties of the native receptor. Science 242:577–579

Block BA, Imagawa T, Campbell KP, Franzini-Armstrong C (1988) Structural evidence for direct interaction between the molecular components of the transverse tubule/sarcoplasmic reticulum junction in skeletal muscle. J Cell Biol 107:2587–2600

Borsotto M, Barhanin J, Norman RI, Lazdunski M (1984a) Purification of the dihydropyridine receptor of the voltage-dependent Ca^{2+} channel from skeletal muscle transverse tubules using (+)-[^{3}H]PN-200-110. Biochem Biophys Res Commun 122:1357–1366

Borsotto M, Norman RI, Fosset M, Lazdunski M (1984b) Solubilization of the nitrendipine receptor from skeletal muscle transverse tubule membranes. Interactions with specific inhibitors of the voltage-dependent Ca^{2+} channel. Eur J Biochem 142:449–455

Borsotto M, Barhanin J, Fosset M, Lazdunski M (1985) The 1,4-dihydropyridine receptor associated with the skeletal muscle voltage-dependent Ca^{2+} channel. Purification and subunit composition. J Biol Chem 260:14255–14263

Bowers CW, Phillips HS, Lee P, Jan YN, Jan LY (1987) Identification and purification of an irreversible presynaptic neurotoxin from the venom of the spider *Hololena curta*. Proc Natl Acad Sci USA 84:3506–3510

Branton WD, Kolton L, Jan YN, Jan LY (1987) Neurotoxins from Plectreurys spider venom are potent presynaptic blockers in *Drosophila*. J Neurosci 7:4195–4200

Brown AM, Birnbaumer L (1988) Direct G-protein gating of ion channels. Am J Physiol 254:401–410

Brown AM, Yatani A, Lacerda AE, Gurrola GB, Possani LD (1987) Neurotoxins that act selectively on voltage-dependent cardiac calcium channels. Circ Res 61:6–9

Brown RH, Hoffmann EP (1988) Molecular biology of Duchenne muscular dystrophy. Trends Neurosci 11:480–484

Brum G, Flockerzi V, Hofmann F, Osterrieder W, Trautwein W (1983) Injection of catalytic subunit of cAMP-dependent protein kinase into isolated cardiac myocytes. Pflugers Arch 398:147–154

Brum G, Stefani E, Rios E (1987) Simultaneous measurements of Ca^{2+} currents and intracellular Ca^{2+} concentrations in single skeletal muscle fibers of the frog. Can J Physiol Pharmacol 65:681–685

Brum G, Fitts R, Pizarro G, Rios E (1988) Voltage sensors of the frog skeletal muscle membrane require calcium to function in excitation-contraction coupling. J Physiol 398:475–505

Brunner J, Spiess M, Aggeler R, Huber P, Semenza G (1983) Hydrophobic labeling of a single leaflet of the human erythrocyte membrane. Biochemistry 22:3812–3820

Bürgisser E, Hancock AA, Lefkowitz RJ, DeLean A (1981) Anomalous equilibrium binding properties of high-affinity racemic radioligands. Mol Pharmacol 19:205–216

Callewaert G, Hanbauer I, Morad M (1989) Modulation of calcium channels in cardiac and neuronal cells by an endogenous peptide. Science 243:663–666

Campbell KP, Lipshutz GM, Denney GH (1984) Direct photoaffinity labelling of the high affinity nitrendipine-binding site in subcellular membrane fractions isolated from canine myocardium. J Biol Chem 259:5384–5387

Campbell KP, Sharp A, Strom M, Kahl SD (1986) High-affinity antibodies to the 1,4-dihydropyridine Ca^{2+}-channel blockers. Proc Natl Acad Sci USA 83:2792–2796

Campbell KP, Knudson CM, Imagawa T, Leung AT, Sutko JL, Kahl SD, Raab CR, Madson L (1987) Identification and characterization of the high-affinity [^{3}H]ryanodine receptor of the junctional sarcoplasmic reticulm Ca^{2+} release channel. J Biol Chem 262:6460–6463

Campbell KP, Leung AT, Sharp AH (1988a) The biochemistry and molecular biology of the dihydropyridine-sensitive calcium channel. Trends Neurosci 11:425–430

Campbell KP, Leung AT, Sharp AH, Imagawa T, Kahl SD (1988b) Calcium channel antibodies: subunit-specific antibodies as probes for structure and function. In: Morad, M et al. (eds) The calcium channel: structure, function and implications. Springer, Berlin Heidelberg New York, pp 586–600

Catterall WA (1986) Molecular properties of voltage-sensitive sodium channels. Annu Rev Biochem 55:953–985

Catterall WA (1988) Structure and function of voltage-sensitive ion channels. Science 242:50–61

Catterall WA, Seagar MJ, Takahashi M (1988) Molecular properties of dihydropyridine-sensitive calcium channels in skeletal muscle. J Biol Chem 263:3535–3538

Catterall WA, Seagar MJ, Takahashi M, Nunoki K (1989) Molecular properties of dihydropyridine-sensitive calcium channels. Ann N Y Acad Sci 560:1–14

Chhatwal GS, Hessler HJ, Habermann E (1983) The action of palytoxin on erythrocytes and resealed ghosts. Formation of small, nonselective pores linked with Na^+, K^+ATPase. Naunyn Schmiedebergs Arch Pharmacol 323:261–268

Conti-Tronconi BM, Raftery MA (1982) The nicotinic receptor: correlation of molecular structure with functional properties. Annu Rev Biochem 51:491–530

Cooper CL, Vandaele S, Barhanin J, Fosset M, Lazdunski M, Hosey MM (1987) Purification and characterization of the dihydropyridine-sensitive voltage-dependent calcium channel from cardiac tissue [published erratum appears in J Biol Chem 1987 Mar 15; 262(8):3927]. J Biol Chem 262:509–512

Crosland RD, Hsiao TH, McClure WO (1984) Purification and characterization of beta-leptinotarsin-h, an activator of presynaptic calcium channels. Biochemistry 23:734–741

Cruz LJ, Olivera BM (1986) Calcium channel antagonists. Omega-conotoxin defines a new high-affinity site. J Biol Chem 261:6230–6233

Cruz LJ, Johnson DS, Olivera BM (1987) Characterization of the omega-conotoxin target. Evidence for tissue-specific heterogeneity in calcium channel types. Biochemistry 26:820–824

Curtis BM, Catterall WA (1983) Solubilization of the calcium antagonist receptor from rat brain. J Biol Chem 258:7280–7283

Curtis BM, Catterall WA (1984) Purification of the calcium antagonist receptor of the voltage-sensitive calcium channel from skeletal muscle transverse tubules. Biochemistry 23:2113–2118

Curtis BM, Catterall WA (1985) Phosphorylation of the calcium antagonist receptor of the voltage-sensitive calcium channel by cAMP-dependent protein kinase. Proc Natl Acad Sci USA 82:2528–2532

Curtis BM, Catterall WA (1986) Reconstitution of the voltage-sensitive calcium channel purified from skeletal muscle transverse tubules. Biochemistry 25:3077–3083

Dascal N, Snutch TP, Lübbert H, Davidson N, Lester HA (1986) Expression and modulation of voltage-gated calcium channels after RNA injection in *Xenopus* oocytes. Science 231:1147–1151

DiVirgilio F, Salviati G, Pozzan T, Volpe P (1986) Is a guanine-nucleotide-binding protein involved in excitation-contraction coupling in skeletal muscle? EMBO J 5:259–262

Doble A, Benavides J, Ferris O, Bertrand P, Menager J, Vaucher N, Burgevin MC, Uzan A, Gueremy C, Le Fur G (1985) Dihydropyridine and peripheral-type benzodiazepine binding sites: subcellular distribution and molecular size determination. Eur J Pharmacol 119:153–167

Dolly JO (1988) Potassium channels – what can protein chemistry contribute? Trends Neurosci 11:186–188

Donaldson SK, Goldberg ND, Walseth TF, Huetteman DA (1988) Voltage-dependence of inositol 1,4,5-trisphosphate-induced Ca^{2+} release in peeled skeletal muscle fibers. Proc Natl Acad Sci USA 85:5749–5753

Dooley DJ, Lupp A, Hertting G (1987) Inhibition of central neurotransmitter release by omega-conotoxin GVIA, a peptide modulator of the N-type voltage-sensitive calcium channel. Naunyn Schmiedebergs Arch Pharmacol 336:467–470

Dooley DJ, Lupp A, Hertting G, Osswald H (1988) Omega-conotoxin GVIA and pharmacological modulation of hippocampal noradrenaline release. Eur J Pharmacol 148:261–267

Ebersole BJ, Molinoff PB (1988) Endogenous ligands for voltage-sensitive calcium channels in extracts of rat and bovine brain. In: Morad M et al. (eds) The calcium channel: structure, function and implications. Springer, Berlin Heidelberg New York, pp 601–610

Edelstein NG, Catterall WA, Moon RT (1988) Identification of a 33-kilodalton cytoskeletal protein with high affinity for the sodium channel. Biochemistry 27:1818–1822

Ehlert FJ (1988) Estimation of the affinities of allosteric ligands using radioligand binding and pharmacological null methods. Mol Pharmacol 33:187–194

Ellis SB, Williams ME, Ways NR, Brenner R, Sharp AH, Leung AT, Campbell KP, McKenna E, Koch WJ, Hui A, Schwartz A, Harpold MM (1988) Structure and expression pattern of mRNAs encoding the alpha$_1$ and alpha$_2$ subunits of a DHP-sensitive calcium channel. Science 241:1661–1664

Ervasti JM, Cleassens MT, Mickelson JR, Louis CF (1989) Altered transverse tubule dihydropyridine receptor binding in malignant hyperthermia. J Biol Chem 264:2711–2717

Fatt P, Ginsborg BL (1958) The ionic requirements for the production of action potentials in crustacean muscle fibres. J Physiol (Lond) 142:516–543

Fatt P, Katz B (1953) The electrical properties of crustacean muscle fibres. J Physiol (Lond) 129:171–204

Feigenbaum P, Garcia ML, Kaczorowski GJ (1988) Evidence for distinct sites coupled to high-affinity omega-conotoxin receptors in rat brain synaptic plasma membrane vesicles. Biochem Biophys Res Commun 154:298–305

Ferguson DG, Schwartz HW, Franzini-Armstrong C (1984) Subunit structure of junctional feet in triads of skeletal muscle: a freeze-drying, rotary-shadowing study. J Cell Biol 99:1735–1742

Ferry DR, Glossmann H (1982) Identification of putative calcium channels in skeletal muscle microsomes. FEBS Lett 148:331–337

Ferry DR, Glossmann H (1984) ^{125}I-iodipine, a new high-affinity ligand for the putative calcium channel. Naunyn Schmiedebergs Arch Pharmacol 325:186–189

Ferry DR, Goll A, Glossmann H (1983a) Calcium channels: evidence for oligomeric nature by target size analysis. EMBO J 2:1729–1732

Ferry DR, Goll A, Glossmann H (1983b) Putative calcium channel molecular weight determination by target size analysis. Naunyn Schmiedebergs Arch Pharmacol 323:292–297

Ferry DR, Goll A, Gadow C, Glossmann H (1984a) (−)-^{3}H-desmethoxyverapamil labelling of putative calcium channels in brain: autoradiographic distribution and allosteric coupling to 1,4-dihydropyridine and diltiazem binding sites. Naunyn Schmiedebergs Arch Pharmacol 327:183–187

Ferry DR, Rombush M, Goll A, Glossmann H (1984b) Photoaffinity labelling of Ca^{2+} channels with [^{3}H]azidopine. FEBS Lett 169:112–118

Ferry DR, Kaempf K, Goll A, Glossmann H (1985) Subunit composition of skeletal muscle transverse tubule calcium channels evaluated with the 1,4-dihydropyridine photoaffinity probe, [^{3}H]AZIdopine. EMBO J 4:1933–1940

Ferry DR, Goll A, Glossmann H (1987) Photoaffinity labelling of the cardiac calcium channel (−)-[^{3}H]azidopine labels a 165-kDa polypeptide, and evidence against a [^{3}H]-1,4-dihydropyridine-isothiocyanate being a calcium-channel-specific affinity ligand. Biochem J 243:127–135

Fill M, Coronado R (1988) Ryanodine receptor channel of sarcoplasmic reticulum. Trends Neurosci 11:453–457

Fitzpatrick LA, Chin H (1988) The role of calcium channels in the regulation of parathyroid hormone release. In: Morad M et al. (eds) The calcium channel: structure, function and implications. Springer, Berlin Heidelberg New York, pp 418–432

Fitzpatrick LA, Brandi ML, Aurbach GD (1986) Control of PTH secretion is mediated through calcium channels and is blocked by pertussis toxin treatment of parathyroid cells. Biochem Biophys Res Commun 138:960–965

Fitzpatrick LA, Chin H, Nirenberg M, Aurbach GD (1988) Antibodies to an alpha subunit of skeletal muscle calcium channels regulate parathyroid cell secretion. Proc Natl Acad Sci USA 85:2115–2119

Fleckenstein A (1983) Calcium antagonism in heart and smooth muscle. Wiley, New York

Fleischer S, Ogunbunmi EM, Dixon MC, Fleer EA (1985) Localization of Ca^{2+} release channels with ryanodine in junctional terminal cisternae of sarcoplasmic reticulum of fast skeletal muscle. Proc Natl Acad Sci USA 82:7256–7259

Flockerzi V, Oeken HJ, Hofmann F, Pelzer D, Cavalie A, Trautwein W (1986a) Purified dihydropyridine-binding site from skeletal muscle T-tubules is a functional calcium channel. Nature 323:66–68

Flockerzi V, Oeken HJ, Hofmann F (1986b) Purification of a functional receptor for calcium-channel blockers from rabbit skeletal-muscle microsomes. Eur J Biochem 161:217–224

Forsayeth JR, Caro JF, Sinha MK, Maddux BA, Goldfine ID (1987) Monoclonal antibodies to the human insulin receptor that activate glucose transport but not insulin receptor kinase activity. Proc Natl Acad Sci USA 84:3448–3451

Fosset M, Jaimovich E, Delpont E, Lazdunski M (1983) [^{3}H]Nitrendipine receptors in skeletal muscle. J Biol Chem 258:6086–6092

Freedman SB, Miller RJ, Miller DM, Tindall DR (1984) Interactions of maitotoxin with voltage-sensitive calcium channels in cultured neuronal cells. Proc Natl Acad Sci USA 81:4582–4585

Galizzi JP, Borsotto M, Barhanin J, Fosset M, Lazdunski M (1986a) Characterization and photoaffinity labeling of receptor sites for the Ca^{2+} channel inhibitors d-cis-diltiazem, (+/−)-bepridil, desmethoxyverapamil, and (+)-PN 200-110 in skeletal muscle transverse tubule membranes. J Biol Chem 261:1393–1397

Galizzi JP, Fosset M, Romey G, Laduron P, Lazdunski M (1986b) Neuroleptics of the diphenyl-butylpiperidine series are potent calcium channel inhibitors. Proc Natl Acad Sci USA 83:7513–7517

Garcia ML, Trumble MJ, Reuben JP, Kaczorowski GJ (1984) Characterization of verapamil binding sites in cardiac membrane vesicles. J Biol Chem 259:15013–15016

Garcia ML, King VF, Siegl PK, Reuben JP, Kaczorowski GJ (1986) Binding of Ca^{2+} entry blockers to cardiac sarcolemmal membrane vesicles. Characterization of diltiazem-binding sites and their interaction with dihydropyridine and aralkylamine receptors. J Biol Chem 261:8146–8157

Gilman A (1987) G proteins: transducers of receptor-generated signals. Annu Rev Biochem 56:615–649

Glossmann H, Ferry DR (1983a) Molecular approach to the calcium channel. Drug Development 9:63–98

Glossmann H, Ferry DR (1983b) Solubilization and partial purification of putative calcium channels labelled with [^{3}H]nimodipine. Naunyn Schmiedebergs Arch Pharmacol 323:279–291

Glossmann H, Ferry DR (1985) Assay for calcium channels. Methods Enzymol 109:513–550

Glossmann H, Striessnig J (1988a) Calcium channels. Vitam Horm 44:155–328

Glossmann H, Striessnig J (1988b) Structure and pharmacology of voltage-dependent calcium channels. ISI Atlas Pharmacol 2:202–210

Glossmann H, Ferry DR, Lübbecke F, Mewes R, Hofmann F (1982) Calcium channels: direct identification with radioligand binding studies. Trends Pharmacol Sci 3:431–437

Glossmann H, Ferry DR, Boschek CB (1983a) Purification of the putative calcium channel from skeletal muscle with the aid of [^{3}H] nimodipine binding. Naunyn Schmiedebergs Arch Pharmacol 323:1–11

Glossmann H, Linn T, Rombusch M, Ferry DR (1983b) Temperature-dependent regulation of d-cis-[^{3}H]diltiazem binding to Ca^{2+} channels by 1,4-dihydropyridine channel agonists and antagonists. FEBS Lett 160:226–232

Glossmann H, Ferry DR, Goll A, Striessnig J, Zernig G (1985) Calcium channels and calcium channel drugs: recent biochemical and biophysical findings. Arzneimittelforschung 35:1917–1935

Glossmann H, Striessnig J, Ferry DR, Goll A, Moosburger K, Schirmer M (1987a) Interaction between calcium channel ligands and calcium channels. Circ Res 61:30–36

Glossmann H, Ferry DR, Striessnig J, Goll A, Moosburger K (1987b) Resolving the structure of the calcium channel by photoaffinity labeling. Trends Pharmacol Sci 8:95–100

Glossmann H, Striessnig J, Hymel L, Schindler H (1988a) Purification and reconstitution of calcium channel drug-receptor sites. Ann N Y Acad Sci 522:150–161

Glossmann H, Striessnig J, Hymel L, Zernig G, Knaus HG, Schindler H (1988b) The structure of the calcium channel: photoaffinity labelling and tissue distribution. In: Morad M, et al. (eds) The calcium channel: structure, function and implications. Springer, Berlin Heidelberg New York, pp 168–192

Glossmann H, Knaus HG, Striessnig J, Marrer S, Hoeltje HD (1988c) Enantioselective interactions of drugs with receptors. Naunyn Schmiedebergs Arch Pharmacol 338:R20 (abstr)

Glossmann H, Striessnig J, Knaus HG, Müller J, Grassegger A, Höltje HD, Marrer S, Hymel L, Schindler H (1989) Structure of calcium channels. Ann NY Acad Sci 560:198–214

Godfraind T, Miller R, Wibo M (1986) Calcium antagonism and calcium entry blockade. Pharmacol Rev 38:321–416

Goldin AI, Snutch T, Lübbert H, Dowsett A, Marshall J, Auld V, Downey W, Fritz LC, Lester HA, Dunn R, Catterall WA, Davidson N (1986) Messenger RNA coding for only the alpha subunit of the rat brain Na channel is sufficient for expression of functional channels in *Xenopus* oocytes. Proc Natl Acad Sci USA 83:7503–7507

Goll A, Ferry DR, Glossmann H (1983a) Target-size analysis reveals subunit composition of calcium channels in brain and skeletal muscle. Naunyn Schmiedebergs Arch Pharmacol 324:R45 (abstr)

Goll A, Ferry DR, Glossmann H (1983b) Target size analysis of skeletal muscle Ca^{2+} channels. Positive allosteric heterotropic regulation by d-*cis*-diltiazem is associated with apparent channel oligomer dissociation. FEBS Lett 157:63–69

Goll A, Ferry DR, Glossmann H (1984a) Target size analysis and molecular properties of Ca^{2+} channels labelled with [^{3}H]verapamil. Eur J Biochem 141:177–186

Goll A, Ferry DR, Striessnig J, Schober M, Glossmann H (1984b) (−)-[^{3}H]Desmethoxyverapamil, a novel Ca^{2+} channel probe. Binding characteristics and target size analysis of its receptor in skeletal muscle. FEBS Lett 176:371–377

Goll A, Glossmann H, Mannhold R (1986) Correlation between the negative inotropic potency and binding parameters of 1,4-dihydropyridine and phenylalkylamine calcium channel blockers in cat heart. Naunyn Schmiedebergs Arch Pharmacol 334:303–312

Gould RJ, Murphy KM, Snyder SH (1982) [^{3}H]Nitrendipine-labelled calcium channels discriminate calcium agonists and antagonists. Proc Natl Acad Sci USA 79:3656–3660

Gould RJ, Murphy KM, Reynolds IJ, Snyder SH (1983) Antischizophrenic drugs of the diphenylbutylpiperidine type act as calcium channel antagonists. Proc Natl Acad Sci USA 80:5122–5125

Gould RJ, Murphy KM, Snyder SH (1984) Tissue heterogeneity of calcium channel antagonist binding sites labeled by [^{3}H]nitrendipine. Mol Pharmacol 25:235–241

Grassegger A, Striessnig J, Weiler M, Knaus HG, Glossmann H (1989) [^{3}H]HOE166 defines a novel calcium antagonist drug receptor – distinct from the 1,4-dihydropyridine binding domain. Naunyn-Schmiedeberg's Arch (in press)

Gray WR, Olivera BM, Cruz LJ (1988) Peptide toxins from venomous conus snails. Annu Rev Biochem 57:665–700

Gredal O, Drejer J, Honore T (1987) Different target sizes of the voltage-dependent Ca^{2+} channel and the [^{3}H]nitrendipine binding site in rat brain. Eur J Pharmacol 136:75–80

Greenberg RM, Striessnig J, Koza A, Devay P, Glossmann H, Hall LM (1989) Native and detergent-solubilized membrane extracts from *Drosophila* heads contain binding sites for phenylalkylamine calcium channel blockers. Insect Biochem 19:309–322

Gu X-H, Casley DJ, Nayler WG (1989) Sarafotoxin S6b displaces specifically bound ^{125}I-endothelin. Eur J Pharmacol 162:509–510

Gusovsky F, Yasumoto T, Daly JW (1987) Maitotoxin stimulates phosphoinositide breakdown in neuroblastoma hybrid NCB-20 cells. Cell Mol Neurobiol 7:317–322

Guy HR, Seetharamulu P (1986) Molecular model of the action potential sodium channel. Proc Natl Acad Sci USA 83:508–512

Habermann E (1984) Apamin. Pharmacol Ther 25:255–270

Habermann E, Fischer K (1979) Bee venom neurotoxin (apamin): iodine labeling and characterization of binding sites. Eur J Biochem 94:355–364

Hamilton SL, Perez M (1987) Toxins that affect voltage-dependent calcium channels. Biochem Pharmacol 36:3325–3329

Hamilton SL, Yatani A, Hawkes MJ, Redding K, Brown AM (1985) Atrotoxin: a specific agonist for calcium currents in heart. Science 229:182–184

Han C, Abel PW, Minneman KP (1987) Alpha$_1$-adrenoceptor subtypes linked to different mechanism for increasing intracellular Ca$_{2+}$ in smooth muscle. Naturems 329:333–335

Hanbauer I, Sanna E, Callewaert G, Morad M (1988) An endogenous purified peptide modulates Ca^{2+} channels in neurons and cardiac myocytes. In: Morad M, et al. (eds) The calcium channel: structure, function and implications. Springer, Berlin Heidelberg New York, pp 611–618

Hescheler J, Trautwein W (1988) Modification of cardiac calcium current by intracellularly applied trypsin. Pfluger's Arch 210:1227 (abstr)

Hescheler J, Kameyama M, Trautwein W, Mieskes G, Soeling HD (1987) Regulation of the cardiac calcium channel by protein phosphatases. Eur J Biochem 165:261–266

Hess P, Lansman JB, Tsien RW (1986) Calcium channel selectivity for divalent and monovalent cations voltage and concentration dependence of single-channel current in ventricular heart cells. J Gen Physiol 88:293–319

Hille B (1984) Ionic channels of excitable membranes. Sinauer, Sunderland, MA

Hirata Y, Yoshimi H, Takata S, Watanabe TX, Kumagai S, Nakajima K, Sakakibara S (1988) Cellular mechanism of action by a novel vasoconstrictor endothelin in cultured rat vascular smooth muscle cells. Biochem Biophys Res Commun 154:868–875

Hoffman EP, Knudson CM, Campbell KP, Kunkel LM (1987) Subcellular fractionation of dystrophin to the triads of skeletal muscle. Nature 330:754–758

Hofmann F, Nastainczyk W, Röhrkasten A, Schneider T, Sieber M (1987) Regulation of L-type calcium channels. Trends Pharmacol Sci 8:393–398

Horne P, Triggle DJ, Venter JC (1984) Nitrendipine and isoproterenol induce phosphorylation of a 42000-dalton protein that co-migrates with the affinity-labeled calcium channel regulatory subunit. Biochem Biophys Res Commun 121:890–898

Horne WA, Weiland GA, Oswald RE (1986) Solubilization and hydrodynamic characterization of the dihydropyridine receptor from rat ventricular muscle. J Biol Chem 261:3588–3594

Hosey MM, Borsotto M, Lazdunski M (1986) Phosphorylation and dephosphorylation of dihydropyridine-sensitive voltage-dependent Ca^{2+} channel in skeletal muscle membranes by cAMP- and Ca^{2+}-dependent processes. Proc Natl Acad Sci USA 83:3733–3737

Hosey MM, Barhanin J, Schmid A, Vandaele S, Ptasienski J, O'Callahan C, Cooper C, Lazdunski M (1987) Photoaffinity labelling and phosphorylation of a 165-kilodalton peptide associated with dihydropyridine- and phenylalkylamine-sensitive calcium channels. Biochem Biophys Res Commun 147:1137–1145

Hymel L, Inui M, Fleischer S, Schindler H (1988a) Purified ryanodine receptor of skeletal muscle sarcoplasmic reticulum forms Ca^{2+}-activated oligomeric Ca^{2+} channels in planar bilayers. Proc Natl Acad Sci USA 85:441–445

Hymel L, Schindler H, Inui M, Fleischer S (1988b) Reconstitution of purified cardiac muscle calcium-release channel (ryanodine receptor) in planar bilayers. Biochem Biophys Res Commun 152:308–314

Hymel L, Striessnig J, Glossmann H, Schindler H (1988c) Purified skeletal muscle 1,4-dihydropyridine receptor forms phosphorylation-dependent oligomeric calcium channels in planar bilayers. Proc Natl Acad Sci USA 85:4290–4294

Hymel L, Schindler H, Inui M, Fleischer S, Striessnig J, Glossmann H (1989) A molecular model of excitation-contraction coupling at the skeletal muscle triad junction via coassociated oligomeric calcium channels. Annu NY Acad Sci 560:185–188

Imagawa T, Leung AT, Campbell KP (1987a) Phosphorylation of the 1,4-dihydropyridine receptor of the voltage-dependent Ca^{2+} channel by an intrinsic protein kinase in isolated triads from rabbit skeletal muscle. J Biol Chem 262:8333–8339

Imagawa T, Smith JS, Coronado R, Campbell KP (1987b) Purified ryanodine receptor from skeletal muscle sarcoplasmic reticulum is the Ca^{2+}-permeable pore of the calcium-release channel. J Biol Chem 262:16636−16643

Inui M, Saito A, Fleischer S (1987a) Isolation of the ryanodine receptor from cardiac sarcoplasmic reticulum and identity with the feet structures. J Biol Chem 262:15637−15642

Inui M, Saito A, Fleischer S (1987b) Purification of the ryanodine receptor and identity with feet structures of junctional terminal cisternae of sarcoplasmic reticulum from fast skeletal muscle. J Biol Chem 262:1740−1747

Itoh Y, Yanagisawa M, Ohkubo S, Kimura C, Kosaka T, Inoue A, Ishida N, Mitsui Y, Onda H, Fujino M, et al. (1988) Cloning and sequence analysis of cDNA encoding the precursor of a human endothelium-derived vasoconstrictor peptide, endothelin: identity of human and porcine endothelin. FEBS Lett 231:440−444

Janis RA, Silver PJ, Triggle DJ (1987) Drug action and cellular calcium regulation. Adv Drug Res 16:309−591

Janis RA, Johnson DE, Shrikhande AV, McCarthy RT, Howard AD, Greguski R, Scriabine A (1988) Endogenous 1,4-dihydropyridine-displacing substances acting on L-type Ca^{2+} channels: isolation and characterization of fractions from brain and stomach. In: Morad M, et al. (eds) The calcium channel: structure, function and implications. Springer, Berlin Heidelberg New York pp 564−574

Johnson DF, Kuo TH, Giacomelli F, Wiener J (1988) Structural analysis of the calcium channel by photoaffinity labelling and limited proteolysis. Biochem Biophys Res Commun 154:455−461

Kamp TJ, Miller RJ (1987) Voltage-dependent nitrendipine binding to cardiac sarcolemmal vesicles. Mol Pharmacol 32:278−285

Kaul PN, Daftari P (1986) Marine pharmacology: bioactive molecules from the sea. Annu Rev Pharmacol Toxicol 26:117−142

Kerr LM, Yoshikami D (1984) A venom peptide with a novel presynaptic blocking action. Nature 308:282−284

Kim YI (1987) Lambert-Eaton myasthenic syndrome: evidence for calcium channel blockade. Annu NY Acad Sci 505:377−379

Kim YI, Neher E (1988) IgG from patients with Lambert-Eaton syndrome blocks voltage-dependent calcium channels. Science 239:405−408

King VF, Garcia ML, Himmel D, Reuben JP, Lam YK, Pan JX, Han GQ, Kaczorowski GJ (1988) Interaction of tetrandrine with slowly inactivating calcium channels. Characterization of calcium channel modulation by an alkaloid of Chinese medicinal herb origin. J Biol Chem 263:2238−2244

King VF, Garcia ML, Shevell JL, Slaughter RS, Kaczorowski GJ (1989) Substituted diphenylbutylpiperidines bind to a unique high affinity site on the L-type calcium channel. J Biol Chem 264:5633−5641

Kirley TL, Schwartz A (1984) Solubilization and affinity labeling of a dihydropyridine binding site from skeletal muscle: effects of temperature and diltiazem on [^{3}H]dihydropyridine binding to transverse tubules. Biochem Biophys Res Commun 123:41−49

Kloog Y, Ambar I, Sokolovsky M, Kochva E, Wollberg Z, Bdolah A (1988) Sarafotoxin, a novel vasoconstrictor peptide: phosphoinositide hydrolysis in rat heart and brain. Science 242:268−270

Knaus HG (1988) Neurotoxic aminoglycoside antibiotics are potent inhibitors of [^{125}I]omega-conotoxin GVIA binding. Naunyn Schmiedebergs Arch Pharmacol 337:R52

Knaus HG, Striessnig J, Koza A, Glossmann H (1987) Neurotoxic aminoglycoside antibiotics are potent inhibitors of [^{125}I]omega-conotoxin GVIA binding to guinea-pig cerebral cortex membranes. Naunyn Schmiedebergs Arch Pharmacol 336:583−586

Knudson CM, Hoffman EP, Kahl SD, Kunkel LM, Campbell KP (1988) Evidence for the association of dystrophin with the transverse tubular system in skeletal muscle. J Biol Chem 263:8480−8484

Knudson CM, Chaudhari N, Sharp AH, Powell JA, Beam KG, Campbell KP (1989) Specific absence of the alpha$_1$ subunit of the dihydropyridine receptor in mice with muscular dysgenesis. J Biol Chem 264:1345−1348

Kobayashi M, Ochi R, Ohizumi Y (1987) Maitotoxin-activated single calcium channels in guinea-pig cardiac cells. Br J Pharmacol 92:665–671

Koenig M, Hoffman EP, Bertelson CJ, Monaco AP, Feener C, Kunkel LM (1987) Complete cloning of the Duchenne muscular dystrophy (DMD) cDNA and preliminary genomic organization of the DMD gene in normal and affected individuals. Cell 50:509–517

Koenig M, Monaco AP, Kunkel LM (1988) The complete sequence of dystrophin predicts a rod-shaped cytoskeletal protein. Cell 53:219–226

Koike K, Judd AM, Login IS, Yasumoto T, MacLeod RM (1986) Maitotoxin, a calcium channel activator, increases prolactin release from rat pituitary tumor 7315a cells by a mechanism that may involve leukotriene production. Neuroendocrinology 43:283–290

Kokubuhn S, Prod'hom B, Becker C, Porzig H, Reuter H (1986) Studies on Ca channels in intact cardiac cells: voltage-dependent effects and cooperative interactions of dihydropyridine enantiomers. Mol Pharmacol 30:571–584

Kosower EM (1985) A structural and dynamic molecular model for the sodium channel of *Electrophorus electricus*. FEBS Lett 182:234–242

Koyano K, Abe T, Nishiuchi Y, Sakakibara S (1987) Effects of synthetic omega-conotoxin on synaptic transmission. Eur J Pharmacol 135:337–343

Kuo TH, Johnson DF, Tsang W, Wiener J (1987) Photoaffinity labeling of the calcium channel antagonist receptor in the heart of the cardiomyopathic hamster. Biochem Biophys Res Commun 148:926–933

Lai FA, Erickson H, Block BA, Meissner G (1987) Evidence for a junctional feet-ryanodine receptor complex from sarcoplasmic reticulum. Biochem Biophys Res Commun 143:704–709

Lai FA, Anderson K, Rousseau E, Liu QY, Meissner G (1988a) Evidence for a Ca^{2+} channel within the ryanodine receptor complex from cardiac sarcoplasmic reticulum. Biochem Biophys Res Commun 151:441–449

Lai FA, Erickson HP, Rousseau E, Liu QY, Meissner G (1988b) Purification and reconstitution of the calcium-release channel from skeletal muscle. Nature 331:315–319

Lang B, Newsom-Davis J, Prior C, Wray D (1983) Antibodies to motor nerve terminals: an electrophysiological study of a human myasthenic syndrome transferred to mouse. J Physiol (Lond) 344:335–345

Lansman JB, Hess P, Tsien RW (1986) Blockade of current through single calcium channels by Cd^{2+}, Mg^{2+}, and Ca^{2+}. Voltage and concentration dependence of calcium entry into the pore. J Gen Physiol 88:321–347

Lazdunski M (1983) Apamin, a neurotoxin specific for one class of Ca^{2+}-dependent K^+ channels. Cell Calcium 4:421–428

Lazdunski M, Frelin C, Barhanin J, Lombet A, Meiri H, Pauron D, Romey G, Schmid A, Schweitz H, Vigne P, et al. (1986a) Polypeptide toxins as tools to study voltage-sensitive Na^+ channels. Annu NY Acad Sci 479:204–220

Lazdunski M, Barhanin J, Borsotto M, Fosset M, Galizzi JP, Renaud JF, Romey G, Schmid A (1986b) Dihydropyridine-sensitive Ca^{2+} channels: molecular properties of interaction with Ca^{2+}-channel blockers, purification, subunit structure, and differentiation. J Cardiovasc Pharmacol 8:S13–S19

Leonhard JP, Nargeot J, Snutch TP, Davidson N, Lester HA (1987) Ca channels induced in *Xenopus* oocytes by rat brain mRNA. J Neurosci 7:875–881

Leonard RJ, Labarca CG, Charnet P, Davidson N, Lester HA (1988) Evidence that the M2 membrane-spanning region lines the ion channel pore of the nicotinic receptor. Science 242:1578–1581

Lester HA (1988) Heterologous expression of excitability proteins: route to more specific drugs. Science 241:1057–1063

Leung AT, Imagawa T, Campbell KP (1987) Structural characterization of the 1,4-dihydropyridine receptor of the voltage-dependent Ca^{2+} channel from rabbit skeletal muscle. Evidence for two distinct high-molecular-weight subunits. J Biol Chem 262:7943–7946

Leung AT, Imagawa T, Block B, Franzini-Armstrong C, Campbell KP (1988) Biochemical and ultrastructural characterization of the 1,4-dihydropyridine receptor from rabbit skeletal muscle. Evidence for a 52000-Da subunit. J Biol Chem 263:994–1001

Lodish HF (1988) Multi-spanning membrane proteins: how accurate are the models? Trends Biochem Sci 13:332–334

Lotan I, Goelet P, Gigi A, Dascal N (1989) Specific block of calcium channel expression by a fragment of dihydropyridine receptor cDNA. Science 243:666–669

Luchowski EM, Yousif F, Triggle DJ, Maurer SC, Sarmiento JG, Janis RA (1984) Effects of metal cations and calmodulin antagonists on [³H]nitrendipine binding in smooth and cardiac muscle. J Pharmacol Exp Ther 230:607–613

Ma J, Fill M, Knudson CM, Campbell KP, Coronado R (1988) Ryanodine receptor of skeletal muscle is a gap junction-type channel. Science 242:99–102

MacLennan DH, Brandl CJ, Korczak B, Green NM (1985) Amino acid sequence of a Ca^{2+}- and Mg^{2+}-dependent ATPase from rabbit muscle sarcoplasmic reticulum, deduced from its complementary DNA sequence. Nature 316:696–700

Madeddu L, Pozzan T, Robello T, Rolandi R, Hsiao TH, Meldolesi J (1985) Leptinotoxin-h action in synaptosomes, neurosecretory cells, and artificial membranes: stimulation of ion fluxes. J Neurochem 45:1708–1718

Maelicke A (1988) Structural similarities between ion channel proteins. Trends Biochem Sci 13:199–202

Malouf NN, Coronado R, McMahon D, Meissner G, Gillespie GY (1987) Monoclonal antibody specific for the transverse tubular membrane of skeletal muscle activates the dihydropyridine-sensitive Ca^{2+} channel. Proc Natl Acad Sci USA 84:5019–5023

Mantione CR, Goldman ME, Martin B, Bolger GT, Lueddens HW, Paul SM, Skolnick P (1988) Purification and characterization of an endogenous protein modulator of radioligand binding to "peripheral-type" benzodiazepine receptors and dihydropyridine Ca^{2+}-channel antagonist binding sites. Biochem Pharmacol 37:339–347

Marqueze B, Martin-Moutot N, Leveque C, Couraud F (1988) Characterization of the omega-conotoxin molecule in rat brain synaptosomes and cultured neurons. Mol Pharmacol 34:87–90

McCleskey EW, Fox AP, Feldman DH, Cruz LJ, Olivera BM, Tsien RW, Yoshikami D (1987) Omega-conotoxin: direct and persistent blockade of specific types of calcium channels in neurons but not muscle. Proc Natl Acad USA 84:4327–4331

McClure WO, Abbott BC, Baxter DE, Hsiao TH, Satin LS, Siger A, Yoshino JE (1980) Leptinotarsin: a presynaptic neurotoxin that stimulates release of acetylcholine. Proc Natl Acad Sci USA 77:1219–1223

McCrea PD, Engelman DM, Popot JL (1988) Topography of integral membrane proteins: hydrophobicity analysis vs immunolocalisation. Trends Biochem Sci 13:289–290

Meiri H, Zeitoun I, Grunhagen HH, Lev-Ram V, Eshhar Z, Schlessinger J (1984) Monoclonal antibodies associated with sodium channel block nerve impulse and stain nodes of Ranvier. Brain Res 310:168–173

Melzer W, Rios E, Schneider MF (1987) A general procedure for determining the rate of calcium release from the sarcoplasmic reticulum in skeletal muscle fibers. Biophys J 51:849–863

McGrew SG, Inui M, Chadwick CC, Boucek RJ, Jung CY, Fleischer S (1989) Comparison of the calcium release channel of cardiac and skeletal muscle sarcoplasmic reticulum by target inactivation analysis. Biochemistry 28:1319–1323

Meyer H, Wehinger E, Bossert F, et al. (1985) Chemistry of 1,4 dihydropyridines. In: Fleckenstein A, Van Breemen C, Gross R, Hoffmeister F (eds) Cardiovascular effects of dihydropyridine-type calcium antagonists and agonists. Springer, Berlin Heidelberg New York, pp 90–103 (Bayer-Symposium, vol 9)

Miasiro N, Yamamoto H, Kanaide H, Nakamura M (1988) Does endothelin mobilize calcium from intracellular store sites in rat aortic vascular smooth muscle cells in primary culture? Biochem Biophys Res Commun 156:312–317

Michalak M, Dupraz P, Shoshan-Barmatz V (1988) Ryanodine binding to sarcoplasmic reticulum membrane; comparison between cardiac and skeletal muscle. Biochim Biophys Acta 939:587−594

Mickelson JR, Gallant EM, Litterer LA, Johnson KM, Rempel WE, Louis CF (1988) Abnormal sarcoplasmic reticulum ryanodine receptor in malignant hyperthermia. J Biol Chem 263:9310−9315

Miljanich GP, Yaeger RE, Hsiao TH (1988) Leptinotarsin-D, a neurotoxic protein, evokes neurotransmitter release from, and calcium flux into, isolated electric organ nerve terminals. J Neurobiol 19:373−386

Miller C (1988) Shaker shakes out potassium channels. Trends Neurosci 1:185−186

Miller RJ (1987) Multiple calcium channels and neuronal function. Science 235:46−52

Mir AK, Spedding M (1987) Calcium antagonist properties of diclofurime isomers. II. Molecular aspects: allosteric interactions with dihydropyridine recognition sites. J Cardiovasc Pharmacol 9:469−477

Moorman JR, Zhou Z, Kirsch GE, Lacerda AE, Caffrey JM, Lam DM, Joho RH, Brown AM (1987) Expression of single calcium channels in Xenopus oocytes after injection of mRNA from rat heart. Am J Physiol 253:985−991

Morton ME, Froehner SC (1987) Monoclonal antibody identifies a 200-kDa subunit of the dihydropyridine-sensitive calcium channel. J Biol Chem 262:11904−11907

Morton ME, Caffrey JM, Brown AM, Froehner SC (1988) Monoclonal antibody to the alpha$_1$-subunit of the dihydropyridine-binding complex inhibits calcium currents in BC3H1 myocytes. J Biol Chem 263:613−616

Nakayama N, Kirley TL, Vaghy PL, Mc Kenna E, Schwartz A (1987) Purification of putative Ca^{2+} channel protein from rabbit skeletal muscle. Determination of the amino-terminal sequence. J Biol Chem 262:6572−6576

Nastainczyk W, Röhrkasten A, Sieber M, Hofmann F (1987) Phosphorylation of the purified receptor for calcium channel blockers by cAMP kinase and protein kinase C. Eur J Biochem 169:137−142

Navarro J (1987) Modulation of [^{3}H]dihydropyridine receptors by activation of protein kinase C in chick muscle cells. J Biol Chem 262:4649−4652

Nilius B (1986) Possible functional significance of a novel type of cardiac Ca channel. Biomed Biochim Acta 45:37−45

Noda M, Shimizu S, Tanabe T, Takai T, Kayano T, Ikeda T, Takahashi H, Nakayama H, Kanaoka Y, Minamino N, Kangawa K, Matsuo H, Raftery MA, Hirose T, Inayama S, Hayashida H, Miyata T, Numa S (1984) Primary structure of Electrophorus electricus sodium channel deduced from cDNA sequence. Nature 312:121−127

Noda M, Ikeda T, Kayano T, Suzuki H, Takeshima H, Kurasaki M, Takahashi H, Numa S (1986a) Existence of distinct sodium channel messenger RNAs in rat brain. Nature 320:188−192

Noda M, Ikeda T, Suzuki H, Takeshima H, Takahashi T, Kuno M, Numa S (1986b) Expression of functional sodium channels from cloned cDNA. Nature 322:826−828

Norman RI, Borsotto M, Fosset M, Lazdunski M, Ellory JC (1983) Determination of the molecular size of the nitrendipine-sensitive Ca^{2+} channel by radiation inactivation. Biochem Biophys Res Commun 111:878−883

O'Callahan CM, Hosey MM (1988) Multiple phosphorylation sites in the 165-kilodalton peptide associated with dihydropyridine calcium channels. Biochemistry 27:6071−6077

Oiki S, Danho W, Madison V, Montal M (1988) M2 delta, a candidate for the structure lining the ionic channel of the nicotinic cholinergic receptor. Proc Natl Acad Sci 85:8703−8707

Olivera BM, Gray WR, Zeikus R, McIntosh JM, Varga J, Rivier J, de Santos V, Cruz LJ (1985) Peptide neurotoxins from fish-hunting cone snails. Science 230:1338−1343

Olivera BM, Cruz LJ, de Santos V, LeCheminant GW, Griffin D, Zeikus R, McIntosh JM, Galyean R, Varga J, Gray WR, et al. (1987) Neuronal calcium-channel antagonists. Discrimination between calcium-channel subtypes using omega-conotoxin from Conus magus venom. Biochemistry 26:2086−2090

Papazian DM, Schwarz TL, Tempel BL, Jan YN, Jan LY (1987) Cloning of genomic and complementary DNA from *Shaker*, a putative potassium channel gene from *Drosophila*. Science 237:749−753

Pastan I, Gottesman M (1987) Multiple-drug resistance in human cancer. N Engl J Med 316:1388−1393

Pauron D, Qar J, Barhanin J, Fournier D, Cuany A, Pralavorio M, Berge JB, Lazdunski M (1987) Identification and affinity labeling of very high affinity binding sites for the phenylalkylamine series of Ca^{2+} channel blockers in the *Drosophila* nervous system. Biochemistry 26:6311−6315

Pelzer D, Grant AO, Cavalie A, Pelzer S, Sieber M, Hofmann F, Trautwein W (1989a) Modulation of calcium channels reconstituted from the skeletal muscle DHP receptor protein complex and its alpha$_1$ peptide subunit in lipid bilayers. Annu NY Acad Sci 560:138−154

Pelzer S, Barhanin J, Pauron D, Trautwein W, Lazdunski M, Pelzer D (1989b) Diversity and novel pharmacological properties of Ca^{2+} channels in Drosophila brain membranes. EMBO J 8:2365−2371

Pessah IN, Francini AO, Scales DJ, Waterhouse AL, Casida JE (1986) Calcium ryanodine receptor complex. J Biol Chem 261:8643−8648

Pessah IN, Stambuk RA, Casida JE (1987) Ca^{2+}-activated ryanodine binding: mechanisms of sensitivity and intensity modulation by Mg^{2+}, caffeine, and adenine nucleotides. Mol Pharmacol 31:232−238

Pin JP, Yasumoto T, Bockaert J (1988) Maitotoxin-evoked gamma-aminobutyric acid release is due not only to the opening of calcium channels. J Neurochem 50:1227−1232

Pincon-Raymond M, Rieger F, Fosset M, Lazdunski M (1985) Abnormal transverse tubule system and abnormal amount of receptors for Ca^{2+} channel inhibitors of the dihydropyridine family in skeletal muscle from mice with embryonic muscular dysgenesis. Dev Biol 112:458−466

Pizarro G, Brum G, Fill R, Rodriguez M, Uribe I, Rios E (1988) The voltage sensor of skeletal muscle excitation contraction coupling: a comparison with calcium channels. In: Morad M, et al. (eds). The calcium channel: structure, function and implications. Springer, Berlin Heidelberg New York, pp 138−158

Porzig H, Becker C (1988) Potential-dependent allosteric modulation of 1,4 dihydropyridine binding by d-*cis*-diltiazem and (±)-verapamil in living cardiac cells. Mol Pharmacol 34:172−179

Powell JA, Fambrough DM (1973) Electrical properties of normal and dysgenic mouse skeletal muscle in culture. J Cell Physiol 82:21−38

Putney JW (1987) Formation and actions of calcium-mobilizing messenger, inositol 1,4,5-trisphosphate. Am J Physiol 252:G149−G157

Qar J, Schweitz H, Schmid A, Lazdunski M (1986) A polypeptide toxin from the coral *Goniopora*. Purification and action on Ca^{2+}-channels. FEBS Lett 202:331−336

Qar J, Galizzi JP, Fosset M, Lazdunski M (1987) Receptors for diphenylbutylpiperidine neuroleptics in brain, cardiac, and smooth muscle membranes. Relationship with receptors for 1,4-dihydropyridines and phenylalkylamines and with Ca^{2+} channel blockade. Eur J Pharmacol 141:261−268

Qar J, Barhanin J, Romey G, Henning R, Lerch U, Oekonomopulos R, Urbach H, Lazdunski M (1988) A novel high-affinity class of Ca^{2+} channel blockers. Mol Pharmacol 33:363−369

Reber BF, Catterall WA (1987) Hydrophobic properties of the beta$_1$ and beta$_2$ subunits of the rat brain sodium channel. J Biol Chem 262:11369−11374

Revel JP (1962) The sarcoplasmic reticulum of the bat crycothyroid muscle. J Cell Biol 12:572−588

Reuter H (1983) Calcium channel modulation by neurotransmitters, enzymes and drugs. Nature 301:569−574

Reynolds IJ, Snowman AD, Snyder SH (1986a) (−)-[^{3}H]Desmethoxyverapamil labels multiple calcium channel modulator receptors in brain and skeletal muscle membranes: differentiation by temperature and dihydropyridines. J Pharmacol Exp Ther 237:731−738

Reynolds IJ, Wagner JA, Snyder SH, Thayer SA, Olivera BM, Miller RJ (1986b) Brain voltage-sensitive calcium-channel subtypes differentiated by omega-conotoxin fraction GVIA. Proc Natl Acad Sci USA 83:8804−8807

Rieger F, Bournaud R, Shimahara T, Garcia L, Pincon-Raymond M, Romey G, Lazdunski M
(1987) Restoration of dysgenic muscle contraction and calcium channel function by co-culture with normal spinal cord neurons. Nature 330:563–566

Rios E, Brum G (1987) Involvement of dihydropyridine receptors in excitation-contraction coupling in skeletal muscle. Nature 325:717–720

Rios E, Pizarro G (1988) The voltage-sensor and calcium channels of excitation-contraction coupling. News Physiol Sci 3:223–227

Rios E, Fitts R, Uribe I, Pizarro G, Brum G (1989) A third role for calcium in excitation-contraction coupling. In: Bacigalupo J (ed) Transduction in biological systems. Plenum, New York (in press)

Rivier J, Galyean R, Gray WR, Azimi-Zonooz A, McIntosh JM, Cruz LJ, Olivera BM (1987) Neuronal calcium channel inhibitors. Synthesis of omega-conotoxin GVIA and effects on ^{45}Ca uptake by synaptosomes. J Biol Chem 262:1194–1198

Röhrkasten A, Meyer HE, Nastainczyk W, Sieber M, Hofmann F (1988) cAMP-dependent protein kinase rapidly phosphorylates Serine-687 of the skeletal muscle receptor for calcium channel blockers. J Biol Chem 263:15325–15329

Romey G, Rieger F, Renaud JF, Pincon-Raymond M, Lazdunski M (1986) The electrophysiological expression of Ca^{2+} channels and of apamin-sensitive Ca^{2+}-activated K^+ channels is abolished in skeletal muscle cells from mice with muscular dysgenesis. Biochem Biophys Res Commun 136:935–940

Romey G, Quast U, Pauron D, Frelin C, Renaud JF, Lazdunski M (1987) Na^+ channels as sites of action of the cardioactive agent DPI 201-106 with agonist and antagonist enantiomers. Proc Natl Acad Sci USA 84:896–900

Rosenthal W, Schultz G (1987) Modulation of voltage-dependent ion channels by extracellular signals. Trends Pharmacol Sci 8:351–354

Ruth P, Flockerzi V, von Nettelbladt E, Oeken J, Hofmann F (1985) Characterization of the binding sites for nimodipine and (−)-desmethoxyverapamil in bovine cardiac sarcolemma. Eur J Biochem 150:313–322

Ruth P, Flockerzi V, Oeken HJ, Hofmann F (1986) Solubilization of the bovine cardiac sarcolemmal binding sites for calcium channel blockers. Eur J Biochem 155:613–620

Saito A, Seiler S, Chu A, Fleischer S (1984) Preparation and morphology of sarcoplasmic reticulum terminal cisternae from rabbit skeletal muscle. J Cell Biol 99:875–885

Saito A, Inui M, Radermacher M, Frank J, Fleischer S (1988) Ultrastructure of the calcium-release channel of sarcoplasmic reticulum. J Cell Biol 107:211–219

Sano K, Enomoto K, Maeno T (1987) Effects of synthetic omega-conotoxin, a new-type Ca^{2+} antagonist, on frog and mouse neuromuscular transmission. Eur J Pharmacol 141:235–241

Sarmiento JG, Epstein PM, Rowe WA, Chester DW, Smilowitz H, Wehinger E, Janis RA (1986) Photoaffinity labelling of a 33000- to 35000-dalton protein in cardiac, skeletal and smooth muscle membranes using a new ^{125}I-labelled 1,4-dihydropyridine calcium channel antagonist. Life Sci 39:2401–2409

Scheffauer F (1989) Immuno-photoaffinity labelling of calcium channel drug receptors. Naunyn Schmiedebergs Arch Pharmacol 339:1245

Schilling WP, Drewe JA (1986) Voltage-sensitive nitrendipine binding in an isolated cardiac sarcolemma preparation. J Biol Chem 261:2750–2758

Schindler H (1989) Planar lipid-protein membranes: strategies of formation and of detecting dependencies of ion-transport functions on membrane conditions. Methods Enzymol (in press)

Schmid A, Barhanin J, Coppola T, Borsotto M, Lazdunski M (1986a) Immunochemical analysis of subunit structures of 1,4-dihydropyridine receptors associated with voltage-dependent Ca^{2+} channels in skeletal, cardiac, and smooth muscles. Biochemistry 25:3492–3495

Schmid A, Barhanin J, Mourre C, Coppola T, Borsotto M, Lazdunski M (1986b) Antibodies reveal the cytolocalization and subunit structure of the 1,4-dihydropyridine component of the neuronal Ca^{2+} channel. Biochem Biophys Res Commun 139:996–1002

Schneider MF (1981) Membrane charge movement and depolarization-contraction coupling. Annu Rev Physiol 43:507–517

Schneider MF, Chandler WK (1973) Voltage-dependent charge movement in skeletal muscle: a possible step in excitation-contraction coupling. Nature 242:244–246

Schneider T, Hofmann F (1988) The bovine cardiac receptor for calcium channel blockers is a 195-kDa protein. Eur J Biochem 174:369–375

Schoemaker H, Langer SZ (1985) [^{3}H]Diltiazem binding to calcium channel-antagonist recognition sites in rat cerebral cortex. Eur J Pharmacol 111:273–277

Schwarz TL, Tempel BL, Papazian DM, Jan YN, Jan LY (1988) Multiple potassium-channel components are produced by alternative splicing at the *Shaker* locus in *Drosophila*. [Published erratum appears in Nature 1988 Apr 21; 332(6166):740]. Nature 331:137–142

Seagar MJ, Labbe-Jullie C, Granier C, Goll A, Glossmann H, Van Rietschoten J, Couraud F (1986) Molecular structure of rat brain apamin receptor: differential photoaffinity labeling of putative K^+ channel subunits and target size analysis. Biochemistry 25:4051–4057

Shalaby IA, Kongsamut S, Miller RJ (1986) Maitotoxin-induced release of gamma-[^{3}H]aminobutyric acid from cultures of striatal neurons. J Neurochem 46:1161–1165

Sharp AH, Campbell KP (1987) Affinity purification of antibodies specific for 1,4-dihydropyridine Ca^{2+} channel blockers. Circ Res 61:37–45

Sharp AH, Campbell KP (1989) Characterization of the 1,4-dihydropyridine receptor using subunit-specific antibodies. J Biol Chem 264:2816–2825

Sharp AH, Imagawa T, Leung AT, Campbell KP (1987) Identification and characterization of the dihydropyridine-binding subunit of the skeletal muscle dihydropyridine receptor. J Biol Chem 262:12309–12315

Sieber M, Nastainczyk W, Röhrkasten A, Hofmann F (1987a) Reconstitution of the purified receptor for calcium channel blockers. Biomed Biochim Acta 46:357–362

Sieber M, Nastainczyk W, Zubor V, Wernet W, Hofmann F (1987b) The 165-kDa peptide of the purified skeletal muscle dihydropyridine receptor contains the known regulatory sites of the calcium channel. Eur J Biochem 167:117–122

Silberberg SD, Poder TC, Lacerda AE (1989) Endothelin increases single-channel calcium currents in coronary arterial smooth muscle. FEBS Lett 247:68–72

Smith JS, McKenna EJ, Ma JJ, Vilven J, Vaghy PL, Schwartz A, Coronado R (1987) Calcium channel activity in a purified dihydropyridine-receptor preparation of skeletal muscle. Biochemistry 26:7182–7188

Soldatov NM (1988) Purification and characterization of dihydropyridine receptor from rabbit skeletal muscle. Eur J Biochem 173:327–338

Srinivasan Y, Elmer L, Davis J, Bennett V, Angelides K (1988) Ankyrin and spectrin associate with voltage-dependent sodium channels in brain. Nature 333:177–180

Striessnig J, Zernig G, Glossmann H (1985a) Ca^{2+} antagonist receptor sites on human red blood cell membranes. Eur J Pharmacol 108:329–330

Striessnig J, Zernig G, Glossmann H (1985b) Human red-blood-cell Ca^{2+}-antagonist binding sites. Evidence for an unusual receptor coupled to the nucleoside transporter. Eur J Biochem 150:67–77

Striessnig J, Goll A, Moosburger K, Glossmann H (1986a) Purified calcium channels have three allosterically coupled drug receptors. FEBS Lett 197:204–210

Striessnig J, Moosburger K, Goll A, Ferry DR, Glossmann H (1986b) Stereoselective photoaffinity labelling of the purified 1,4-dihydropyridine receptor of the voltage-dependent calcium channel. Eur J Biochem 161:603–609

Striessnig J, Knaus HG, Grabner M, Moosburger K, Seitz W, Lietz H, Glossmann H (1987) Photoaffinity labelling of the phenylalkylamine receptor of the skeletal muscle transverse-tubule calcium channel. FEBS Lett 212:247–253

Striessnig J, Meusburger E, Grabner M, Knaus HG, Glossmann H, Kaiser J, Schölkens B, Becker R, Linz W, Henning R (1988a) Evidence for a distinct Ca^{2+}-antagonist receptor for a novel benzothiazinone compound HOE 166. Naunyn Schmiedebergs Arch Pharmacol 337:331–340

Striessnig J, Knaus HG, Glossmann H (1988b) Photoaffinity labelling of the calcium channel-associated 1,4-dihydropyridine and phenylalkylamine receptor in guinea-pig hippocampus. Biochem J 253:37–46

Striessnig J, Scheffauer F, Mitterdorfer J, Schirmer M, Glossmann H (1990) Identification of the benzothiazepine-binding polypeptide of skeletal muscle calcium channels with (+)-cis-azidodiltiazem and antiligand antibodies. J Biol Chem (in press)

Suszkiw JB, Murawsky MM, Fortner RC (1987) Heterogeneity of presynaptic calcium channels revealed by species differences in the sensitivity of synaptosomal ^{45}Ca entry to omega-conotoxin. Biochem Biophys Res Commun 145:1283–1286

Takahashi M, Catterall WA (1987a) Dihydropyridine-sensitive calcium channels in cardiac and skeletal muscle membranes: studies with antibodies against the alpha subunits. Biochemistry 26:5518–5526

Takahashi M, Catterall WA (1987b) Identification of an alpha subunit of dihydropyridine-sensitive brain calcium channels. Science 236:88–91

Takahashi M, Tatsumi M, Ohizumi Y, Yasumoto T (1983) Ca^{2+} channel-activating function of maitotoxin, the most potent marine toxin known, in clonal rat pheochromocytoma cells. J Biol Chem 258:10944–10949

Takahashi M, Seagar MJ, Jones JF, Reber BF, Catterall WA (1987) Subunit structure of dihydropyridine-sensitive calcium channels from skeletal muscle. Proc Natl Acad Sci USA 84:5478–5482

Tanabe T, Takeshima H, Mikami A, Flockerzi V, Takahashi H, Kangawa K, Kojima M, Matsuo H, Hirose T, Numa S (1987) Primary structure of the receptor for calcium channel blockers from skeletal muscle. Nature 328:313–318

Tanabe T, Beam KG, Powell JA, Numa S (1988) Restoration of excitation-contraction coupling and slow calcium current in dysgenic muscle by dihydropyridine receptor cDNA. Nature 336:134–139

Tang CM, Presser F, Morad M (1988) Amiloride selectively blocks the low-threshold (T) calcium channel. Science 240:213–215

Tempel BL, Papazian DM, Schwarz TL, Jan YN, Jan LY (1987) Sequence of a probable potassium-channel component encoded at Shaker locus of Drosophila. Science 237:770–775

Timpe LC, Schwarz TL, Tempel BL, Papazian DM, Jan YN, Jan LY (1988) Expression of functional potassium channels from Shaker cDNA in Xenopus oocytes. Nature 331:143–145

Toutant M, Barhanin J, Bockaert J, Rouot B (1988) G-proteins in skeletal muscle. Biochem J 254:405–409

Triggle DJ (1988) Endogenous ligands for the calcium channel – mythos and realities. In: Morad M, et al. (eds). The calcium channel: structure, function and implications. Springer, Berlin Heidelberg New York, pp 549–563

Triggle DJ, Janis RA (1987) Calcium channel ligands. Annu Rev Pharmacol Toxicol 27:347–369

Tsien RW, Lipscombe DV, Madison KR, Bley KR, Fox AP (1988) Multiple types of neuronal calcium channels and their selective modulation. Trends Neurosci 11:431–438

Turner TJ, Goldin SM (1985) Calcium channels in rat brain synaptosomes: identification and pharmacological characterization. High-affinity blockade by organic Ca^{2+} channel blockers. J Neurosci 5:841–849

Umbach JA, Gundersen CB (1987) Expression of an omega-conotoxin-sensitive calcium channel in Xenopus oocytes injected with mRNA from Torpedo electric lobe. Proc Natl Acad Sci USA 84:5464–5468

Vaghy PL, Striessnig J, Miwa K, Knaus HG, Itagaki K, McKenna E, Glossmann H, Schwartz A (1987) Identification of a novel 1,4-dihydropyridine- and phenylalkylamine-binding polypeptide in calcium channel preparations. J Biol Chem 262:14337–14342

Vaghy PL, McKenna E, Itagaki K, Schwartz A (1988) Resolution of the identity of the Ca^{2+}-antagonist receptor in skeletal muscle. Trends Pharmacol Sci 9:398–402

Vandaele S, Fosset M, Galizzi JP, Lazdunski M (1987) Monoclonal antibodies that coimmunoprecipitate the 1,4-dihydropyridine and phenylalkylamine receptors and reveal the Ca^{2+} channel structure. Biochemistry 26:5–9

Venter JC, Fraser CM, Schaber JS, Jung CY, Bolger G, Triggle DJ (1983) Molecular properties of the slow inward calcium channel. Molecular weight determinations by radiation inactivation and covalent affinity labeling. J Biol Chem 258:9344–9348

Vergara J, Tsien RY, Delay M (1985) Inositol 1,4,5-trisphosphate: a possible chemical link in excitation-contraction coupling in muscle. Proc Natl Acad Sci USA 82:6352–6356

Volpe P, Salviati G, Di Virgilio F, Pozzan T (1985) Inositol 1,4,5-trisphosphate induces calcium release from sarcoplasmic reticulum of skeletal muscle. Nature 316:347–349

Volpe P, Di Virgilio F, Pozzan T, Salviati G (1986) Role of inositol 1,4,5-trisphosphate in excitation-contraction coupling in skeletal muscle. FEBS Lett 197:1–4

Wagner JA, Snowman AM, Olivera BM, Snyder SH (1987) Aminoglycoside effects on voltage-sensitive calcium channels and neurotoxicity (letter). N Engl J Med 317:1669

Wagner JA, Snowman AM, Biswas A, Olivera BM, Snyder SH (1988) Omega-conotoxin GVIA binding to a high-affinity receptor in brain: characterization, calcium sensitivity, and solubilization. J Neurosci 8:3354–3359

Waser PG (1986) The cholinergic receptor. In: Parnham MJ, Bruinvels J (eds) Discoveries in pharmacology. Elsevier, Amsterdam, pp 157–202

Watkins SC, Hoffman EP, Slayter HS, Kunkel LM (1988) Immunoelectron-microscopic localization of dystrophin in myofibres. Nature 333:863–866

Wu CH, Narahashi T (1988) Mechanism of action of novel marine neurotoxins on ion channels. Annu Rev Pharmacol Toxicol 28:141–161

Yamaguchi T, Saisu H, Mitsui H, Abe T (1988) Solubilization of the omega-conotoxin receptor associated with voltage-sensitive calcium channels from bovine brain. J Biol Chem 263:9491–9498

Yanagisawa M, Kurihara H, Kimura S, Tomobe Y, Kobayashi M, Mitsui Y, Yazaki Y, Goto K, Masaki T (1988) A novel potent vasoconstrictor peptide produced by vascular endothelial cells. Nature 332:411–415

Yang CP, Mellado W, Horwitz SB (1988) Azidopine photoaffinity labeling of multidrug resistance-associated glycoproteins. Biochem Pharmacol 37:1416–1421

Yatani A, Codina J, Imoto Y, Reeves JP, Birnbaumer L, Brown AM (1987) A G-protein directly regulates mammalian cardiac calcium channels. Science 238:1288–1292

Yatani A, Kunze DL, Brown AM (1988) Effects of dihydropyridine calcium channel modulators on cardiac sodium channels. Am J Physiol 254:H140–H147

Yeager RE, Yoshikami D, Rivier J, Cruz LJ, Miljanich GP (1987) Transmitter release from presynaptic terminals of electric organ: inhibition by the calcium channel antagonist omega *Conus* toxin. J Neurosci 7:2390–2396

Yoshii M, Tsunoo A, Kuroda Y, Wu CH, Narahashi T (1987) Maitotoxin-induced membrane current in neuroblastoma cells. Brain Res 424:119–125

Zech C, Greenberg RM, Hall L (1989) Very high affinity interaction of sodium channel ligands with *Drosophila* head membranes. Naunyn Schmiedebergs Arch Pharmacol 339:1245

Zernig G, Glossmann H (1988) A novel 1,4-dihydropyridine binding site on mitochondrial membranes from guinea-pig heart, liver and kidney. Biochem J 253:49–58

Zernig G, Moshammer T, Graziadei I, Glossmann H (1988) The mitochondrial high-capacity, low-affinity [^{3}H]nitrendipine binding site is regulated by nucleotides. Eur J Pharmacol 157:67–73

Note Added in Proof

1. The complementary DNA of the rabbit skeletal muscle ryanodine receptor (Ca^{2+}-release channel) has been sequenced and expressed by Takeshima et al. (1989). The calculated M_r of the protein is 565 223 (5037 amino acids). Compared to the enormous total length, the four predicted transmembrane segments (located at the C-terminal end) are a surprisingly small fraction (approximately one tenth) of the entire polypeptide. Sequence similarities to the

M2–M3 transmembrane segments of the nicotinic acetylcholine receptor exist. Calmodulin-, nucleotide- and Ca^{2+}-binding domains are found ("modulator-binding sites") between the sequences forming the foot region and the "channel"-forming C-terminal end. An expression plasmid which carried the entire protein-coding sequence was used to transfect CHO-cells, and high-affinity ($K_d = 19\,nM$) sites for [^{3}H]ryanodine were found in transfected cell membranes, with densities up to 6 pmol/mg protein.

2. The primary structure and functional expression of alpha$_1$ from rabbit cardiac muscle has been reported by Mikami et al. (1989). This alpha$_1$-subunit is composed of 2171 amino acids (M_r 242 771). The degree of sequence homology between cardiac and skeletal muscle alpha$_1$-subunits is 66%. Most important, the cAMP-dependent phosphorylation site which was suggested to play an important regulatory role (Ser 687 between repeats II and III of skeletal muscle alpha$_1$, see Röhrkasten et al. 1988) is not conserved. This is in line with observations that purified L-type channel complexes from heart cannot be phosphorylated in digitonin buffer. The N- and C-terminal regions of the cardiac alpha$_1$ are larger than for the skeletal muscle isochannel. When mRNA specific for transcription of the cardiac alpha$_1$ subunit was injected into *Xenopus* oocytes a 1,4 DHP-sensitive Ca^{2+} current was expressed. The co-injection of alpha$_2$ mRNA resulted in a larger Ca^{2+} current, probably reflecting the increased expression of alpha$_1$ by facilitation of its localisation in the membrane. It is as yet unknown whether alpha$_1$ from cardiac muscle can restore contraction in myoblasts from dysgenic mice even when flux of Ca^{2+} ions is blocked, e.g. by Cd^{2+}. Chimeric alpha$_1$ subunits may give clues about the regions which are necessary to couple alpha$_1$ to the ryanodine receptor.

3. When murine L-cells are transfected with a plasmid containing the complete open reading frame of the rabbit skeletal muscle alpha$_1$ subunit, low but significant binding of a radiolabelled 1,4 DHP [(+) PN 200–110] with K_d of 0.4 nM is expressed in membranes (Perez-Reyes et al. 1989). With the patch clamp method Ca^{2+} currents (which are activated by Bay K 8644 and blocked by Cd^{2+}) could be monitored. The half-times for activation and inactivation were considerably longer than for skeletal muscle or phenotypic skeletal muscle cell lines. The alpha$_1$ subunit (by immunoblot analysis) has an M_r considerably larger (by 20000) in this heterologous expression system than in native skeletal muscle membranes. Taken together, these results indicate that alpha$_1$ (from either heart or skeletal muscle) is sufficient to form voltage-dependent Ca^{2+} channels and to bind 1,4 DHP with high affinity even in the absence of the alpha$_2$ subunit. Proteolytic processing of the skeletal muscle alpha$_1$ appears to be a property of homologous systems, indicating tissue-specific channel tailoring.

References

Mikami A, Imoto K, Tanabe T, Niidome T, Mori Y, Takeshima H, Narumiya S, Numa S (1989) Primary structure and functional expression of the cardiac dihydropyridine-sensitive calcium channel. Nature 340:231–233

Perez-Reyes E, Kim HS, Lacerda AE, Horne W, Wei X, Rampe D, Campbell KP, Brown AM, Birnbaumer L (1989) Induction of calcium currents by the expression of the alpha$_1$-subunit of the dihydropyridine receptor from skeletal muscle. Nature 340:233–236

Takeshima H, NIshimura S, Matsumoto T, Ishida H, Kangawa K, Minamino N, Matsuo H, Ueda M, Hanaoka M, Hirose T, Numa S (1989) Primary structure and expression from complementary DNA of skeletal muscle ryanodine receptor. Nature 339:439–445

Rev. Physiol. Biochem. Pharmacol., Vol. 114
© Springer-Verlag 1990

Properties and Regulation of Calcium Channels in Muscle Cells

DIETER PELZER, SIEGRIED PELZER[1] and TERENCE F. McDONALD[2]

Contents

[1] II. Physiologisches Institut, Medizinische Fakultät der Universität des Saarlandes, D-6650 Homburg/Saar, FRG
[2] Department of Physiology and Biophysics, Dalhousie University, Halifax, Nova Scotia B3H 4H7, Canada

1 Introduction

With hindsight, and for the reasons mentioned below, it is clear that the modern era in the study of Ca channels began to draw to a close in the early 1980s. The reviews published around that time detail the areas of consensus and of contention (Bolton 1979; Carmeliet and Vereecke 1979; Reuter 1979, 1983; van Breemen et al. 1979; Coraboeuf 1980; Hagiwara and Byerly 1981; McDonald 1982; Tsien 1983), which are summarized in the following list to provide a perspective on the more current work that is given primary attention in this article.

1. Channel Types. With rare exceptions (e.g., starfish eggs), the Ca current (I_{Ca}) was assumed to flow through a single homogenous set of Ca channels.

2. Current-Voltage Relation. At physiological external Ca concentration (Ca_o), the current-voltage relation was bell-shaped with the threshold near -40 mV, a maximum near 0 mV, and an extrapolated reversal potential (E_{rev}) of $+50$ to $+70$ mV.

3. Conductance. The open channel conductance was considered to be ohmic.

4. Selectivity. The selectivity of Ca channels for Ca over Na was large, but not that large. In fact, the results of experiments in which Ca_o and Na_o were varied suggested that Na might carry as much or more of the current than Ca under physiological conditions. This was the major reason for the widespread use of the annotations I_s or I_{si} (slow current or slow inward current) instead of I_{Ca}, at least by cardiac electrophysiologists. In regard to divalent cations, external Ba and Sr, but not Mg, were effective charge carriers when substituted for Ca.

5. Kinetics. Ca channel kinetics were assumed to be purely voltage dependent with activation and inactivation governed by first-order processes. Steady-state activation (d_∞) and steady-state inactivation (f_∞) were sigmoidal functions of voltage with the voltages at half maximum (V_h), being near -20 mV.

6. Block. I_{Ca} was blocked by a number of multivalent ions such as La, Mn, Ni, Co, and Cd. In addition, it was blocked by organic compounds (so-called Ca antagonists) in a manner which depended on voltage and frequency of stimulation ("use-dependent" block).

7. Modulation. Neurotransmitters were thought to regulate I_{Ca} by occlusive changes or by changes in channel density with no effect on channel kinetics.

There were clear signs that the foregoing was a rough, far from definitive framework, particularly in regard to Ca channel permeability/selectivity and Ca channel kinetics. For example, the relative contribution of Na to I_{Ca} was difficult to assess from experiments in which Na_o was varied because this manipulation affected the internal Ca concentration (Ca_i) via the Na-Ca exchange mechanism. Clear-cut demonstrations of E_{rev}, and changes in E_{rev}, were rarely, if ever, achieved due to the presence of large overlapping capacitive currents and other currents. The latter currents also made it difficult to ascertain whether activation followed a monoexponential time course or some other time course. Times to peak I_{Ca} and time constants of current decay had an unreasonably wide scatter, the former ranging from about 5 to 50 ms and the latter from about 50 to 500 ms. Finally, there were indications that inactivation was not entirely voltage dependent.

Investigators involved in that generation of studies on Ca channels labored under two major handicaps: firstly, borderline voltage control due to large membrane areas and complex tissue geometry, and secondly, lack of experimental control over the intracellular solution. The solution to the first problem arrived in the form of viable, enzymatically dissociated single cells (e.g., Isenberg and Klöckner 1980; Trube 1983). The solution to the second problem was the use of large bore (microns) suction pipettes (e.g., Lee et al. 1980; Hamill et al. 1981). With gentle suction these could be attached to the surface membrane of single cells, and with further suction a "breakthrough" of the patch of membrane under the pipette lumen provided direct access of the pipette-filling solution to the cell interior. In this way, the intracellular solution could be dialyzed against the pipette-filling solution, thereby permitting the intracellular application of ions, channel blockers, second messengers, nucleotides, enzymes, and so on.

Aside from providing a means of recording "whole-cell" currents, the suction pipette was also instrumental in opening up a new field of study, that of single-channel currents (Hamill et al. 1981; Sakmann and Neher 1983, 1984). Suction applied to the pipette leads to a tight sealing of the open glass tip to the surface membrane of the cell. This effectively isolates the patch of membrane under the tip lumen from the rest of the cell membrane. The cell-attached membrane patch may contain a number of types of channels in addition to Ca channels, but these can be blocked by including the appropriate agents (e.g., tetrodotoxin, TTX) in the pipette-filling solution. The voltage across the patch can be clamped by applying voltage to the intrapipette solution, and current flow due to the opening of Ca channels can be recorded. On some occasions the patch may contain a number of Ca channels, on others there may be only one functional channel present. In a further refinement of the tight-seal technique, the membrane patch can be excised from the cell and single channel activity recorded during experimental control of solutions bathing the two faces of the membrane.

Coincident with the development of patch-clamp studies on single Ca channels was the arrival of radioactively labeled chemical probes of high affinity and specifity for Ca channel binding sites (see Janis et al. 1987; Triggle and Janis 1987). These have led to the identification and isolation of the protein components of the Ca channel (see Lazdunski et al. 1987; Campbell et al. 1988d; Catterall 1988; Glossmann and Striessnig 1988, 1990; Hosey and Lazdunski 1988). Since isolated channel proteins can be reconstituted into functional single Ca channels in lipid bilayers (see Flockerzi et al. 1986; Rosenberg et al. 1986), they offer the opportunity for investigating Ca channel properties without interference from other cellular moieties.

Studies using the whole-cell and single-channel approaches have provided complementary information that has greatly refined the picture of the Ca channel that had emerged from the multicellular era. For example, it is now established that in addition to the well-studied predominant species of Ca channel (the low-threshold, slow-inactivating, large-conductance, dihydropyridine-sensitive L-type) there is at least one other set of Ca channels (the high-threshold, fast-inactivating, small-conductance, dihydropyridine-insensitive T-type) which is found along with L-type channels in heart, smooth muscle, and skeletal muscle, as well as in a wide variety of neuronal, endocrine, and exocrine cells. A third kind of Ca channel (the intermediate N-type) seems to coexist with the two other types in neuronal membranes (see Sect. 7 for details on the properties of Ca channel types).

The conductance properties and kinetics of the classic L-type channel are more complicated than previously thought, and it turns out that L-type channel selectivity for Ca is remarkably high. Channel permeation appears to be a multistep, single-file process, and block by poorly permeant divalent cations can be better appreciated in this context. There have been great strides in understanding the regulation of L-type Ca channels by intracellular factors, as well as the modulation exerted by Ca channel stimulants and inhibitors.

The objective of this review is to present an update on the properties and regulation of the classic L-type Ca channel in heart, smooth muscle, and skeletal muscle cells (see Sects. 1–6). We do this in a manner that is not encyclopedic and where pertinent, we draw on our own material and views. We give only a sketchy account of the other types of Ca channels (see Sect. 7) and of the properties and regulation of Ca channels in nonmuscle cells. For information on the latter, the reader is referred to excellent recent reviews (Miller 1987a, b; Tsien 1987; Levitan 1988; Tsien et al. 1988). Finally, we are aware that the studies cited here constitute only a part of the work in this area and apologize for any serious omissions.

2 Kinetics of L-Type Calcium Channels

2.1 Overview

When muscle preparations are depolarized by a voltage-clamp pulse from, say, -50 to 0 mV for several hundred milliseconds, macroscopic Ca channel current quickly turns on (activation), reaches a peak, and then (usually) decays (inactivation) with a distinctly slower time course. When the membrane is then repolarized to -50 mV, a subsequent depolarization elicits a smaller peak current unless sufficient time has been granted at -50 mV to permit removal of the depolarization-induced inactivation. This transition from the inactivated state to the closed (resting, available) state is termed reactivation (restoration, repriming). If, instead, the depolarization from -50 to 0 mV is only applied for tens of milliseconds, such that a substantial fraction of the Ca channels have not yet entered the inactivated state, repolarization does not instantaneously close the channels; the potential change is accompanied by a tail current whose time course of decay reflects the deactivation of channels.

As mentioned in the Introduction (Sect. 1), the processes of activation, in-activation, and reactivation in multicellular muscle tissues were assumed to obey first-order kinetics governed by voltage (e.g., Bassingthwaighte and Reuter 1972; Bolton 1979; Reuter 1979; van Breemen et al. 1979; McDonald 1982; Sánchez and Stefani 1983; Tsien 1983). Studies on macroscopic and microscopic Ca channel currents in more suitable muscle preparations have led to modifications of these earlier views, and they will be reviewed here. The section on activation (Sect. 2.2) begins with an overview of macroscopic and microscopic currents and covers the topics of voltage dependence, maximal activation, and current-voltage relations. Inactivation (Sect. 2.3) is discussed in terms of apparent relations between voltage and inactivation parameters, Ca-dependent inactivation, and voltage-dependent inactivation. This is fol-lowed by the section on reactivation (Sect. 2.4).

The reader is forewarned that most of this material is drawn from the litera-ture on cardiac cells. One reason for this is that information on the kinetic properties of Ca channels in skeletal and smooth muscle is relatively sparse. Another reason is that in heart the contribution of T-type channels to whole-cell or single-channel currents can be easily identified and eliminated. In fact, T-channel activity is often completely absent, even on depolarization from very negative holding potentials (e.g., Mitra and Morad 1986; Campbell et al. 1988a; Hadley and Hume 1988). By contrast, macroscopic Ca current kinet-ics are difficult to ascertain in noncardiac muscle cells due to seemingly vari-able proportions of T- and L-like (low and high threshold or fast and slow) Ca channels. In some cases the separation of components appears to have been satisfactory (e.g., Bean et al. 1986; Benham et al. 1987; Beam and Knud-son 1988a, b), but in others the properties of two or more suspected channel populations (see Sect. 7) were not easily distinguished from one another (e.g., Aaronson et al. 1988; Nakazawa et al. 1988). Although we have attempted to focus on L-type data, these constraints need to be kept in mind.

2.2 Activation

2.2.1 Overview of Activation of Macroscopic and Microscopic Currents

The first voltage-clamp studies on I_{Ca} in single heart cells (Isenberg and Klöckner 1980, 1982) indicated that the time to peak current was considerably shorter than that usually found in multicellular cardiac tissue, which in turn was about ten times shorter than, for example, in frog skeletal muscle fibers at similar temperatures (cf. Stanfield 1977; Sánchez and Stefani 1978, 1983; Almers and Palade 1981). Isenberg and Klöckner (1980, 1982) reported that I_{Ca} triggered by a step depolarization to 0 mV peaked within 2–4 ms in rat and bovine ventricular myocytes at 35 °C. In isolated smooth muscle cells, the activation of I_{Ca} was also shown to be much quicker than previously registered from smooth muscle tissue (Walsh and Singer 1981). The turn-on of the current in cardiac myocytes appeared to follow a monoexponential time course with a time constant τ_d of about 1.1 ms close to threshold and of about 0.5 ms at $+10$ mV (35 °C ; Isenberg and Klöckner 1982). In frog twitch muscle fibers at room temperature, the time constant of activation was about 180 ms at -30 mV and declined to about 45 ms at $+20$ mV (Sánchez and Stefani 1983). After a short depolarization, the decay of cardiac I_{Ca} tail current in bovine ventricular myocytes at -50 mV was fitted with a single exponential ($\tau = 0.4-0.8$ ms, 35 °C; Isenberg and Klöckner 1982). Josephson et al. (1984) also found that the decay of I_{Ca} tail current could be approximated by a single exponential with $\tau = 1.7-1.9$ ms in rat ventricular myocytes at 22 °C.

In these early studies on single cells, the initial phases of I_{Ca} activation and deactivation were obscured by the flow of large capacitive currents. In addition, the voltage-clamp techniques that were used (one- or two-microelectrode methods) in rather large cells may not have been optimal. When tight-seal pipettes were used to record whole-cell currents in smaller cardiac myocytes (Lee and Tsien 1982, 1984; Pelzer et al. 1986b), the turn-on of current had a distinctly sigmoidal time course, indicating that the activation of Ca channel current involves more than a simple transition from a single closed state to an open one. The detection and state assignment of additional transitions related to I_{Ca} activation emerged from patch-clamp recordings of single Ca channels in cardiac myocytes (Reuter et al. 1982; Cavalié et al. 1983, 1986), and a step by step analysis is warranted at this point in this review.

The currents elicited by activating depolarizations applied to a cardiac membrane patch containing a single Ca channel can be divided into two types: those that do not exhibit channel openings (blanks, nulls) and those that do. For now, we focus on the features of records (sweeps) with openings since these give information on the time course of the activation process. Although short openings of the channel are usually followed by short closings, they are sometimes followed by markedly longer ones. A sequence of short openings punctuated by short closings is called a burst, and the subsequent long closing prior to the next burst is the interburst interval (Reuter et al. 1982; Cavalié et al. 1983, 1986).

During bursting activity at a given voltage, the lifetime of a particular state will be determined by the reciprocal of the sum of the rate constants leaving that state. If there is just one open state, the distribution of the open times should be exponential. This distribution can be determined by measuring the duration of a large number of openings such as might occur in an ensemble of current records generated by a train of 100–200 depolarizations. From the same ensemble of records, one may enquire about the distribution of closed states and whether their lifetimes can be described by a single exponential.

A monoexponential distribution of open times and a double exponential distribution of closed times have been the norm in studies of single Ca channel currents in cardiac membrane patches (e.g., Reuter et al. 1982; Cachelin et al. 1983; Cavalié et al. 1983, 1986; Hess et al. 1984). At potentials corresponding to near maximal macroscopic Ca channel current (V_{peak}) in guinea pig ventricular myocytes, τ_o ranged from 0.44 to 1.27 ms (mean 0.68 ms), τ_{C1} from 0.1 to 0.35 ms (mean 0.19 ms), and τ_{C2} from 0.72 to 3.75 ms (mean 1.8 ms) (Cavalié et al. 1986). The interpretation of these findings is that Ca channels carrying Ca or Ba ions have only one conducting open state (but see below), that τ_{C1} corresponds to the mean lifetime of the closures between openings within bursts, and that τ_{C2} corresponds to the mean lifetime of the closures between bursts.

The clear separation of the average lifetimes in the two latter groups of millisecond range closings points to the participation of two separate closed states in channel bursting activity. The relation between these closed states and the open state has been examined by analyzing the time lags (first latencies) between the onset of a depolarizing jump and the first opening of a cardiac single Ca channel (Trautwein and Pelzer 1985b, 1986; Cavalié et al. 1986; Pelzer et al. 1986a). In ensembles collected from repetitive depolarization to V_{peak} potentials, the distribution of first latencies has a biphasic shape. It rises from zero to a peak at 0.25–0.5 ms, and has very few latencies longer than 6 ms. This finding indicates that the two closed states (C_1, C_2) precede the open state (O) of the Ca channel and that the open state is entered after transit through the two closed states. An alternative arrangement, O between C_1 and C_2, can be ruled out since the probability-density function was biphasic rather than monotonically declining.

The evidence suggests that activation kinetics can be explained by a scheme of sequential states with four adjustable rate constants (k) as follows:

$$C_1 \underset{k_2}{\overset{k_1}{\rightleftarrows}} C_2 \underset{k_4}{\overset{k_3}{\rightleftarrows}} O$$

There are several important considerations that arise out of this kinetic model. (a) A major difference from a Hodgkin-Huxley-like m^2 activation description (with rate constants such that $k_1 = 2k_3$, and $k_4 = 2k_2$ and where m denotes the activation variable) is that k_1 is much smaller than the other rate constants and therefore rate-limits activation (Cachelin et al. 1983; Cavalié et al. 1983, 1986; Tsien 1983; Brum et al. 1984). (b) The cumulative probability-density distribution of first latencies, obtained by integration of the first-latency distribution, estimates the time course of the first entrance of the single Ca channel into the open state. After suitable scaling, the cumulative function superimposes on the activation phase of the corresponding ensemble mean current (Cavalié et al. 1986; Pelzer et al. 1986a; Trautwein and Pelzer 1988). (c) Simulations with the model scheme produce rising phases compatible with the turn-on of whole-cell Ca channel currents (Cavalié et al. 1986; Pelzer et al. 1986a; Tsien et al. 1986; Trautwein and Pelzer 1988).

The kinetics of single L-type (or high threshold) Ca channel currents in cell-attached patches of smooth muscle cells have not yet been analyzed in similar detail. However, there are indications that they share features with cardiac microscopic Ca channel currents. (a) Successive depolarizations draw a mixture of sweeps with channel openings and blanks in mammalian vascular (Isenberg and Klöckner 1985c; Worley et al. 1986; Benham et al. 1987; Yatani et al. 1987c; Nelson et al. 1988) and intestinal cells (Yoshino and Yabu 1985). (b) Rapid bursting activity with long interburst intervals characterized records obtained in the foregoing studies. Finally, (c) the turn-on time course of single Ca channel currents resembled that observed in macroscopic currents recorded from similar smooth muscle cells (Yoshino and Yabu 1985; Benham et al. 1987).

Information on single skeletal muscle L-like Ca channel kinetics comes from bilayer recordings of reconstituted skeletal muscle T-tubular channels (Affolter and Coronado 1985; Rosenberg et al. 1986; Coronado 1987; Yatani et al. 1988) and of solubilized and purified dihydropyridine (DHP)-binding sites reconstituted as functional Ca channels (Pelzer et al. 1988, 1989b). The outcome of these studies can be summarized as follows. Openings occur in bursts and clusters of bursts which are stochastic in nature. Open time frequency histograms are best fit by the sum of two first-order decay functions with two well-separated time constants − both of which were considerably shorter in those studies, which did not use DHP agonists to promote Ca channel opening (Pelzer et al. 1988, 1989b). Closed time frequency histograms are best fit by a two-exponential probability distribution function. All in all, it appears that skeletal muscle Ca channels in lipid bilayers gate somewhat slower than cardiac or smooth muscle Ca channels. However, a constraint to be kept in mind with kinetic data from lipid bilayer experiments is that the bilayer lipid composition may affect single-channel kinetics (Coronado 1987).

2.2.2 Voltage Dependence of Channel Opening

Studies on macroscopic and microscopic Ca channel currents have produced complementary findings on the voltage dependence of Ca channel opening. It is therefore useful to begin this section by describing how macroscopic Ca channel currents (I) in a cell (or tissue) relate to elementary current (i) through an open single pore. As formulated by Tsien et al. (1986)

$$I = N_f \times p_o \times i$$

where N_f is the number of channels in the available pool, and p_o is the probability that the channel will be open, given that it is available. N_f is defined as

$$N_f = N_T \times p_f$$

where N_T is the total number of channels in the cell membrane of the preparation (available plus unavailable), and p_f is the probability that a given channel is available. The addition of these two equations gives

$$I = N_T \times p_f \times p_o \times i \ .$$

The two open-state probabilities, p_f and p_o, are easily distinguished because they reflect processes operating on very different time scales. The factor p_f (probability that a single channel is available) is indicated by the fraction of sweeps that are blank within an ensemble. Blanks often cluster together to form "runs" within an ensemble generated by pulsing at 0.5 Hz; this suggests that they are governed by very slow kinetics (cardiac muscle: Fox et al. 1984; Cavalié et al. 1986; Pelzer et al. 1986a; Kawashima and Ochi 1988; Trautwein and Pelzer 1988; smooth muscle: Nelson et al. 1988). By contrast, the open-state probability p_o is determined by the millisecond kinetics of channel opening and closing in nonblank sweeps (cardiac muscle: Cavalié et al. 1986; Pelzer et al. 1986a; Tsien et al. 1986; Trautwein and Pelzer 1988; smooth muscle: Nelson et al. 1988). This probability can be designated for a particular time after the onset of depolarizing steps (e.g., 5 ms), or it can be designated as a time-averaged probability over the duration of the pulse. Finally, it is important to note that p_o and p_f are sometimes lumped together and called p.

When muscle preparations are depolarized by pulses applied from a negative holding potential, the degree of L-type Ca channel activation depends on the potential of the depolarizing pulse. When the external bathing solution contains a low millimolar concentration of Ca or Ba, the threshold for activation in cardiac myocytes is around -40 mV (Beeler and Reuter 1970; Isenberg and Klöckner 1980, 1982; Mitchell et al. 1983; Campbell et al. 1988a). Within a Hodgkin-Huxley framework, the steady-state activation variable, d_∞, ranges from 0 at threshold to 1 near $+10$ mV; it has a sigmoidal shape with a slope factor of about 7 mV (Bassingthwaighte and Reuter 1972; Trautwein et al. 1975; Isenberg and Klöckner 1982; Kass and Sanguinetti 1984; Cohen and Lederer 1987).

The activation range of L-type Ca channels in smooth muscle cells does not appear to be very different from that of cardiac cells. In guinea pig urinary bladder cells bathed with 1.2 mM Ca solution, Klöckner and Isenberg (1985) reported a threshold around -40 mV and V_{peak} near -5 mV. The V_h of the steady-state activation curve was -14 mV and $k = 6$ mV, compared to $V_h = -27$ mV and a slope factor of 8 mV in guinea pig ileum cells (Droogmans and Callewaert 1986). Thresholds ranging from -40 to -20 mV have been measured in vascular smooth muscle cells (Friedman et al. 1986; Benham et al. 1987; Pacaud et al. 1987; Toro and Stefani 1987; Aaronson et al. 1988; Ohya et al. 1988) as well as in intestinal and other smooth muscle cells (Amédée et al. 1987; Ganitkevich et al. 1988; Nakazawa et al. 1988).

The position of the activation range of L-type (slow) Ca channel current in skeletal muscle appears to be more variable than that in heart and smooth muscle cells. For example, V_h of the steady-state activation relation in rat fast twitch fibers bathed in 10 mM Ca solution was $+16$ mV with $k = 7.1$ mV (Beam and Knudson 1988b). Activation threshold was near 0 mV, compared with thresholds near -30 mV in cultured rat myoballs (but with 2.5 mM Ba_o; Cognard et al. 1986), near -30 mV in frog skeletal muscle fibers (10 mM Ca solution; Almers et al. 1984; Cota and Stefani 1986), and between -40 and -60 mV in one study on frog twitch muscle fibers with 10 mM Ca_o ($V_h = -35.2$ mV, $k = 9.9$ mV; Sánchez and Stefani 1983).

The position of the activation curve, but not its general shape, is affected by alterations in the membrane surface potential. Divalent cations can effectively shield or neutralize the external surface charge and thereby affect the voltage dependence of Ca channel opening (e.g., Ohmori and Yoshii 1977; Wilson et al. 1983; Kass and Krafte 1987; Ganitkevich et al. 1988). Thus, an increase in Ca_o from approximately 2 mM to 5 or 7.5 mM shifts activation in the positive direction by $5-10$ mV in mammalian and amphibian cardiac myocytes (Tseng et al. 1987; Campbell et al. 1988a), an increase from 2 to 10 mM shifts it by 15 mV in rat skeletal muscle fibers (Donaldson and Beam 1983), and an increase from 0.5 to 30 mM shifts relations by nearly 30 mV in guinea pig taenia caeci cells (Ganitkevich et al. 1988). Compared to the threshold in millimolar Ca_o or Ba_o, the threshold in $90-110$ mM Ba_o is shifted in the positive direction by about 30 mV in smooth muscle myocytes (Bean et al. 1986; Benham et al. 1987; Aaronson et al. 1988). Similarly, the voltage dependence of Ca channel activation in guinea pig ventricular cardiomyocytes is shifted by about 30 mV in the positive direction when a 90 mM Ba solution replaces 3.6 mM Ba solution (McDonald et al. 1986; Pelzer et al. 1986b).

The dependence of Ca channel activation on voltage can also be ascertained in single-channel experiments. There is a distinct threshold potential negative to which there are no channel openings and positive to which there is an increasing open-state probability which reaches a maximal value (cardiac muscle: Reuter et al. 1982; Cavalié et al. 1983; Trautwein and Pelzer 1985a; smooth muscle: Yoshino and Yabu 1985; Worley et al. 1986; Nelson et al. 1988). The increase in open-state probability with depolarizing voltage is due to two factors. Firstly, the fraction of nonblank records in an ensemble (p_f) rises from zero at potentials below threshold to some maximal value at more positive potentials. Secondly, there is an increase in the open-state probability within nonblank records (p_o), as estimated in several different ways: (a) from single-channel currents time-averaged over the duration of the pulse and then divided by the unitary current amplitude (Reuter et al. 1982; Nelson et al. 1988), (b) from the fraction of records in an ensemble that have an open channel event at a selected time (e.g., 10 ms) during depolarization (Cavalié et al.

1983; Trautwein and Pelzer 1985a), and (c) from the peak amplitude of the ensemble average current I* divided by unitary current amplitude i (McDonald et al. 1986). In heart, the voltage dependencies of p_f and p_o are quite similar, and have a sigmoidal shape with a slope factor of about 7 mV (McDonald et al. 1986). In fact, after account is taken of the surface potential change due to high Ba_o, the voltage dependence of single Ca channel opening overlays the steady-state activation variable (d_∞) of the global Ca channel current determined in cardiac myocytes and tissues superfused with solutions containing millimolar Ca (Pelzer et al. 1986b). In arterial smooth muscle cells, V_h and k of single Ca channel activation curves (80 mM Ba) range from -15 to -23 mV and from 7 to 8 mV, respectively (Nelson et al. 1988). In lipid bilayer recordings, p_o and open times of reconstituted single L-like skeletal muscle Ca channels also increase with increasing depolarization (Pelzer et al. 1988, 1989b; Yatani et al. 1988), as expected for a voltage-dependent activation process.

Methodological problems have thus far precluded accurate measurements of the relations between activation time constants of macroscopic Ca channel current and voltage. However, some direction can be gleaned from the voltage dependence of the times to peak current. In guinea pig ventricular myocytes bathed in 3.6 mM Ca or Ba solutions at 36°C, times to peak were about $10-15$ ms at potentials near -20 mV and declined to $2-3$ ms at potentials near $+30$ to $+40$ mV (McDonald et al. 1986). After correction for a surface potential shift of approximately 30 mV, the times to peak during superfusion with 90 mM Ba solution were similar to the above. From single-channel studies, it is clear that the increase in p_o with voltage is due to an acceleration of the forward rate constants, leading to the open state and a slowing of the backward rate constants, leading away from the open state. A direct consequence of these kinetic changes, particularly the increase in k_1, is that the average latency to first channel opening declines with depolarization (Cavalié et al. 1983). This response by the single Ca channel to applied voltage corresponds with the shortening of time to peak macroscopic current. Times to peak currents in skeletal muscle cells, as long as several hundred milliseconds near threshold at room temperature, shorten by about fivefold at more positive potentials (Almers et al. 1984; Cota and Stefani 1986; Beam and Knudson 1988b). In smooth muscle the times to peak are more like those in cardiac cells, and they also shorten with increasing depolarization (e.g., Bean et al. 1986; Benham et al. 1987).

2.2.3 Maximal Probability of Channel Opening

A general assumption prior to single-channel studies was that the current elicited by depolarization to a maximally activating potential from a noninactivating holding potential reflected the contribution of almost all of the functional Ca channels in the cell membrane. This assumption was wrong on two counts. Firstly, the nature of the bursting activity of single Ca channels means that only a fraction of the channels which open will be open at any given instant after depolarization, i.e., p_o will normally be far less than unity (cardiac muscle: Reuter et al. 1982; Cavalié et al. 1983, 1986; Hess et al. 1984; Kawashima and Ochi 1988; smooth muscle: Worley et al. 1986; Nelson et al. 1988; skeletal muscle: Pelzer et al. 1988, 1989b; Yatani et al. 1988). Secondly, functional channels seem to flit in and out of the available state (i.e., $p_f < 1$) even when voltage-dependent inactivation is maximally removed by large hyperpolarization (cardiac muscle: Reuter et al. 1982; Cavalié et al. 1983, 1986; Hess et al. 1984; Kawashima and Ochi 1988; smooth muscle: Worley et al. 1986; Nelson et al. 1988).

2.2.4 Current-Voltage Relations

The voltage dependence of Ca channel activation is a major influence on the current-voltage relations of peak current. One reason for this is that, at the time of peak current, activation is nearly complete whereas inactivation is poorly developed.

There are other factors that determine the current-voltage relations of peak current and these include channel selectivity and driving force. The latter, along with the single-channel conductance, determines the current-voltage relation of the elementary current (i-V). In cardiac cell-attached patches, the i-V relation is linear over most of the voltage range, whether the charge carrier is Ca, Ba (Reuter et al. 1982; Cavalié et al. 1983; Hess et al. 1986; McDonald et al. 1986), or Na (Matsuda 1986), i.e., the conductance over this range is constant and the amplitude varies with driving force. A similar statement applies to single Ca channels in smooth muscle patches (Yoshino and Yabu 1985; Worley et al. 1986; Benham et al. 1987; Nelson et al. 1988) and to skeletal muscle T-tubular Ca channels incorporated into lipid bilayers (Affolter and Coronado 1985; Flockerzi et al. 1986; Talvenheimo et al. 1987; Trautwein et al. 1987; Pelzer et al. 1988, 1989b; Yatani et al. 1988). Lipid bilayer Ca channel recordings from cardiac and skeletal muscle membrane preparations also revealed that the nonlinearity of i-V relations at very positive membrane potentials is related to the asymmetry of external and internal solutions, since i-V curves are almost linear in symmetrical solutions (Rosenberg et al. 1986). The product of the voltage-dependent elementary current amplitude (i(V)) and the voltage-dependent open-state probability (p(V)) determines the voltage dependence of the ensemble average single Ca channel current (I*(V)). Since, as noted earlier, p(V) rises sigmoidally with voltage (slope factor ≈ 7 mV) and saturates at about 50 mV positive to threshold, the product of p(V) and linear i(V) is bell-shaped.

As expected from the foregoing, the I-V relation of macroscopic Ca channel current is bell-shaped in multicellular cardiac tissues (Carmeliet and Vereecke 1979; Coraboeuf 1980), independent of whether the current is carried by Ca, Ba, or Sr (Kass and Sanguinetti 1984). Similar I-V relations have been measured in a variety of single cardiac and smooth muscle myocytes superfused with solutions containing permeant divalent cations (e.g., Isenberg and Klöckner 1982; Mitchell et al. 1983; Klöckner and Isenberg 1985; Fischmeister and Hartzell 1986; Tseng et al. 1987; Nakazawa et al. 1988) or permeant univalent cations (e.g., Isenberg and Klöckner 1985c; Matsuda 1986; Hadley and Hume 1987; Campbell et al. 1988a).

One way of comparing I-V relations determined under different experimental conditions is to plot I against a "normalized" V axis (V_{peak} + V) (McDonald et al. 1986). When global Ca channel currents are compared in this way, relations from multicellular and single-cell preparations superimpose on each other, independent of the extracellular charge carrier and its concentration (Pelzer et al. 1986b). The threshold is then around V_{peak} −40 mV, the voltage at one-half peak current amplitude is near V_{peak} −15 mV, and the reversal potential (E_{rev}) is about 55−70 mV positive to V_{peak}. In fact, the ensemble average single Ca channel current, I*(V), superimposes satisfactorily on global I-V data (McDonald et al. 1986).

2.3 Inactivation

Since the discovery that Ca channel inactivation in *Paramecium* is dependent on Ca entry rather than on voltage (Brehm and Eckert 1978), considerable attention has been paid to this topic in studies on a wide variety of other cell types (Eckert and Chad 1984). In regard to cardiac cells, there is now a large body of evidence suggesting that both Ca-dependent and voltage-dependent mechanisms govern Ca channel inactivation. We review this area in cardiac cells by outlining the apparent relations between voltage and inactivation, examining the evidence of favor of Ca-dependent inactivation and voltage-dependent inactivation respectively, and briefly summarizing the outcome. We then turn to inactivation in smooth muscle and skeletal muscle.

2.3.1 Apparent Relations Between Voltage and Inactivation Parameters

During a maintained depolarization to a plateau-phase potential, the rapid activation of I_{Ca} is followed by a slower decay (inactivation). In some studies on multicellular cardiac preparations, the decay phase was well described by a single exponential; in others at least two exponentials were required (see McDonald 1982; Mentrard et al. 1984). The time course of inactivation in cardiac myocytes is not less complicated. In some cases the decay has been well described by a single exponential (e.g., Campbell et al. 1988 c), and in others, a double-exponential fit was the most suitable (Isenberg and Klöckner 1982; Irisawa 1984; Imoto et al. 1985; Tseng et al. 1987). However, there have also been studies in which neither of the above descriptions seemed appropriate (Bean 1985; Bechem and Pott 1985; McDonald et al. 1986; Hadley and Hume 1987), and the authors resorted to a more generic parameter, time to one-half decay $(t_{1/2})$, to describe their findings. The foregoing applies to large, relatively fast, decay phases. There may also be a small ultraslow component of I_{Ca} inactivation in cardiac tissue (Kass and Scheuer 1982) and myocytes (Hume and Giles 1983; Noble 1984; Isenberg and Klöckner 1985 a).

When I_{Ca} inactivation in myocytes has been described by the sum of two exponentials, the first of these usually has a time constant that is five to ten times shorter than the second one. There is not complete agreement on the shape of the voltage dependence of the fast phase time constant; it may be U-shaped with minimum values around 0 mV (Josephson et al. 1984; Uehara and Hume 1985; Tseng et al. 1987), or curvilinear increasing with positive voltage (Isenberg and Klöckner 1982; Irisawa 1984). There is better agreement on the shape of the voltage dependence of the slow phase time constant; it is U-shaped with a minimum near 0 mV and several-fold longer values at -30 mV and $+30$ mV (Isenberg and Klöckner 1982; Irisawa 1984; Josephson et al. 1984; Iijima et al. 1985). The actual time constants and the relative amplitudes of the two phases show considerable scatter. However, rough averages for guinea pig myocytes at 35 °C are $3-7$ ms and $30-80$ ms for the two time constants, with the amplitude of the fast phase being somewhat smaller than that of the slow phase. In two studies where I_{Ca} decay was fitted with monoexponential functions, the voltage dependence was U-shaped with a minimum near 0 mV (Tseng et al. 1987; Campbell et al. 1988 c).

Does cardiac I_{Ca} fully inactivate during a maintained depolarization? In multicellular preparations, inactivation appeared to go to completion within $300-500$ ms in some studies, but not quite to completion in others (McDonald 1982). A similar spread of results applies to myocytes. For example, I_{Ca} near V_{peak} was found to inactivate completely in some studies (Imoto et al. 1985; Fischmeister et al. 1986) and to inactivate to 93% completion (Hadley and Hume 1987) or to 80%–90% completion in others (Uehara and Hume 1985; Wendt-Gallitelli and Isenberg 1985).

A prepulse to potentials more positive than -50 mV reduces the amplitude of I_{Ca} on a subsequent test pulse. When the degree of this reduction is plotted against prepulse potential, the resulting relation estimates the voltage dependence of channel availability and is termed steady-state inactivation f_∞ (Reuter 1979; McDonald 1982). The dependence of I_{Ca} availability on prepulse voltage, at least up to the range $+10$ to $+20$ mV (see Sect. 2.3.2), has a sigmoidal shape with values approaching 1.0 at -50 mV and 0 at $+10$ mV. The relation can often be well described by the equation:

$$I_{Ca}/I_{Ca} \text{ (max)} = 1/\{1 + \exp\ [(V-V_h)/k]\}$$

where V_h is the prepulse voltage that reduces test I_{Ca} to one-half of maximum test I_{Ca}, and k is the slope of the relation. In bovine ventricular myocytes (Isenberg and Klöckner 1982), rat ventricular myocytes (Josephson et al. 1984), canine ventricular myocytes (Tseng et al. 1987), frog ventricular myocytes (Fischmeister and Hartzell 1986), and frog atrial myocytes (Uehara and Hume 1985; Campbell et al. 1988c), V_h ranged from -20 to -30 mV, and k from -4 to -11 mV. These values are in good accord with determinations on multicellular tissue (Coraboeuf 1980; McDonald 1982; Kass and Sanguinetti 1984) and also agree with measurements on single Ca channels with due regard for shifts related to screening (Pelzer et al. 1986b).

2.3.2 Calcium-Dependent Inactivation

A number of experimental expectations spring from the hypothesis (Brehm and Eckert 1978) that the inactivation of Ca channels occurs by a mechanism related to Ca entry. These are discussed in order below (also, see Sect. 5.3).

The first expectation is that the rate of inactivation during a depolarization should increase when I_{Ca} amplitude is increased by raising Ca_o or by enhancing activation via voltage clamp protocols. Affirmative results have come from studies on cat ventricular tissue (Kohlhardt et al. 1975), calf Purkinje fibers (Lee et al. 1985), frog atrium (Mentrard et al. 1984; Nilius and Benndorf 1986), guinea pig ventricular myocytes (Lee et al. 1985), frog atrial myocytes (Hume and Giles 1983), frog S-A nodal myocytes (Shibata and Giles 1985), and guinea pig atrial cardioballs (Bechem and Pott 1985). In addition, I_{Ca} inactivation rates at V_{peak} were faster in cells which by change had larger densities of I_{Ca} than other similar cells (guinea pig cardioballs: Bechem and Pott 1985; frog ventricular myocytes: Fischmeister et al. 1987).

The second expectation is that the rate of inactivation during a depolarization should decrease when cations other than Ca carry the Ca channel current. Slower inactivation has been observed when external Ca was replaced by Ba or Sr in cat ventricular tissue (Kohlhardt et al. 1973); calf Purkinje fibers (Kass and Sanguinetti 1984; Lee et al. 1985), frog atrium (Mentrard et al. 1984), guinea pig ventricular myocytes (Kokubun and Irisawa 1984; McDonald et al. 1986), rat ventricular myocytes (Mitchell et al. 1983; Josephson et al. 1984), frog atrial myocytes (Campbell et al. 1988c), and dog atrial myocytes (Bean 1985). Similarly, monovalent ionic currents through Ca channels inactivate slower than I_{Ca} in frog atrial tissue (Chesnais et al. 1975) and guinea pig ventricular myocytes (Imoto et al. 1985; Matsuda 1986; Hadley and Hume 1987).

The third expectation is that the rate of I_{Ca} inactivation should decrease when Ca entry is buffered by intracellular ethylene glycol-bis(β-aminoethyl ether)N,N,N',N'-tetraacetic acid (EGTA). This has been observed in guinea pig ventricular myocytes (Kurachi 1982; Fedida et al. 1988a), rat ventricular myocytes (Josephson et al. 1984; Mitchell et al. 1985), and guinea pig atrial cardioballs (Bechem and Pott 1985). The converse forms the fourth expectation, that inactivation should speed up when Ca_i is higher. This was observed when sheep Purkinje fibers were injected with Ca (Isenberg 1977) and when Ca_i was increased by low Na_o treatment in frog atrium (Mentrard et al. 1984), rabbit S-A node (Brown et al. 1984), and rabbit A-V node (Kokubun et al. 1982).

The fifth expectation is that the steady-state inactivation curve determined by the two-pulse method will not be a smooth function of voltage, but instead will be dictated by the amplitude of I_{Ca} on the first (P1) of the two pulses. When I_{Ca} is large (P1 to V_{peak} potentials), I_{Ca} on the second pulse (P2) should be small. However, when I_{Ca} is small (P1 to threshold or E_{rev} regions), I_{Ca} on subsequent P2 tests should be large. Thus, when $I_{Ca}/I_{Ca(max)}$ measured on P2 is plotted against P1 voltage, the curve should be U-shaped, i.e., the fraction should be near unity at threshold, fall to a minimum near V_{peak}, and climb to near unity again at more positive potentials. U-shaped curves have been determined in studies on frog atrium (Mentrard et al. 1984; Nilius and Benndorf 1986), Purkinje fiber strands (Lee et al. 1985), rat ventricular myocytes (Josephson et al. 1984), guinea pig ventricular myocytes (Josephson et al. 1984; Lee et al. 1985; Hadley and Hume 1987), guinea pig atrial cardioballs (Bechem and Pott 1985), frog atrial cells (Campbell et al. 1988c), and frog ventricular cells (Fischmeister and Hartzell 1986). In addition, a near linear relation between the estimated Ca charge transferred on P1 and the degree of inactivation has been observed on P2 in frog atrium (Mentrard et al. 1984; Nilius and Benndorf 1986) and in guinea pig and rat myocytes (Josephson et al. 1984).

The final expectation is that the rate of I_{Ca} inactivation should be increased when I_{Ca} is increased by stimulating agents, and decreased when I_{Ca} is decreased by inhibitory agents. There are diverse experimental results on this point. For example, when β-adrenergic stimulation increased I_{Ca} amplitude by two to ten times control, the inactivation rate was unchanged in bovine ventricular tissue (Reuter and Scholz 1977b), frog atrial tissue (Mentrard et al. 1984), rat ventricular myocytes (Mitchell et al. 1983), bovine ventricular myocytes (Isenberg and Klöckner 1982), and frog ventricular myocytes (Fischmeister and Hartzell 1986), and actually slowed in guinea pig ventricular myocytes (Tsien et al. 1986). On the other hand, partial block of I_{Ca} by Co or Cd has in some cases been shown to slow inactivation (Mentrard et al. 1984; Bechem and Pott 1985), but partial block by D600 does not seem to do so (Pelzer et al. 1982; Lee and Tsien 1983; McDonald et al. 1984; Mentrard et al. 1984; Uehara and Hume 1985). A comment on this expectation is that it may by unrealistic to expect anything other than diverse results. For exam-

ple, part of a D600 reduction of I_{Ca} amplitude will be due to the stabilization of inactivation in a fraction of the channels (McDonald et al. 1984;
Pelzer et al. 1985; Trautwein and Pelzer 1985a). The remaining fraction of
channels, during partial block, may be allowing nearly the same Ca entry as
channels in nontreated preparations (McDonald et al. 1989). With regard to
β-adrenergic stimulation, one should keep in mind that isoproterenol, aside
from increasing I_{Ca}, may also slow inactivation and thereby mask enhanced
inactivation related to Ca entry. Thus, one can compare the lack of effect of
β-adrenergic stimulation on the decay of I_{Ca} (see above) with the slowing of
the inactivation of Ba currents through Ca channels (Tsien et al. 1986; Trautwein and Pelzer 1988).

2.3.3 Voltage-Dependent Inactivation

The abundant evidence in favor of Ca-dependent inactivation not withstand-
ing, there is good reason to believe that a voltage-dependent inactivation
mechanism is intrinsic to cardiac Ca channels. Arguments for the latter are
based on results obtained on both the macroscopic and microscopic levels,
and we deal with the former first.

Ca channel currents carried by Ba or Sr undergo inactivation during step
depolarization of cardiac tissue (Kass and Sanguinetti 1984; Lee et al. 1985)
and myocytes (Mitchell et al. 1983; Josephson et al. 1984; Lee et al. 1985; Mc-
Donald et al. 1986). Since the rate of inactivation was slower than when Ca
carried the current, one could argue that Ba and Sr are simply less effective
than Ca in eliciting Ca-dependent inactivation. The counter arguments are
that (a) inward-going Ca channel currents carried by Na also inactivate (Hess
and Tsien 1984; Hadley and Hume 1987), as do outward-going ones carried
by K or Cs (Lee and Tsien 1982; Lee et al. 1985; McDonald et al. 1986; Hadley
and Hume 1987; Campbell et al. 1988c), (b) the voltage dependency of the
inactivation time constants of current carried by Ba, Sr, or Na is not U-
shaped (e.g., Kass and Sanguinetti 1984), (c) large increases in the amplitude
of current carried by Sr (Kass and Sanguinetti 1984) or Ba (McDonald et al.
1986) to not cause more rapid inactivation, and (d) prepulses that fail to trigger measurable I_{Ca} nevertheless produce large inactivation (Nilius and Benndorf 1986; Hadley and Hume 1987; Campbell et al. 1988c).

On the single Ca channel level, inactivation (increased number of blanks)
following prepulses that failed to open the channel has been observed in neonatal rat heart myocytes (Reuter et al. 1982) and guinea pig ventricular
myocytes (Cavalié et al. 1983). At depolarization levels that elicit channel
openings in guinea pig ventricular myocytes, the ensemble average current inactivates whether the charge is carried by Ba or Ca (Cavalié et al. 1983). Currents through single cardiac Ca channels incorporated into lipid bilayers also
have a relaxation phase (Rosenberg et al. 1986, 1988; Imoto et al. 1988). As

with whole-cell Ba currents, the rate of inactivation of average Ba currents in guinea pig ventricular myocyte patches does not have a U-shaped dependence on voltage (McDonald et al. 1986).

The foregoing summarizes the evidence indicating that a voltage-dependent inactivation mechanism is an inherent property of the cardiac Ca channel. Further insight into this mechanism, as viewed from the single channel perspective, has emerged from the study by Cavalié et al. (1986) on cell-attached patches of guinea pig ventricular myocytes. During maintained depolarizations to partially activating potentials, channel activity was characterized by short closings within bursts ($\tau_{C1} = 0.4$ ms), longer closings between bursts ($\tau_{C2} = 2.1$ ms), and even longer closings (>21 ms) between "clusters" of bursts. Records with openings had clusters (average lifetime 275 ms) that were preferentially located at the beginning of 900 ms-long depolarizations (0.2 Hz pulsing). Corresponding to the average cluster lifetime, the inactivation of ensemble average current (I*) was described by a monoexponential curve with $\tau = 275$ ms. Thus, there seemed to be a connection between entry of the channel into the cluster-ending, long-lived closed (inactivated?) state, and the decay of I* during depolarization. To determine whether the long-lived closed state terminating a cluster was an "unavailable" state, step depolarizations were applied from a relatively low-p prepulse potential to a more positive, higher-p potential. Ca channel opening was never triggered by the second step when the latter was applied during an intercluster closing. Conversely, channel opening was always triggered by the second step when the latter was applied during a cluster. A pertinent finding is that a blank is twice as likely to be drawn after a sweep which ends after cluster termination than after a sweep which apparently ends during a cluster (unpublished).

In a given ensemble generated by step depolarizations to a positive potential, the number of blank sweeps usually far exceeds the number that could be accounted for by channel bursting kinetics (Reuter et al. 1982; Cavalié et al. 1983; Hess et al. 1984). In the Cavalié et al. (1986) study, the incidence of blanks ranged from 7% to 62% in ensembles generated by 300 ms depolarizations to V_{peak} potentials at 0.5 Hz. Large hyperpolarizations from apparently noninactivating holding potentials did not reduce the basal incidence of blanks in this or other studies (e.g., Reuter et al. 1982; Hess et al. 1984). Extra blanks occur after depolarizing prepulses, and the fraction of these per ensemble rises sigmoidally with voltage in a manner that closely resembles steady-state inactivation of whole-cell Ba currents (McDonald et al. 1986). This behavior is also observed in single Ca channels incorporated into lipid bilayers (Yatani et al. 1987a; Imoto et al. 1988; Rosenberg et al. 1988). In fact, when surface charge screening is taken into account, steady-state inactivation curves of Ca channel currents from cardiac tissues, myocytes, and single channels are practically indistinguishable (Pelzer et al. 1986b).

2.3.4 Summary: A Dual Mechanism of Inactivation in Heart Cells

The conclusion that can be drawn from the review of Ca channel inactivation in the heart is that it has two components, a Ca-dependent one and a voltage-dependent one. Lee et al. (1985) considered ways in which two such mechanisms could interact. In the first scheme, gates were controlled by Ca_i and potential, i.e., inactivation was the product of two variables, $h_1(V)$ and $h_2(Ca)$; in the second scheme, Ca_i modulation of inactivation time constants (V, Ca_i) was explored. In the final analysis, Lee and colleagues were unhappy with both of these schemes; whether the two mechanisms are tightly coupled or operate independently of each other remains an open question. One aspect of the dual mechanism is clear: Ca-dependent inactivation has a considerably faster onset at negative and moderately positive potentials than voltage-dependent inactivation; it may or may not produce more complete inactivation at these potentials. Hadley and Hume (1987) made a direct comparison of the inactivation of Ca channel current carried by Ca and that carried by Na in the same guinea pig ventricular myocyte. They observed that the Ca-dependent mechanism can produce nearly complete inactivation during 500 ms prepulses to potentials between -20 and $+20$ mV. At more positive potentials, the voltage-dependent mechanism assumed dominance. Its superimposition affects the U-shaped Ca-dependent relation and explains why the ascending limb of the U-shaped curve expected for a pure Ca-dependent mechanism tails off at 0.4 relative current.

2.3.5 Inactivation in Smooth Muscle and Skeletal Muscle Cells

The shape of the steady-state inactivation curve in smooth muscle cells is similar to that in heart cells, with slope factors ranging from 6 to 10 mV (e.g., Klöckner and Isenberg 1985; Benham et al. 1987). A major difference from heart is the location of the curve on the voltage axis. With external solutions containing millimolar Ca or Ba, the V_h of the relation lies between -55 and -38 mV in vascular and other smooth muscle myocytes (Klöckner and Isenberg 1985; Droogmans and Callewaert 1986; Aaronson et al. 1988; Ohya et al. 1988). When vascular smooth muscle cells were bathed in $110-115$ mM Ba solution, V_h was between -23 and -15 mV (Bean et al. 1986; Benham et al. 1987; Aaronson et al. 1988). Thus, V_h of steady-state inactivation in smooth muscle cells is displaced by about 20 mV in the negative potential direction compared to V_h in heart cells.

The decay of I_{Ca} during a depolarizing pulse has usually been described in terms of a multiexponential process. At room temperature with millimolar external Ca, two exponentials ($\tau = 15-20$ and $80-215$ ms) fitted the decay of I_{Ca} in cells from guinea pig ileum (Droogmans and Callewaert 1986) and rat vas deferens (Nakazawa et al. 1987), although the latter also noted a small, slowly inactivating component with $\tau > 1$ s. These values fit reasonably well with those determined at higher temperature (35 °C) in guinea pig urinary bladder cells ($\tau = 5$ and 40 ms, and a small slow phase, $\tau = 200$ ms) (Klöckner and Isenberg 1985).

The voltage dependencies of the fast and the ultraslow time constants are unclear, but a number of studies indicate that the intermediate time constant is U-shaped with a minimum near 0 mV (Klöckner and Isenberg 1985; Amédée et al. 1987; Ganitkevich et al. 1987; Aaronson et al. 1988; also see below). Partial U-shaped steady-state inactivation curves have also been documented (Ganitkevich et al. 1986, 1987; Amédée et al. 1987) suggesting the presence of Ca-induced inactivation. Additional evidence for this mechanism in smooth muscle cells is that, first, I_{Ca} inactivates faster when current amplitude is increased by raising Ca_o, and inactivates slower when I_{Ca} is partially blocked by Co (Ganitkevich et al. 1987); second, replacement of Ca_o by Ba, Sr, or Na slows the rate of inactivation (Droogmans and Callewaert 1986; Jmari et al. 1986, 1987; Amédée et al. 1987; Ganitkevich et al. 1987); and third, the amplitude of I_{Ca} evoked by P2 of a double pulse protocol is inversely related to Ca entry on P1 (Ganitkevich et al. 1987; Ohya et al. 1988).

There is also evidence for voltage-dependent inactivation in smooth muscle cells. For example, increasing the EGTA concentration of the dialysing solution had no effect on the decay of I_{Ca} in guinea pig urinary bladder cells (Klöckner and Isenberg 1985). Ba-carried currents were inactivated faster and to a larger degree at $+30$ mV than at $+10$ mV in rabbit ear artery myocytes (Benham et al. 1987), and a similar observation has been made on I_{Ca} in taenia caeci cells (Ganitkevich et al. 1986). Walsh and Singer (1987) noted that significant inactivation developed during depolarizations to subthreshold potentials for I_{Ca} in toad stomach cells.

Additional evidence in favor of a voltage-regulated inactivation mechanism is that single Ca channel currents in isolated patches from vascular smooth muscle cells also undergo inactivation that is similar to that seen in cell-attached patches (Benham et al. 1987). In cell-attached patches, single Ca channel currents carried by Na in urinary bladder cells (Isenberg and Klöckner 1985 c) and by Ba in taenia caeci cells (Yoshino and Yabu 1985) also undergo inactivation. In conclusion, there is every indication that Ca channel inactivation in smooth muscle cells is similar to that in heart cells; both Ca-induced and voltage-induced inactivation appear to be present under physiological conditions.

The inactivation of L-type (or slow) Ca channel current during depolarization of skeletal muscle fibers is much slower than the inactivation of L-type Ca channel current in most other cell types (e.g., Beaty and Stefani 1976; Bernard et al. 1976; Stanfield 1977; Sánchez and Stefani 1978; Almers and Palde 1981). Almers et al. (1981) concluded that the seconds-long process in frog fibers was due to time-dependent depletion of Ca_o in the T-tubules rather than to a voltage-dependent or Ca-dependent mechanism (also see Palade and Almers 1985). More recently, Beam and Knudson (1988a, b) found that embryonic rat skeletal muscle myotubes with sparse T-tubules possessed slow Ca currents that hardly inactivated at all during activating depolarizing pulses. However, slow inactivation was apparent in similar adult preparations with well-developed T-tubular systems.

A further indication that T-tubular depletion may be responsible for inactivation is that steady-state inactivation in frog skeletal muscle fibers can occur in direct proportion to the magnitude of I_{Ca} on the conditioning pulse (Almers et al. 1981; also see Stanfield 1977). In line with the foregoing, single mammalian skeletal muscle Ca channels in lipid bilayers do not inactivate at all (Affolter and Coronado 1985; Coronado and Affolter 1986; Flockerzi et al. 1986; Rosenberg et al. 1986; Talvenheimo et al. 1987; Trautwein et al. 1987; Pelzer et al. 1988, 1989b; Yatani et al. 1988).

There is also evidence that a more classical mode of inactivation can be present in skeletal muscle. First, Stefani and colleagues (Sánchez and Stefani 1983; Cota et al. 1984) showed that I_{Ca} in intact frog fibers bathed in hypertonic solution inactivated in a voltage-dependent manner ($V_h = -44$ to -33 mV, $k = 6-9.5$ mV) that was independent of I_{Ca} amplitude on the conditioning pulse. In fact, 7-s conditioning pulses to potentials that did not activate I_{Ca} produced significant ($50\% - 80\%$) inactivation. Similarly, Beam and Knudson (1988a) reported that in addition to inactivation that could be induced by fairly short (about 2 s) depolarizations to threshold potentials (about 0 mV) and above, 60-s depolarizations to much more negative potentials also caused inactivation of I_{Ca}. When these investigators applied the latter protocol to embryonic and neonatal rat muscle preparations, V_h was between -40 and -20 mV.

Cognard et al. (1986) investigated the steady-state inactivation of L-type Ca channels in newborn rat thigh muscle cells that were cultured and converted into myoballs by colchicine treatment. When myoballs were bathed in 2.5 mM Ba or Ca solution, Ca channel currents inactivated more slowly with Ba_o than with Ca_o. The threshold for activation of Ba-carried current was around -30 mV. As in the Beam and Knudson (1988a) study, long (30 s) conditioning pulses to subthreshold potentials produced marked inactivation. The steady-state inactivation relation had a $V_h = -72$ mV and $k = 5.4$ mV.

2.4 Reactivation

Ca channels inactivated by depolarization can be restored to an available status by repolarization of the membrane to negative potentials. The reactivation (restoration, repriming) process proceeds faster at more negative potentials in both amphibian and mammalian heart tissue (Trautwein et al. 1975; Kass and Sanguinetti 1984; Mentrard et al. 1984) and myocytes (Isenberg and Klöckner 1982; Lee et al. 1985; Fischmeister and Hartzell 1986). For example, the $t_{1/2}$ for reactivation of I_{Ca} at room temperature is several hundred milliseconds near $-50\,mV$ and only about 100 ms at -80 to $-100\,mV$ in myocytes from frog ventricle (Fischmeister and Hartzell 1986), frog atrium (Campbell et al. 1988c) and guinea pig ventricle (Hadley and Hume 1987, 1988). Faster rates of I_{Ca} reactivation are evident at higher temperatures. For example, a $t_{1/2}$ of $30-50$ ms near $-50\,mV$ has been measured in bovine ventricular myocytes (Isenberg and Klöckner 1982) and canine ventricular myocytes (Tseng 1988) at $35\,°-37\,°C$. These data are quite compatible with the values for $t_{1/2}$ determined in the other studies at room temperature (≈ 300 ms) when account is taken of temperature coefficient per $10\,°C$ (Q_{10}) ≈ 3 for Ca channel kinetics (Mitchell et al. 1983; Cavalié et al. 1985).

The foregoing indicates that the reactivation rate of I_{Ca} is voltage-dependent, temperature-sensitive, and relatively independent of myocyte type, although Josephson et al. (1984) reported that reactivation was at least twice as slow in rat myocytes as in guinea pig myocytes. It also seems certain that the reactivation process is not simply the reciprocal of inactivation as in the Hodgkin-Huxley framework (Hodgkin and Huxley 1952); reactivation is a considerably slower process when measured at the same potential as inactivation (see Kohlhardt et al. 1975; Trautwein et al. 1975; Isenberg and Klöckner 1982). However, what is most unclear is the actual time course of I_{Ca} reactivation. It has been described as (a) monoexponential in a number of cardiac tissues (cat ventricle, calf Purkinje fibers) and myocytes (bovine ventricle, bullfrog atrium, guinea pig ventricle; see Trautwein et al. 1975; Isenberg and Klöckner 1982; Kass and Sanguinetti 1984; Campbell et al. 1988c; Hadley and Hume 1988), (b) biexponential in calf Purkinje fibers (Kass and Sanguinetti 1984), guinea pig ventricular myocytes (Josephson et al. 1984), and rat ventricular myocytes (Josephson et al. 1984), (c) oscillatory in calf Purkinje fibers (Weingart et al. 1978), dog ventricular trabeculae (Hiraoka and Sano 1978), and ventricular myocytes from guinea pig (Fedida et al. 1987; Tseng 1988) and dog (Tseng 1988), and (d) sigmoidal in frog atrial strands (Mentrard et al. 1984).

The reason for this rich diversity has not been established, but unusual influences of voltage clamp protocols (Shimoni 1981; Kass and Sanguinetti 1984; Campbell et al. 1988c) and extracellular Ca concentration (Noble and Shimoni 1981; Tseng 1988) suggest that fluctuations in Ca_i may be an important variable.

Tseng (1988) investigated the role of Ca_i on the reactivation of I_{Ca} in dialyzed guinea pig and canine ventricular myocytes. In both types of cells, reactivation during superfusion with $5\,mM\ Ca_o$ was monoexponential at $-30\,mV$ but oscillatory with an overshoot at potentials negative to $-50\,mV$.

The overshoot on the test depolarization was large when the test pulse was applied 50–300 ms after the inactivating prepulse, and much smaller when it was placed 1–3 s after it. The size of the overshoot was about 120% of control I_{Ca} amplitude when the inactivating prepulse to +10 mV was 50 ms long, around 20% when the prepulse was 100 ms long, and less than 5% when the prepulse was 500 ms long. The overshoot was essentially abolished when the dialysate contained 40 mM EGTA instead of 10 mM, or when it contained 10 mM 1,2-bis(2-aminophenoxy)ethane N,N,N'-tetraacetic acid (BAPTA) instead of 10 mM EGTA. It was also abolished when Ca_o was reduced from 5 to 2 mM, when Ba replaced Ca as the external divalent cation, or when sarcoplasmic reticulum (SR) function was modified by pretreatment with 10 mM caffeine or 1–2 µM ryanodine.

On the basis of these results, and proposals that during depolarization in heart cells it is primarily the Ca released from the SR which causes channel inactivation (Marban and Wier 1985; Mitchell et al. 1985), Tseng (1988) put forward two hypotheses that might explain oscillatory reactivation. The first was that the strong inactivation of SR Ca-release channels at early times after depolarizing prepulses (cf. Fabiato 1985) would attenuate Ca release and permit a larger than normal I_{Ca} (overshoot) to flow into cells; at later times, reactivation of SR Ca-release channels would restore normal Ca release and Ca_i influence on I_{Ca}. The second hypothesis was that gradual, time-dependent depletion of Ca_i after a depolarization results in a time window during which the moderately elevated concentrations of Ca in the myoplasm exert a facilatory effect on I_{Ca}. The hypothesis that Ca_i might have an up-regulatory as well as a down-regulatory action has also emerged from other studies, and is discussed further in Sect. 5.

The dependence of the reactivation rate on the preceding influx of Ca is far from resolved. When I_{Ca} was increased by increasing Ca_o, reactivation was faster in cat ventricular muscle (Kohlhardt et al. 1975). In frog atrial strands, it was faster in one study (Shimoni 1981) and slower in another (Mentrard et al. 1984), whereas in myocytes from this tissue it was unchanged (Campbell et al. 1988c). When I_{Ca} is greatly increased by β-adrenergic stimulation, there are equally divergent results on reactivation rates (see Sect. 4.1). When Ca influx was varied by voltage protocol or low concentrations of Ca channel blockers, smaller amplitudes of I_{Ca} on inactivating prepulses did not affect reactivation rate in guinea pig ventricular myocytes (Hadley and Hume 1987) or frog atrial myocytes (Campbell et al. 1988c).

A final area of contention concerns the influence of external charge carrier on the rate of reactivation. The replacement of Ca by Ba or Sr had no effect on reactivation in calf Purkinje fibers (Kass and Sanguinetti 1984), slowed it in frog atrial strands (Noble and Shimoni 1981) and dog ventricular myocytes (Tseng 1988), and speeded it up in frog atrial strands (Mentrard et al. 1984). Thus, these contrasting results with replacement of Ca by Ba or Sr provide little direction on the relative importance of Ca-induced inactivation in heart cells. One might expect that the larger the Ca influx on the prepulse, the slower Ca_i in the vicinity of the channel would be restored to resting level, and the slower would be the removal of inactivation (Mentrard et al. 1984). Replacement of Ca_o by Ba_o or Sr_o would therefore result in a significant increase in the rate of reactivation. When the external charge carrier is Na, rather than a divalent cation, an even greater change in reactivation rate can be expected. In fact, Hadley and Hume (1987) found that the reactivation of Ca channel current carried by external Na ions was not faster than that carried by external Ca.

Reactivation has not yet been well studied in skeletal muscle or smooth muscle cells. In skeletal muscle fibers with T-tubules, the possibility that T-tubular Ca_o depletion may occur on depolarization that induces large inactivation may confound the measurements. In rat uterine myocytes at room temperature, the reactivation of I_{Ca} at -60 mV was described by a two-exponential process with $\tau = 380$ and 5190 ms (Amédée et al. 1987). The replacement of 10 mM Ca_o by Ba shortened the time to near complete reactivation from 5 to 2 s, mainly by speeding up the slower phase. In rat uterine tissue at $30\,°C$, the reactivation of Na-carried Ca channel current at -50 mV was monoexponential with $\tau = 280$ ms (Jmari et al. 1987).

Recovery of I_{Ca} from inactivation in isolated guinea pig taenia caeci cells has been examined at both -50 and -90 mV (Ganitkevich et al. 1987). At $23\,°C$, the half-time of reactivation was about 300 ms at -50 mV, and 100 ms at -90 mV. These values are in good agreement with the I_{Ca} reactivation data of Amédée et al. (1987) (half-time near 300 ms at -50 mV), and the speeding up of reactivation at more negative potentials is in accord with the observations on heart cells.

3 Permeation, Selectivity, and Block

3.1 Overview

The permeation of ions through Ca channels can best be studied when conditions permit unambiguous resolution of the channel current, variation of the intracellular and extracellular ionic composition over a wide range, and good control of the membrane potential. Thus it is not surprising that the early groundwork of the subject, built from studies on multicellular muscle preparations during the 1970s and early 1980s, has been refined and greatly supplemented by studies of macroscopic Ca channel currents in dialyzed muscle cells and microscopic Ca channel currents in cell-attached myocyte patches and planar lipid bilayers.

We begin this section by listing the species of ions that have been shown to carry inward and/or outward current through Ca channels, and discuss their permeability in terms of current-carrying ability. We then consider channel selectivity as determined by E_{rev} measurements and changes in Ca channel activity when external solutions contain mixtures of ions (including classical inorganic blocking cations). Finally, we consider evidence suggesting that permeating ions may be able to influence the state of the Ca channel. (The block of channel current by intracellular inorganic cations such as Ca, Mg, and H is discussed with reference to cell metabolism in Sect. 5).

3.2 Permeability (Current-Carrying Ability)

3.2.1 Channel-Permeating Cations

L-type Ca channels in muscle cells are permeable to a large number of inorganic cations. External solutions containing millimolar Ca, Sr, or Ba support large macroscopic Ca channel currents in cardiac tissues (Kohlhardt et al. 1975; Kass and Sanguinetti 1984), cardiac myocytes (Mitchell et al. 1983; Lee et al. 1985; Campbell et al. 1988a), skeletal muscle fibers (Stanfield 1977; Cota and Stefani 1984; Palade and Almers 1985), smooth muscle tissue (Bolton 1979; Jmari et al. 1986; Bülbring and Tomita 1987), and smooth muscle myocytes (Klöckner and Isenberg 1985; Ganitkevich et al. 1988). Currents carried by these three ions have also been resolved at the single-channel level in cardiac myocytes (Hess et al. 1986).

Mn, although better known as a blocker of I_{Ca}, has also been shown to carry significant current through Ca channels in cardiac tissue (Ochi 1970, 1975) and skeletal muscle fibers (Fukuda and Kawa 1977; Almers and Palade 1981; also see Akaike et al. 1983; Anderson 1983). Mg at concentrations up to 100 mM did not have detectable charge-carrying ability in cardiac (Matsuda and Noma 1984; Hess et al. 1986) or smooth muscle (Ganitkevich et al. 1988) myocytes, but small Mg currents through Ca channels have been measured in frog atrium (Mentrard et al. 1984) and frog skeletal muscle (Almers and Palade 1981). Other divalent cations such as Cd, Co, and Ni have no detectable permeability, but may very well dribble through in the same way as relatively impermeable La ions (see Sect. 3.4.1). Protons can probably be placed in the same category.

In the absence of external divalent cations, external monovalent cations can carry inward-going Ca channel current. For example, external Na supports large currents in cardiac tissue (Garnier et al. 1969; Chesnais et al. 1975), cardiac myocytes (Hess and Tsien 1984; Imoto et al. 1985; Matsuda 1986; Hadley and Hume 1987), skeletal muscle fibers (Potreau and Raymond 1982; Almers and McCleskey 1984; Almers et al. 1984), smooth muscle tissue (Jmari et al.

1987), and smooth muscle myocytes (Isenberg and Klöckner 1985b). Single Ca channel currents carried by Na have also been recorded from cell-attached patches in guinea pig ventricular myocytes (Matsuda 1986; Hess et al. 1986; Lacerda et al. 1988) and guinea pig urinary bladder myocytes (Isenberg and Klöckner 1985c), and from single channels (bovine ventricular membrane) incorporated into lipid bilayers (Rosenberg et al. 1986, 1988). External Na as charge carrier can be substituted by Li in cardiac tissue (Chesnais et al. 1975) and by Li, Rb, and Cs in frog skeletal muscle fibers (Almers et al. 1984). Li, Cs, and K have been shown to carry inward-going current in cardiac cell-attached patches (Lansman et al. 1986; Pietrobon et al. 1988), and internal K and Cs can carry outward-going whole-cell Ca channel current in cardiac myocytes (Lee and Tsien 1984; Lee et al. 1985; McDonald et al. 1986; Hadley and Hume 1987) and smooth muscle myocytes (Klöckner and Isenberg 1985; Droogmans and Callewaert 1986; Amédée et al. 1987).

Aside from the divalent and monovalent cations mentioned above, there is also evidence that Ca channels are permeable so some organic cations. Based on E_{rev} measurements, the tetramethylammonium ion was found to be the largest methylated derivative of ammonium that was permeant through Ca channels in skeletal muscle fibers and cardiac ventricular myocytes (McCleskey and Almers 1985; McCleskey et al. 1985). This fixed the minimal pore diameter at 6 Å, a patency that was confirmed by Coronado and Smith (1987) in permeability studies on single Ca channels from rat muscle T-tubules incorporated into planar lipid bilayers. Coronado and Smith (1987) also reported that ammonium ions had a larger single-channel conductance than Na, Li, K, and Cs.

3.2.2 Current-Carrying Ability

The current-carrying ability of the physiological charge carrier, Ca, has been compared with that of the two other highly permeant divalent cations, Ba and Sr, in a wide variety of muscle cells. The conclusion from many, though not all, of the studies on macroscopic Ca channel current has been that Ba- or Sr-carried currents are larger than Ca-carried currents when the comparison is made at either physiological or higher concentrations.

Ba and Sr currents have been reported to be larger than Ca currents in multicellular preparations from heart (Kohlhardt et al. 1973; Mentrard et al. 1984) and uterus (Jmari et al. 1987), as well as in single skeletal muscle fibers from frog (Almers et al. 1984; Cota and Stefani 1984) and barnacle (Hagiwara et al. 1974). A similar conclusion has been reached in a number of studies on dissociated myocytes. For example, Lee and Tsien (1984) and Hess and Tsien (1984) reported that Ba currents were about 50% larger than Ca currents in guinea pig ventricular myocytes bathed in solutions containing 10 mM divalent cation. In frog atrial myocytes, the ratio of Sr to Ba and Ca currents was 4:3:2 at 2.5 mM concentration (Campbell et al. 1988a), and in myocytes from rabbit ear artery (Droogmans et al. 1987) Ba currents were twice as large as Ca currents at external concentrations of 10 mM.

Despite the foregoing, there have been a number of careful studies in which Ba or Sr currents have been either smaller or not larger than Ca currents. For example, Fedida et al. (1987) reported that the current generated by 2.5 mM Sr$_o$ was smaller than that generated by 2.5 mM Ca$_o$ in guinea pig ventricular myocytes, and Amédée et al. (1987) calculated that Ba- and Sr-carried currents (10 mM external concentration) were 30% smaller than I$_{Ca}$ in rat myometrial cells. With external concentrations between 1.8 and 3.6 mM, Ba currents and Ca currents were of similar amplitude in rat ventricular myocytes (Mitchell et al. 1983), guinea pig ventricular myocytes (McDonald et al. 1986), and rat skeletal muscle myoballs (Cognard et al. 1986). At 10 mM external concentration, Ba- and Ca-carried currents were of similar size in neonatal (Beam and Knudson 1988a) and adult (Donaldson and Beam 1983) rat skeletal muscle.

There are three plausible explanations for these divergent results. (a) A likely partial block by Mg$_o$ of Ba or Sr currents (see Sect. 3.4.2 below) may have skewed the results in a number of studies. (b) Differences in surface charge which screen the effects of Ca, Ba, and Sr (see McDonald et al. 1986; Amédée et al. 1987; Ganitkevich et al. 1988) have often not been considered. Finally, (c) the relative amplitudes of currents will depend on the concentration chosen for the comparison (see Sect. 3.2.3 below).

Nevertheless, it is certain that the limiting conductance for Ba is much greater than for Ca. For external ion concentrations in the range 20–110 mM, unitary Ba currents through single Ca channels in guinea pig myocytes are much larger than Ca currents, and the limiting conductances are of the order of 23 pS and 9 pS, respectively (Cavalié et al. 1983; Hess et al. 1986). Differences in limiting conductances may be even larger in smooth muscle cells since (a) single-channel conductance for Ba seems to be higher than in cardiac cells (see Yoshino and Yabu 1985; Benham et al. 1987), and (b) ratios of maximum current-carrying ability in guinea pig taenia caeci cells were estimated as 8:1.5:1 for Ba, Sr, and Ca respectively (Ganitkevich et al. 1988).

In the absence of divalent cations, Ca channels are highly permeable to monovalent cations. Compared with the 20–23 pS single-channel conductance observed with 110 mM Ba, the conductance measured in the presence of 110–150 mM Na was 75–110 pS in cardiac myocytes from guinea pig (Hess et al. 1986; Matsuda 1986) and embryonic chick (Levi and De Felice 1986). The cardiac single Ca channel conductance to Li is also very high, about 45 pS (Hess et al. 1986). In single Ca channels of rat T-tubular membrane incorporated into lipid bilayers, Coronado and Smith (1987) demonstrated that monovalent cations have high conductances that follow the sequence $Cs \approx K > Na > Li$, with $Cs/Li \approx 1.7$ (compare with Almers et al. 1984).

Monovalent cation current through Ca channels has also been documented at the whole-cell and tissue levels. Identification criteria have included an insensitivity to TTX and block by inorganic and organic Ca channel blockers (e.g., Chesnais et al. 1975; Almers and McCleskey 1984; Almers et al. 1984; Matsuda 1986; Jmari et al. 1987). The usual findings in both cardiac and smooth muscle cells bathed in divalent cation-free, $100-150$ mM Na solution are that the current activation threshold is more negative than for I_{Ca}, the decay of Na-carried current is slower, and the reversal potential is $30-50$ mV lower (e.g., Chesnais et al. 1975; Matsuda 1986; Hadley and Hume 1987; Jmari et al. 1987). These alterations explain the marked lengthening of the action potential duration and the lowering of the plateau potential in preparations from heart (e.g., Rougier et al. 1969; Chesnais et al. 1975) and smooth muscle (e.g., Prosser et al. 1977; Mironneau et al. 1982). The reactivation of Na-carried current was shown to be monoexponential in studies on guinea pig ventricular myocytes (Hadley and Hume 1987) and rat uterine strips (Jmari et al. 1987).

In keeping with the almost twofold smaller conductance of Li than Na, determined in single Ca channel studies on cardiac myocytes (Hess et al. 1986), replacement of Na by Li reduced whole-cell Ca channel current by nearly 70% in guinea pig ventricular myocytes (Matsuda 1986). A similar result was obtained with frog atrial strands (Chesnais et al. 1975), while Na-supported long action potentials in smooth muscle disappear with Li substitution (Prosser et al. 1977; Mironneau et al. 1982). By contrast, Almers et al. (1984) did not find any diminution of monovalent current when Li replaced Na in the solution bathing skeletal muscle fibers, and Kostyuk et al. (1983) reported that Li was nearly as permeant as Na in molluscan neurones.

It is important to keep in mind that the very large macroscopic currents and single Ca channel conductances referred to above were measured when preparations were bathed in divalent cation-free solutions. When even micromolar concentrations of divalent cations are present, the permeability of Ca channels to monovalent cations is drastically reduced (see Sect. 3.4 below).

3.2.3 Saturation

The amplitude of whole-cell I_{Ca} depends on the extracellular Ca concentration. Although the relationship is reasonably linear for Ca_o between about 0.1 and 3 mM in cardiac myocytes (Hume and Giles 1983; Hess and Tsien 1984) and frog skeletal muscle (Almers et al. 1984), higher concentrations produce proportionally smaller increases in current. For example, Hume and Giles (1983) found that a threefold increase in Ca_o from 2.5 to 7.5 mM only increased I_{Ca} by twofold. A similar increase in rat skeletal muscle I_{Ca} occurred when Ca_o was raised from 2 to 10 mM (Donaldson and Beam 1983). In smooth muscle preparations, Nakazawa et al. (1987) noted that a doubling of Ca_o from 1.8 to 3.6 mM only increased I_{Ca} by 30%, and Jmari et al. (1987) found that a doubling of I_{Ca} required a five fold increase in Ca_o from 2 to 10 mM.

When Ca_o is increased into the 30−50 mM range, the amplitude of I_{Ca} reaches a plateau. Half-saturation was reached with 13.8 mM Ca_o in frog atrial myocytes (Campbell et al. 1988 b), 20−30 mM in frog skeletal muscle (Almers et al. 1984), and a very low 1.2 mM in guinea pig taenia caeci cells (Ganitkevich et al. 1988). In the latter cells, the concentration of Ba_o for half-saturation of Ba-carried current was 9.6 mM. The higher value for Ba than Ca half-saturation is a common finding (e.g., Almers et al. 1984) that has been verified in single-channel studies on cardiac myocytes (28 mM Ba_o versus 14 mM Ca_o: Hess et al. 1986).

Monovalent cation currents through Ca channels also saturate but at a much higher external concentration than divalent cations. In rat T-tubular Ca channels incorporated into lipid bilayers, the external concentration of Na required for half-saturation was 200−300 mM (Coronado and Affolter 1986). In the same study, Ba-carried current half-saturated at 40 mM.

3.3 Selectivity as Determined by Reversal Potential Measurements

A traditional method of determining the relative channel permeabilities of two ions is to measure E_{rev} in the presence of known extracellular and intracellular concentrations of the two ions (Meves and Vogel 1973; Hille 1984). If ion a is infinitely more permeable than ion b, E_{rev} will be at the Nernst equilibrium potential for a (i.e., E_a), if ion a is only somewhat more permeable than ion b, E_{rev} will be between E_a and E_b. Even though early measurements of E_{rev} in cardiac tissue were beset by difficulties such as large overlapping capacitive transients and outward ionic currents, they were sufficient to establish that cardiac Ca channels are much more permeable to Ca than to monovalent cations. The best estimates placed Ca permeability at about 100 times that of Na or K (Reuter and Scholz 1977 a; Coraboeuf 1980; McDonald 1982). This degree of selectivity greatly exceeds that of the Na channel for Na over K (roughly 12:1; see Chandler and Meves 1965; Meves and Vogel 1973).

Improvements in methodology have produced results which indicate that the channel is enormously more selective for divalent cations than previously thought. Estimates of relative permeability can be obtained from the equation $E_{rev} = RT/2F \ln (4P_D D_o / P_M M_i)$ where R, T, and F have their usual meanings (R gas constant, T absolute temperature, F Faraday's constant), P is permeability, D_o refers to external divalent ion activity, M_i refers to internal monovalent ion activity and subscripts D and M refer to divalent and monovalent ions. The equation is an approximation of Eqs. 2 and 4 in Meves and Vogel (1973) that holds when $D_i \approx 0$ mM, M_o is low, V' (surface potential difference) is set at 0 mV, and E_{rev} is a positive potential (Lee and Tsien 1984; McDonald et al. 1986; also see Fatt and Ginsborg 1958). When this or a similar equation is applied (see Tsien et al. 1987; Campbell et al. 1988b), P_{Ca}/P_{Na} and P_{Ca}/P_k are of the order of 1000–11000 in frog atrial myocytes (Hume and Giles 1983; Campbell et al. 1988b) and guinea pig ventricular myocytes (Lee and Tsien 1982, 1984; Hess et al. 1986). P_{Ca}/P_{Cs} in guinea pig ventricular myocytes has been placed at 4200–10000 (Matsuda and Noma 1984; Hess et al. 1986), and P_{Ba}/P_{Cs} at 1356–1700 (Lee and Tsien 1984; Hess et al. 1986; McDonald et al. 1986). These estimates of very high permeability for divalent cations over monovalent cations, calculated from cardiac whole-cell E_{rev} data, are also supported by E_{rev} estimates (McDonald et al. 1986) and measurements (Hess et al. 1986) from single Ca channel experiments in guinea pig ventricular myocytes.

Ca channels in noncardiac muscle cells also have a very high selectivity for divalent cations over monovalent cations. In external solutions containing millimolar Ca or Ba, E_{rev} in skeletal muscle cells is around +50 mV (Almers and McCleskey 1984; Cognard et al. 1986). Similar E_{rev} values have been determined in voltage-clamp experiments on smooth muscle tissue (e.g., Mironneau 1974; Inomata and Kao 1976; Jmari et al. 1986). In a variety of isolated smooth muscle cells dialyzed with ca. 130 mM Cs solution, E_{rev} was between +40 and +60 mV when Ca_o was 1.2 to 1.8 mM (Klöckner and Isenberg 1985; Droogmans and Callewaert 1986; Benham et al. 1987; Nakazawa et al. 1987) and +70 to +80 mV when Ca_o was 2.5 to 5 mM (Amédée et al. 1987; Pacaud et al. 1987). Calculations with these E_{rev} values (see the above equation) lead to P_{Ca}/P_{Cs} and P_{Ba}/P_{Cs} ratios similar to those estimated for heart cells. When bathing solutions contain millimolar Ca, Sr, or Ba, E_{rev} is more positive in Ca solution than in either of the other two (Amédée et al. 1987; Tsien et al. 1987).

Bi-ionic potential measurements have also been used to assess the permeability of univalent cations under divalent cation-free conditions. In frog skeletal muscle fibers loaded with Cs, E_{rev} was more positive when external Cs was replaced by Rb, Li, or Na, suggesting a selectivity sequence of Li ≈ Na > Rb > Cs, with P_{Na} about 1.8 times larger than P_{Cs} (Almers et al. 1984). This sequence is corroborated by the results of Coronado and Smith (1987) on rat T-tubular Ca channels incorporated into lipid bilayers: P_{Li} (1.4) > P_{Na} (1.0) > P_K (0.7) ≈ P_{Cs} (0.8). In addition, cardiac Ca channels have a similar relative permeability to these ions. In whole-cell and cell-attached patch experiments, P_{Li} (9.9) > P_{Na} (3.6) > P_K (1.4) > P_{Cs} (1.0) (Hess et al. 1986; Tsien et al. 1987). One piece of evidence indicating that the cardiac Ca channel clearly selects Li over Na was that the open channel i-V relation extrapolated to a $10-15$ mV more positive E_{rev} with external Li than with external K in K-loaded cells (Hess et al. 1986). In agreement with this finding, Matsuda (1986) observed that macroscopic Ca channel current in Cs-dialyzed guinea pig ventricular cells reversed at a potential $10-15$ mV more positive with external Li than with external Na.

In summary, the experimental results on the selectivity of muscle Ca channels are quite consistent. The order of preference is $Ca > Sr > Ba \gg Li > Na > K > Cs$. This sequence is the direct opposite of that determined on the basis of current-carrying ability. Thus, the argument has been made that selectivity is determined by the *affinity* of an ion for a binding site within the channel, whereas current-carrying ability is based on the *mobility* of the ion through the pore (Almers and McCleskey 1984; Hess and Tsien 1984; Hess et al. 1986). For permeant ions of the same charge, the ionic radius seems to have a bearing on mobility: the more mobile Ba is smaller than Ca, and the more mobile Cs is smaller than Li (Hess et al. 1986; Tsien et al. 1987; see also McCleskey and Almers 1985). In fact, Coronado and Smith (1987) have made the point that the relative conductance of alkali cations through single skeletal muscle Ca channels closely matches their relative mobility in dilute solutions.

3.4 Block by External Inorganic Cations

A number of inorganic cations block Ca (and Ba)-dependent action potentials and Ca channel current in muscle tissues (e.g., Reuter 1973). The actions of these classic inorganic blockers, including the trivalent La and the divalent Cd, Co, Mn, and Ni, are reviewed in Sect. 3.4.1. The relatively impermeant Mg, as well as the highly permeant Ca and Ba, can also exert a blocking action; their effects are covered in Sects. 3.4.2, 3.4.3. The blocking action of external protons has some unusual aspects and these are outlined in Sect. 3.4.4.

3.4.1 Block by Classical Inorganic Cation Blockers

Cd has almost become the divalent cation blocker of choice for Ca- and Ba-carried currents, perhaps because it is a potent blocker and because the block produced is often reversible. Its potency is such that a concentration of 0.5 mM can completely block I_{Ca} generated from ca. 2 mM Ca solution in cardiac myocytes (e.g., Hume and Giles 1983; Mitchell et al. 1983; Lee and Tsien 1984; Fischmeister and Hartzell 1986). Complete suppression of cardiac I_{Ca} by Cd concentrations up to 0.5 mM suggests that the half-blocking concentration could be in the order of $10-50$ μM. However, in frog skeletal muscle fibers bathed with 10 mM Ca solution, the EC_{50} for I_{Ca} inhibition is $0.3-0.4$ mM (Palade and Almers 1985; Cota and Stefani 1986). The fact that half-block in frog skeletal muscle fibers required a tenfold higher concentration of Cd than in heart cells can be explained, at least in part, by the fivefold higher Ca_o in the muscle cell experiments, since there is competition between the permeating Ca ions and the blocking cations for a site within the channel pore (see below). A secondary explanation is that frog skeletal muscle Ca channels require higher concentrations of Cd for block than cardiac Ca channels. In this regard, skeletal muscle from rat appears to be more sensitive than skeletal muscle from frog: I_{Ca} from fibers in 10 mM Ca_o was completely blocked by $0.5-1.0$ mM Cd (Donaldson and Beam 1983).

EC_{50} values for block of I_{Ca} (10 mM Ca_o) in frog skeletal muscle fibers by other divalent cations are as follows (Palade and Almers 1985): Ni, 0.68 mM; Co, 1.28 mM; and Mn, 13.5 mM. Comparable data are not available for heart and smooth muscle cells, although we note that (a) full-blocking concentration of these ions on I_{Ca} (ca. 2 mM Ca_o) are all of the order of $3-5$ mM in heart cells (Table 4, Pelzer et al. 1989a), and (b) 5 mM Ni and 5 mM Co completely blocked I_{Ca} ($1.8-2.5$ mM Ca_o) in smooth muscle cells (Klöckner and Isenberg 1985; Ganitkevich et al. 1987).

Ca channel block by these inorganic blockers has been studied at the single-channel level. In cell-attached patches of guinea pig ventricular myocytes bathed by 50 mM Ba solution, channel open time was reduced to about one-half control by 20 μM Cd, 100 μM Mn, and 250 μM Co (Lansman et al. 1986). By resolving individual steps of block and unblock at the single-channel level, Lansman et al. (1986) were able to rule out the possibility that block was due to a local reduction of charge-carrying ions at the channel's external entrance. Rather, the reduced flux of permeating ions appeared to be due to competition with blocking ions. The competition seems to take place within the channel itself, an idea originally proposed by Hagiwara et al. (1974) to explain block of Ca channel currents in barnacle muscle. Lansman et al. (1986) observed that when blocking ion concentration was increased, unitary Ba current amplitude was unchanged, indicating no reduction of charge-carrier concentration. However, there was an increase in the number of discrete blocking and unblocking events with increasing blocker concentration, as expected with intrachannel competition.

Lansman et al. (1986) also detailed another important aspect of block by multivalent cations: a dependence on voltage. Membrane hyperpolarization produced steep increases in unblocking rates (i.e., relief from block) as if blocking ions were ejected from the pore into the myoplasm by the applied electric field. For example, Cd block was less severe at negative potentials, a result also obtained by Byerly et al. (1985) on snail neurons. The blocking rate constant was unaffected by voltage, as if the approach of Cd to the blocking site was unaffected by the membrane field (Lansman et al. 1986).

Lansman et al. (1986) also enquired whether there was less block by Cd ions when the concentration of the permeant Ba ions was raised (a result that is expected if there is competition for binding to a common site in the channel). When they applied a fixed 20 μM concentration of Cd, and varied Ba$_o$ between 20 and 110 mM, the rate constant of unblock increased twofold and the rate constant of block declined by about 2.5-fold (i.e., relief from block). They explained the increased rate of unblock as being due to a Ba$_o$-dependent increase in channel occupancy by Ba ions with resultant quicker expulsion of Cd into the cell. The decreased rate of block was explained by an increase in Ba$_o$-dependent occupancy of a channel site near the external mouth.

La is an even more potent blocker of Ca- and Ba-carried current than Cd. At the single cardiac Ca channel level, it curtails channel openings like Cd, but it prolongs interburst closings much more than Cd (Lansman et al. 1986). In whole-cell experiments on frog atrial myocytes, Campbell et al. (1988a) have shown that I$_{Ca}$ generated by 2.5 mM Ca$_o$ was completely blocked by 10 μM La; comparable block by Cd required a 100 μM concentration. Nathan et al. (1988) have also reported that I$_{Ca}$ (2.5 mM Ca$_o$) in frog atrial myocytes is completely blocked by 10 μM La (a much lower concentration than had previously been used for this purpose on cardiac tissues; cf. Katzung et al. 1973; Kass and Tsien 1975). This concentration of La did not shift Ca channel kinetics, and assuming a 1:1 binding ratio of La to a channel site, the dissociation constant was estimated to be slightly less than 1 μM (Nathan et al. 1988). Competition between Ca and La for a binding site was indicated by the near 15-fold increase in half-blocking concentration of La when Ca$_o$ was raised from 2.5 to 7.5 mM (Nathan et al. 1988).

La has also been shown to be a potent blocker of I$_{Ca}$ in uterine smooth muscle (Anderson et al. 1971), and Bean et al. (1986) noted that 30 μM La was sufficient for complete block of Ca channel current generated by 110 mM Ba$_o$ in cells isolated from rat mesenteric artery. Like Cd block, La block is relieved with membrane hyperpolarization, presumably because the electric field enhances the ejection of blocking ions from the pore into the myoplasm (Lansman et al. 1986). Thus, even ions like Cd and La that bind in ultrastrong fashion may have a limited permeability through the channel. In fact, Wendt-Galitelli and Isenberg (1985) have detected intracellular La after prolonged stimulation of guinea pig ventricular myocytes in La-containing solution.

We noted above that there appears to be competition between the permeating ion and the blocking ion for a Ca channel binding site near the outer mouth. Thus, increasing Ba$_o$ reduced the block by a given concentration of Cd, and increasing Ca$_o$ reduced the block by La. An additional test of the competition hypothesis is to determine whether current carried by low-affinity permeating ions is blocked by lower concentrations of blockers than current carried by high-affinity permeating ions. This certainly appears to be the case when La block of Ca-carried current is compared with La block of Ba-carried current. Ba is the lower affinity permeator of this pair (see below) and concentrations of La which only block 50% of cardiac I$_{Ca}$ generated by

7.5 mM Ca_o exert a stronger block on cardiac Ba current generated by 50–110 mM Ba_o (see above). Likewise, 0.9 mM Co_o blocks about 70% of I_{Ca} generated by 0.9 mM Ca_o (Matsuda and Noma 1984) and 2 mM Co_o blocks a similar fraction of I_{Ca} generated by 1.8 mM Ca_o (McDonald et al. 1981), but 1–3 mM Co_o blocks 90 mM Ba_o-generated current through cardiac (unpublished) and reconstituted skeletal muscle Ca channels (Pelzer et al. 1988, 1989b).

The same test can be applied to block of Ca channel current carried by a relatively high-affinity divalent cation versus a low-affinity monovalent cation such as Na. In rat uterine strips, Jmari et al. (1987) determined that I_{Ca} generated from 2.1 mM Ca solution was half-blocked by the same concentrations of Co (0.5 mM), Mn (0.5 mM) and Ni (1 mM) required to block Na-carried current (130 mM Na_o). In frog skeletal muscle fibers, the results were equivocal on comparing 10 mM Ca_o with 32 mM Na_o conditions. The EC_{50} for block of I_{Ca} was significantly higher than the EC_{50} for block of Na-carried current when the blocker was Mn or Co, but not when it was Ni or Cd (Almers et al. 1984; Palade and Almers 1985). By contrast, Hadley and Hume (1987) required 500 µM Cd to block I_{Ca} in guinea pig ventricular myocytes bathed in 2.5 mM Ca solution, but only 20 µM to block Na-carried current (124 mM Na_o). Further comparisons of block of divalent cation currents versus monovalent ones are cited below.

3.4.2 Block by Magnesium

External Mg can block both divalent and monovalent cation currents through muscle Ca channels, though the blocking action is not nearly as effective as that exerted by the classical inorganic blockers. The blocking action of Mg on I_{Ca} is particularly weak. For example, 5 mM Mg_o had no effect on I_{Ca} of frog atrial cells in 2.5 mM Ca_o (Campbell et al. 1988a). I_{Ca} in some smooth muscle cells may be more susceptible to Mg_o block: Ganitkevich et al. (1988) have reported that I_{Ca} in guinea pig taenia caeci myocytes bathed in 2.5 mM Ca solution is reduced by 25% with 5 mM Mg_o and 50% with 10 mM Mg_o. This appears to be in good agreement with the blocking activity in skeletal muscle fibers where the EC_{50} was about 33 mM for the block of I_{Ca} (10 mM Ca_o) (Almers et al. 1984).

Ba- and Sr-carried currents are more susceptible than Ca-carried currents to block by Mg_o. In frog atrial cells, 5 mM Mg_o almost completely inhibited currents generated from 2.5 mM Ba or Sr solutions (Campbell et al. 1988a). This result agrees with that of Lansman et al. (1986) who determined that 10 mM Mg_o half-blocked single Ca channel current carried by Ba when cell-attached patches in guinea pig ventricular myocytes were bathed in 50 mM Ba solution. In further experiments, Lansman et al. (1986) found that increases in Ba_o between 20 and 110 mM reduced the Mg block. Surprisingly, and unlike the situation with Cd and La blocks, Mg block was enhanced at more negative potentials as though these potentials do not push Mg out into the cytoplasm but simply wedge the ion deeper into the pore (also see Fukushima and Hagiwara 1985).

Ca channel currents carried by monovalent cations in the absence of divalent charge carriers are blocked by Mg_o at millimolar or lower concentration. In frog skeletal muscle bathed in 32 mM Na solution, Na-carried current was half-blocked by 2.8 mM Mg_o (Almers et al. 1984), whereas in guinea pig ventricular myocytes bathed in 145 mM Na solution, the half-blocking concentration was about 60 μM (Matsuda 1986). Thus, these results with Mg_o suggest a Ca channel affinity sequence of Ca ≫ Mg ≈ Sr, Ba ≫ Na.

3.4.3 Block by Calcium and Barium

Although both Ca and Ba ions move easily through Ca channels, their affinity for channel binding is nevertheless significant, especially when measured against monovalent cations (see E_{rev} data Sect. 3.3). Therefore, they are expected to exert a considerable block against permeating monovalent cations. We will deal with this topic shortly, but first we focus on experimental findings related to the fact that the channel affinity for Ca ions appears to be five to ten times larger than for Ba ions (e.g., Tsien et al. 1987; Campbell et al. 1988 b).

3.4.3.1 Anomalous Mole Fraction Effect

The difference in affinity between the highly permeant ions, Ca and Ba, raises the interesting question of what happens to Ca channel current when the external solution contains varying proportions of the two species. In K channels where ions are thought to traverse multisite pores in single file, mixtures of charge carriers produce the phenomenon known as the "anomalous mole fraction effect" (Hille and Schwarz 1978; cf. Eisenman and Horn 1983), and a similar effect is observed with muscle Ca channels. When the external solution bathing heart myocytes contained mixtures of Ca and Ba such that the total (Ca + Ba) concentration was 2.5, 3.6, or 10 mM, the Ca channel current was up to 50% smaller than when the external solution contained either Ca or Ba alone at 2.5, 3.6, or 10 mM concentration (Hess and Tsien 1984; McDonald et al. 1986; Campbell et al. 1988 a). When the proportions of Ca and Ba are varied, the current goes through a broad minimum that is inconsistent with the prediction of a simple one-site pore (Tsien et al. 1987). Similar anomalous mole fraction effects have been found in frog skeletal muscle fibers (Almers and McCleskey 1984) and rat uterine smooth muscle tissue (Jmari et al. 1987). In general, the block of predominantly Ba current by small concentrations of Ca is much more pronounced than the block of predominantly Ca current by small concentrations of Ba.

Employing fluctuation analysis of whole-cell Ca channel current from frog ventricular myocytes, and single-channel recordings in guinea pig ventricular myocytes, Hess and Tsien (1984) and Lansman et al. (1986) established that the inhibition of Ba currents by Ca occurs at an intrapore site. When 10 mM Ca was added to 50 mM Ba solution there was a 50% block manifested as a reduction in unitary current amplitude without detectable flickering. The authors concluded that the apparent reduction in unitary current was the consequence of very rapid blocking and unblocking transitions that were not resolved with their recording system.

3.4.3.2 Block of Monovalent Current

When muscle cells are bathed in divalent cation-free solutions, external Na can carry large currents through Ca channels. These inward currents can be completely blocked by the addition of micromolar Ca to the bathing solution. In cardiac myocytes, monovalent Na current is reduced to one-half by 1 μM Ca (Hess and Tsien 1984; Hess et al. 1986; Matsuda 1986; Hadley and Hume 1987). In skeletal muscle fibers, the half-blocking concentration is about the same (Almers et al. 1984), and in smooth muscle cells (Isenberg and Klöckner 1985 b) and tissue (Jmari et al. 1987) it may be even lower. This high-affinity channel site for Ca contrasts with the low-affinity site determining I_{Ca} amplitude (K_D in the millimolar range as reviewed earlier) (see Sect. 3.2.3) and provides a strong argument in favor of multisite channel permeation (Almers and McCleskey 1984; Hess and Tsien 1984).

Lansman et al. (1986) have studied the block of inward monovalent Ca channel current by micromolar Ca_o at the single-channel level. Inward current generated from 150 mM Li bathing solution was blocked by micromolar Ca_o. As Ca_o was increased from 0.7 up to 5.5 μM, long channel openings grew sparser and current records were instead punctuated by series of rapid spike-like events. Histograms of open and closed times indicated that the blocking rate increased linearly with Ca_o, while the unblocking rate was not changed. Hyperpolarization shortened Ca blocking events by increasing the rate of unblock, suggesting that the blocking site is in the permeation path, and that the change in electric field drives blocking Ca ions out of the pore into the myoplasm.

3.4.4 Block by Extracellular Protons

I_{Ca} in cardiac ventricular and atrial tissue is depressed when the extracellular solution is made acidic (e.g., Chesnais et al. 1975; Kohlhardt et al. 1976). This observation has been confirmed by experiments on isolated rabbit nodal cells (Satoh et al. 1982) and rat ventricular cells (Yatani and Goto 1983). In these studies, significant block was only observed when pH_o was lowered by 1.5 – 3 units, which raised the possibility that the effect might, at least in part, be mediated by intracellular acidosis (see Sect. 5.6). In addition, it was evident that H ions affected the voltage dependence of I_{Ca}, perhaps by screening membrane surface charges (see Yatani and Goto 1983).

Irisawa and Sato (1986) considered both of these "side effects" in their study on I_{Ca} in dialyzed guinea pig ventricular muscles bathed with 1.8 mM Ca solution. When pH_o was lowered from 7.4 to 6, I_{Ca} was depressed by nearly 50%. However, when the intracellular dialysis solution contained 50 mM N-(2-Hydroxyethyl)piperazine-N'-(2-ethanesulfonic acid) (HEPES) instead of the standard 5 mM, half-block of I_{Ca} was not achieved until pH_o was lowered to 5. Since the positive shift of 10 mV in channel gating induced by lowering pH_o was accounted for in their measurements, Irisawa and Sato (1986) concluded that Ca channel activity was more susceptible to inhibition by changes in pH_i (see Sect. 5.6) than by changes in pH_o.

An examination of the graphs in the study by Yatani and Goto (1983) indicates that the proton block of Ba- and Sr-carried currents was more pronounced than the block of I_{Ca}. This result is therefore in keeping with the competition for a channel binding site as discussed in the preceding sections. In single Ca channel work on guinea pig ventricular myocytes, Hess et al. (1986) conducted some of their experiments on Na-carried currents at pH 9 because they felt that rapid chopping in currents recorded at pH 7.6 was related to block by protons. This seems to argue for sharper competition against low-affinity Na ions than against Ca, Sr, or Ba. In fact, there was less chopping of open channel records when, instead of Na, the higher affinity Li ion carried the Ca channel current (Hess et al. 1986).

Hess and his colleagues have pursued the question of how external protons depress Ca channel opening in cardiac cells (Prod'hom et al. 1987; Pietrobon et al. 1988). By varying external pH between 6 and 9, they found that protonation of a site located at the external surface of the channel greatly reduced single-channel conductance (i reduced to one-third) when the external charge carrier was Na. External Deuterium ions (D) also caused flickery transitions between high- and low-conductance states, and block by protons or D was not influenced by membrane potential, suggesting that the binding site was not in the channel itself. In contrast to the potent block of Na-carried current by external H, Ca channel currents generated by 110 mM Ba_o were not affected by pH_o as low as 6. This led the investigators to conclude that the higher affinity divalent cation could bias the protonation equilibrium to the nonprotonated state. In fact, they found that the apparent pK of the external protonation site was 8.6 when Cs was the charge carrier, but 8.2 and 7.4 respectively with K and Na as the charge carriers. Based on i-V relations extrapolated to zero current, they found that channel selectivity by affinity was in the order K > Na > Cs, i.e., the same order as for the potency of destabilization of the protonated state.

Destabilization was primarily due to an increase in the H dissociation rate constant ($k_{off}(H)$) and thus they concluded that there was a direct link between channel occupancy and $k_{off}(H)$. They supported their hypothesis by showing that the block of K current by micromolar external Ca created a marked increase in the number of transient closings from the low-conductance (protonated) state. Further, openings after these closings were invariably to the high-conductance level, indicating that channel-blocking sojourns by individual Ca ions (transient closings) shifted the protonation equilibrium towards the unprotonated state. Analysis of the average current preceding a blocking event indicated that Ca entry into the channel was not determined by the protonated state of the channel. This allowed the authors to conclude that the protonation site was far enough removed from the permeation pathway that protonation-dependent surface potential could not affect the local concentration of permeant ions. They argued that the results exclude the notion of direct competition between protons and permeant cations for the same site because (a) a Ca ion would then only be able to enter the unprotonated state (whereas block frequently ensued from the low-level conductance state), (b) protons do not seem to block the channel completely (in contrast to Ca, for example), and (c) there were only minor effects of permeant ions on the H association rate constant $k_{on}(H)$ of the protonation kinetics. The conclusion reached by Hess and colleagues was that protons and permeant cations interact allosterically at different sites. Occupancy of the cation permeation site changes the protein conformation in a way that favors deprotonation of the proton site. Protonation triggers a conformational change which reduces the channel conductance for monovalent ions with little effect on selectivity. Conversely, occupancy of the cation permeation site (high with millimolar divalent cation) changes the channel protein conformation in a way that favors deprotonation of the proton site. It follows that short-lived conformational changes induced by a permeating ion could influence the permeation of the next ion entering the pore. In this regard it is interesting that the time course of activation of Ba-carried current in cardiac myocytes appears to be different from that of Ca-carried current (McDonald et al. 1986; Campbell et al. 1988a).

3.5 Summary: Hypothesis on Calcium Channel Permeation

The review of permeability, selectivity, and block presented above emphasizes the concept that permeant ions have a relatively high mobility through the Ca channel and low affinity for multiple binding sites within it, whereas blocking ions have relatively low mobility and high affinity. Neither mobility nor affinity is determined by ionic size alone. Two examples suffice. First, although Ca, Na, and Cd have almost identical Pauling radii, they have enormously different mobilities and affinities for cardiac Ca channels (Hess et al. 1986; Lansman et al. 1986; Tsien et al. 1987). Second, rat T-tubular Ca channels incorporated into lipid bilayers had a very high conductance for ammonium ions (60 pS) and a fourfold lower conductance for hydrazinium ions (15 pS) which are almost identical in size (Coronado and Smith 1987).

By way of a summary, we present the hypothesis on the mechanism of Ca channel permeation formulated by Almers and McCleskey (1984) and Hess and Tsien (1984; see also Hagiwara and Byerly 1981; Kostyuk et al. 1983; Tsien et al. 1987). The hypothesis explains how, under physiological conditions, Ca channels can on the one hand exercise a high selectivity for Ca over Na, and on the other hand, allow the passage of ions at a rate near millions per second.

The key points of this hypothesis are:

1. Ions pass through the pore in single-file, interacting with multiple binding sites along the way.

2. Rapid permeation by Ca ions depends upon simultaneous ionic occupancy of two intrapore binding sites.

3. Double occupancy becomes significant at millimolar Ca_o and produces electrostatic repulsion or other ion-ion interaction which helps promote a quick exit of Ca ions from the pore into the cell.

4. Selectivity is largely determined by ion affinity to the binding site rather than by filter-like exclusion.

5. Divalent cations with lower binding affinities will move more quickly through the pore, and carry a larger current, than divalent cations with higher affinity.

6. Divalent cations with very high binding site affinities stick in the pore and may block permeation by other species.

7. Since monovalent cations have a much lower binding affinity than divalent cations, their open-channel flux is large but very sensitive to block by divalent cations.

The potent block of Na movement by low Ca_o explains why under physiological conditions, Na ions make a negligible contribution to inward Ca channel current. Thus, the complete removal of Na_o does not affect the amplitude of I_{Ca} in mammalian cardiac myocytes (Isenberg and Klöckner 1982; Mitchell et al. 1983; Matsuda and Noma 1984), amphibian cardiac myocytes (Hume and Giles 1983; Campbell et al. 1988a), frog skeletal muscle (Almers et al. 1984), or smooth muscle myocytes (Klöckner and Isenberg 1985).

4 Extracellular Signalling and Intracellular Regulation

4.1 Overview

The expansion of investigation into extracellular signalling and intracellular regulation of Ca channels has been nothing less than phenomenal over the past 5 years. Here, we have chosen to review the subject by presenting at first an outline of receptor-activated intracellular enzymatic cascades that are currently thought to impinge on Ca channel activity in heart. Next, we tackle β-adrenergic and muscarinic stimulation, beginning with the responses to agonists, and proceeding to the effects of intracellular perturbations of the cascades. We then briefly review the actions of several other important neurohumoral agents (adenosine, atrial natriuretic factor, and angiotensin II). Finally, we review evidence indicating that regulatory G proteins may have direct actions on Ca channels. In each of the above sections, we focus first on heart cells and then proceed to other muscle cells as the topic develops.

4.2 Brief Overview of Intracellular Enzymatic Responses to Receptor Activation

The major steps between agonist binding to the β-receptor and the increase in cardiac Ca channel current are outlined below. The agonist-receptor complex catalyses the conversion of the inactive guanosine diphosphate (GDP)-associated regulatory protein, G_s, to the active guanosine triphosphate (GTP)-associated form, G_s^*. G_s^* (or more correctly its active subunit, α_s^*) associates with the catalytic subunit of adenyl cyclase. This triggers an increase in the production of cyclic adenosine monophosphate (cAMP) which enhances the activity of the catalytic (C) subunit of protein kinase A (PKA). In the presence of adenosine triphosphate (ATP), the C subunit phosphorylates Ca channel protein. Other agents which cause stimulation of this pathway include (a) exogenous histamine (H_2 receptor binding stimulates adenyl cyclase, by activating G_s; (Watanabe et al. 1984), (b) membrane-permeable forskolin (increase in cAMP activity after direct activation of C subunit; Daly 1984; Seamon and Wetzel 1984), (c) permeable phosphodiesterase inhibitors such as theophylline and various xanthines (increase in cAMP after inhibition of the conversion of cAMP to 5'-AMP; Watanabe et al. 1984), and (d) cholera toxin (CTX; increase in cAMP by adenosine diphosphate (ADP)-ribosylation mediated activation of G_s; Birnbaumer et al. 1987; Graziano and Gilman 1987).

Agonist binding to muscarinic receptors can affect the cAMP cascade in a number of ways. First, it is well established that acetylcholine (ACh) inhibits adenyl cyclase activity in heart and reduces cAMP levels (Murad et al. 1962;

Watanabe and Besch 1975; Biegon and Pappano 1980). The current view is that muscarinic agonists stimulate the conversion of a second regulatory protein, G_i, to the active form G_i^*. G_i^* depresses the activity of adenyl cyclase and can thereby curtail the production of cAMP (Fleming et al. 1987). Adenosine appears to activate G_i in a manner similar to ACh (Isenberg et al. 1987). Pertussis toxin (PTX) ADP-ribosylates G_i and suppresses G_i activation by receptor agonists (Ui 1984; Wolff et al. 1984). Secondly, muscarinic receptor occupation also increases intracellular cyclic guanosine monophosphate (cGMP) (George et al. 1970; Kuo et al. 1972) by activating guanyl cyclase (Goldberg and Haddox 1977; Lincoln and Keely 1981). Atrial natriuretic factor (ANF) also increases cGMP (Ballermann and Brenner 1986; Huang et al. 1986), presumably due to activation of a guanyl cyclase (Rapoport et al. 1986), and therefore may also cause depression of adenyl cyclase activity (Anand-Srivastava et al. 1984). Nitroprusside application can also increase cGMP levels in tissues (Diamond et al. 1977). cGMP itself appears to stimulate cAMP hydrolysis (Beavo et al. 1971). Thirdly, the fact that muscarinic activation of G_i and its subunits can directly affect G_s and its subunits should not be excluded.

There is evidence that muscarinic receptor occupation activates yet another regulatory protein, $G_p(?)$, that differs from G_i in that it is PTX-insensitive (Tajima et al. 1987; Jones et al. 1988). The putative G_p activates phospholipase C and causes the breakdown of phosphatidylinositol 4,5-bisphosphate (PIP_2) into two second messengers, inositol trisphosphate (IP_3) and diacylglycerol (DAG) (Nishizuka 1988). The former releases Ca from intracellular storage sites while the latter activates cytosolic protein kinase C (PKC) which translocates to the membrane where it may phosphorylate the Ca channel (see below). A number of DAG analogues, as well as tumor-promoting phorbol esters, are potent activators of PKC (Nishizuka 1988).

4.3 Responses to β-Adrenergic Receptor Stimulation

Reuter provided early evidence that β-adrenergic stimulation increases the Ca permeability of cardiac membranes when he showed that epinephrine increased the uptake of ^{45}Ca and the amplitude of Ca-dependent regenerative responses in cardiac tissue (Reuter 1965, 1966). Voltage-clamp experiments on various cardiac tissues established that these effects of β-adrenergic catecholamines were due to enhanced I_{Ca}, and that the latter played an important role in the positive inotropic and chronotropic effects of these agents on cardiac muscle and pacemaker tissue (Reuter 1967, 1974; Vassort et al. 1969; Noma et al. 1980). An involvement of cAMP in the β-adrenergic responses of cardiac Purkinje fibers was demonstrated by Tsien et al. (1972), and the effects of cAMP injection into these tissues led to the proposal that cAMP-dependent protein phosphorylation could mediate β-adrenergic modulation of cardiac Ca channels (Tsien 1973). Shortly afterwards, Sperelakis and Schneider (1976) proposed that in the presence of sufficient ATP, cAMP-dependent protein kinase was required to phosphorylate channel protein and thereby make the channel available for voltage-dependent activation. Dephosphorylation by a phosphatase might then return the channel to a nonfunctional pool (Reuter and Scholz 1977 b). Thus, β-adrenergic stimulation of cardiac cells was postulated to increase I_{Ca} by increasing the number of functional Ca channels available for voltage activation (Reuter and Scholz 1977 b).

From this vantage point, investigation into the mechanism of β-adrenergic stimulation proceeded on two fronts: (a) a step-by-step probing of the intracellular enzymatic cascade between receptor activation and enhancement of Ca influx, and (b) an analysis of the resultant changes in channel gating and channel availability.

4.3.1 The cAMP Cascade and Calcium Channel Modulation

There is overwhelming evidence that the mechanism of β-adrenergic stimulation of I_{Ca} in heart cells is due to Ca channel phosphorylation by enhanced PKA activity. The evidence can be divided into (a) the outcome of interventions that ultimately should stimulate phosphorylation, and (b) the outcome of interventions that ultimately should suppress phosphorylation.

4.3.1.1 Stimulation

cAMP activity in heart cells can be increased by (a) external application of membrane-permeable forskolin (a direct activator of the catalytic subunit of adenyl cyclase; Seamon and Daly 1983), (b) external application of permeable cAMP analogues that can activate PKA, (c) external or internal application of phosphodiesterase inhibitors to suppress cAMP conversion, (d) direct intracellular dialysis with cAMP-containing solutions, and (e) photochemical conversion of an intracellular inactive analogue to active cAMP.

Each of these procedures has produced the expected increment in whole-cell or single-channel current. Forskolin ($0.15-1\ \mu M$) produced a 0.5- to 3-fold increase in I_{Ca} of guinea pig ventricular myocytes (Trautwein et al. 1986; Isenberg et al. 1987), and the phosphodiesterase inhibitor IBMX increased the current by nearly 40% (Trautwein et al. 1986). Cachelin et al. (1983) observed an increase in the opening of single Ca channels in neonatal rat heart myocytes after extracellular application of 8-bromo-cAMP. Dialysis of guinea pig myocytes with cAMP-containing solutions enhances I_{Ca} (Irisawa and Kokubun 1983): $10-30\ \mu M$ cAMP increased I_{Ca} by three- to fourfold, a stimulation similar to that observed by supramaximal concentrations of isoproterenol (Kameyama et al. 1985). In frog ventricular cells where the stimulatory effect of isoproterenol is particularly pronounced (Bean et al. 1984), the increase in I_{Ca} is duplicated by intracellular dialysis with cAMP solution (Fischmeister and Hartzell 1986). Amphibian heart cells also respond to a flash-induced step increase in cAMP with a step increase in macroscopic I_{Ca} (Nargeot et al. 1983).

Stimulation of the cAMP cascade has also been simulated by intracellular application of the C subunit of PKA to guinea pig ventricular myocytes. Injection of C subunit increased whole-cell I_{Ca} (Osterrieder et al. 1982; Brum et al. 1983), as did intracellular dialysis, in a manner that was quantitatively consistent with I_{Ca} stimulation by isoproterenol and cAMP (Kameyama et al. 1985). Kameyama et al. (1985, 1986a,b) and Hescheler et al. (1987) also showed that whole-cell I_{Ca} in guinea pig ventricular myocytes gradually increased when ATP-γ-S, the dephosphorylation-resistant analogue of ATP, was present in the cell. When cells were loaded with ATP-γ-S, a submaximal concentration of isoproterenol (10^{-8} M) elicited maximal I_{Ca} stimulation. In addition, the phosphatase inhibitor, PPase-inhibitor-2 (cf. Cohen 1982) slowly increased I_{Ca}, and accentuated stimulation by isoproterenol, as though an endogenous phosphatase that normally kept Ca channel activity in check were suppressed.

4.3.1.2 Inhibition

The effects of interventions that inhibit the cAMP cascade have also been investigated. Muscarinic receptor activation is a classical method of blunting β-adrenergic stimulation but this topic is dealt with in detail below. A more direct test is to inhibit PKA with cAMP-dependent protein kinase inhibitor (PKI; Ashby and Walsh 1972). When guinea pig ventricular myocytes were dialyzed with 1 μM PKI, the enhancement of I_{Ca} by 5×10^{-8} M isoproterenol was completely blocked (Kameyama et al. 1986a). The fact that ATP must be involved in the phosphorylation process was indicated by the strong suppression of I_{Ca} stimulation (10^{-7} M isoproterenol or 3 μM C subunit) after cell dialysis with AMP-PNP (adenylimidodiphosphate), a nonhydrolysable ATP analogue (Kameyama et al. 1985). When normal phosphorylation processes were intact, the intracellular application of dephosphorylating agents (phosphatase-1 and -2A; cf. Cohen 1982; Ingebritsen and Cohen 1983) abolished the increment in I_{Ca} caused 5×10^{-8} M isoproterenol (Kameyama et al. 1986b; Hescheler et al. 1987).

Under control conditions in the absence of β-adrenergic stimulation, none of the cascade-inhibitory treatments above depressed I_{Ca} by more than 30%. This suggests that cAMP-dependent phosphorylation is not an absolute requirement for I_{Ca} in guinea pig ventricular cells. The same conclusion can be derived from experiments on Ca channels reconstituted into bilayers. After tissue homogenization procedures lasting for 6–12 h, single cardiac Ca channels incorporated from sarcolemmal vesicles are quite capable of opening in the apparent absence of phosphorylating enzymes and ATP (Rosenberg et al. 1986; Imoto et al. 1988). The same holds true for cardiac Ca channels reconstituted into liposomes (Kameyama and Nakayama 1988) and for single skeletal muscle Ca channels (DHP-binding sites) reconstituted into lipid bilayers (Flockerzi et al. 1986; Trautwein et al. 1987; Pelzer et al. 1988; Yatani et al. 1988).

The activity of Ca channels in skeletal muscle T-tubules may also be regulated by cAMP-mediated phosphorylation (Schmid et al. 1985). Indeed, it has been shown that epinephrine and isoproterenol have strong stimulatory effects on two types of I_{Ca} (fast, slow) in frog skeletal muscle fibers (Arreola et al. 1987; Stefani et al. 1987). In these studies, the stimulation of I_{Ca} by catecholamines was duplicated by the application of dibutyryl cAMP. When cAMP, PKA, and ATP were applied to the cytoplasmic face of single Ca channels reconstituted from rabbit skeletal muscle DHP-binding sites, there was a large increase in the open-state probability (Flockerzi et al. 1986; Trautwein et al. 1987; Pelzer et al. 1988; Yatani et al. 1988). Further, it was shown that the purified DHP-binding sites reconstituted into 20 pS channels (90 mM Ba; Flockerzi et al. 1986) can be directly phosphorylated by incubation in solutions containing the C subunit of PKA and ATP (Nastainczyk et al. 1987; Trautwein et al. 1987).

4.3.2 Investigation of Calcium Channel Gating and Availability

A more precise description of the changes in Ca channel activity elicited by cAMP-cascade stimulation has emerged from studies on single Ca channel activity in cell-attached cardiac membrane patches. The first important observation was that the unitary current amplitude is not increased by β-adrenergic agonists (Reuter et al. 1983; Brum et al. 1984) nor by cAMP analogues (Cachelin et al. 1983). However, there are clearcut changes in both the fast and slow gating of the channel. Isoproterenol produces a moderate lengthening of millisecond-long openings, and a moderate shortening of millisecond-long closings, in single Ca channels of cardiac myocytes from neonatal rat (Reuter et al. 1983) and adult guinea pig ventricular myocytes (Brum et al. 1984). A similar prolongation of open times and abbreviation of closed times is observed in cell-attached patches of guinea pig ventricular myocytes treated with adrenaline (Trautwein and Pelzer 1985a, 1988; McDonald et al. 1989) and neonatal rat heart cells superfused with 8-bromo-cAMP solution (Cachelin et al. 1983).

These changes in millisecond-kinetics only account for part of the increase in P_o induced by β-adrenergic stimulation; the remainder is due to marked changes in the slow (seconds) gating kinetics of cardiac Ca channels (see analysis by Tsien et al. 1986). This is most apparent in the increase of the proportion of nonblank sweeps within ensembles of single Ca channel currents. For example, the percentage of nonblank sweeps in guinea pig ventricular myocytes increased from 24% to 95% after the application of $14\,\mu M$ isoproterenol (Tsien et al. 1986), from 63% to 91% after $1\,\mu M$ adrenaline (Trautwein and Pelzer 1985a, 1988), and from 38% to 63% after $2\,\mu M$ epinephrine (Ochi et al. 1986).

Nonblank records tend to occur in clusters (Cavalié et al. 1986; Tsien et al. 1986), and Ochi et al. (1986) noted that there was a threefold increase in the mean duration of clusters after $2\,\mu M$ epinephrine. In terms of an intact cell, the more frequent opening of Ca channels on depolarizations after β-adrenergic stimulation should be observed in noise fluctuation studies as an increase in the functional number of channels (N_f) (Tsien et al. 1986). Fluctuation analysis of I_{Ca} in frog ventricular myocytes indicated that N_f was increased nearly threefold after β-adrenergic stimulation with $0.5\,\mu M$ isoproterenol (Bean et al. 1984).

The usual explanation for blank sweeps in control ensembles of single Ca channel currents is that the channel is in an inactivated state at the time of these sweeps (Cavalié et al. 1986; McDonald et al. 1986; Pelzer et al. 1986a; Tsien et al. 1986; Trautwein and Pelzer 1988). Therefore, the increase in the proportion of nonblank sweeps caused by β-adrenergic agonists can be interpreted as a reduction in the average sojourn of the channel in the inactivated state. In most single channel studies, the pulses are short in comparison to the interpulse intervals, and so the attenuation of sojourns in the inactivated state can be viewed as an agonist action that is exerted at negative interpulse holding potentials. This action should translate into a speeding up of the reactivation of whole-cell I_{Ca}. A faster reactivation of I_{Ca} in calf Purkinje fibers treated with adrenaline or isoproterenol has been described by Weingart et al. (1978), and similar observations have been made on other cardiac tissues (Shimoni et al. 1984; Tsuji et al. 1985) and on frog (Bean et al. 1984) and guinea pig (Shimoni et al. 1987) ventricular myocytes. On the other hand, the very opposite effect, a slowing of I_{Ca} reactivation, has also been reported in catecholamine-treated tissues and myocytes (Josephson et al. 1984; Mentrard et al. 1984; Mitchell et al. 1985; Fischmeister and Hartzell 1986; Fischmeister et al. 1987), as has the absence of any effect (Reuter and Scholz 1977b; Isenberg and Klöckner 1982). This lack of agreement on changes in reactivation rate may be due in part to an influence of Ca_i on cardiac Ca channel inactivation and reactivation (see Sect. 2).

One can also examine the flip side of the reduction in blank sweeps by β-adrenergic agonists, i.e., the lengthening of nonblank sweep clusters. This action can be viewed as a hindering of the voltage-dependent inactivation (Sect. 2) that normally occurs at depolarized potentials or, stated another way, a tilt in equilibrium towards the activated state. This alteration should be reflected in a slowing of the decay phase of ensemble average single Ca channel currents elicited by step depolarizations, a similar slowing of whole-cell current decay, and possibly a shift of the I-V relation to more negative potentials. When single Ca channel current is carried by Ba, the time course of ensemble average current is slowed by a factor of 2 in both frog (Tsien et al. 1986) and guinea pig (Brum et al. 1984; Trautwein and Pelzer 1985a, 1988) ventricular myocytes. A slowing of macroscopic Ca channel current has been difficult to discern, perhaps because there is an overlaying influence of Ca_i-induced inactivation. With Ba as charge carrier, a marked slowing has been detected in frog ventricular myocytes (Bean et al. 1984) and in guinea pig ventricular myocytes pretreated with 3 μM Bay K8644 (Tsien et al. 1986). A conclusion from most studies on β-adrenergic stimulation, and cAMP-cascade stimulation, is that the intervention simply scales the I-V relation of Ca channel current. However, closer examination of some of these results (e.g., Fischmeister and Hartzell 1986, Fig. 6, frog ventricular myocytes) suggests that stimulation produces a negative shift in the peak of the I_{Ca}-V relation. We have also observed a distinct shift after stimulation of guinea pig ventricular myocytes when whole-cell Ca channel currents were carried by either Ba (unpublished) or Ca (Y.M. Shuba, unpublished). Where a definite shift in V_{peak} is difficult to determine, a larger stimulation at negative potentials than at positive ones can be quite obvious (unpublished). Recently, Walsh and Kass (1988) reported that a membrane permeable cAMP analogue, 8-chlorphenylthio cAMP, increased I_{Ca} by nearly 600% at −10 mV, but only by 80% at +20 mV in guinea pig ventricular myocytes. Bean (1989) found a similar voltage-dependent response of L-type Ca channel current in frog ventricular and rabbit atrial myocytes treated with β-adrenergic agonists or dialyzed with cAMP. These observations were substantiated by recordings of shifts in Ca channel gating currents to more negative potentials. Bean (1989) also investigated the action of norepinephrine on sympathetic neurons and reported that the well-known inhibition of N-type Ca channel current in that preparation was more pronounced at negative activating potentials than at positive ones.

As in cardiac cells, stimulation of cAMP-dependent phosphorylation lengthens the open time and curtails the closed time of reconstituted T-tubular Ca channels (Flockerzi et al. 1986; Trautwein et al. 1987; Pelzer et al. 1988; Yatani et al. 1988). In smooth muscle cells, there is no evidence of a cAMP-mediated response by I_{Ca} to β-adrenergic agonists (Bülbring and Tomita 1987). However, recent studies indicate that L-type Ca channels in vascular smooth muscle cells do respond to norepinephrine. I_{Ca} in rat portal vein was both increased and decreased by the agonist (Pacaud et al. 1987), whereas I_{Ca} in rabbit ear artery cells was either unaffected or reversibly depressed by the neurotransmitter (Droogmans et al. 1987). However, in similar ear artery cells, Benham and Tsien (1988a) found no depressant effect due to norepinephrine on DHP-sensitive L-type Ca channel current. With 110 mM Ba as charge carrier, whole-cell currents were increased up to threefold by 20 μM

agonist. Inclusion of 200 µM GTP in the pipette dialysate enhanced the stimulation. Epinephrine was as effective as norepinephrine, and stimulation was
not blocked by either α- or β-adrenoreceptor antagonists.

The fact that norepinephrine stimulation of contraction in rabbit mesenteric artery is voltage-dependent and extremely sensitive to DHP antagonism
suggested the participation of L-type voltage-sensitive Ca channels to Nelson
et al. (1988). In single-channel experiments with 80 mM Ba as charge carrier
and Bay K8644 added to promote channel opening, the addition of 10 µM
norepinephrine to the bath solution caused a marked increase in open-state
probability. The response following the bath (rather than pipette) application
of agonist led to the conclusion that a second messenger was involved,
perhaps via signal transduction from α-adrenergic receptors.

4.4 Responses to Muscarinic Receptor Stimulation

4.4.1 Effect of ACh on Basal Calcium Current

The effects of extracellular application of ACh on cardiac Ca channels have
long been controversial. By the early 1980s, there was a body of evidence suggesting that ACh has an inhibitory effect on "basal" (i.e., control) I_{Ca} in cardiac tissues. For example, ACh had been shown to suppress I_{Ca} in amphibian
atrial tissue (Giles and Noble 1976; Ikemoto and Goto 1977; Garnier et al.
1978; Nargeot and Garnier 1982), mammalian atrial tissue (Ten Eick et al.
1976), mammalian ventricular tissue (Inui and Imamura 1977; Hino and Ochi
1980), chick ventricular tissue (Josephson and Sperelakis 1982; Inoue et al.
1983), and sheep Purkinje fibers (Carmeliet and Ramon 1980). Despite this
evidence, it was disturbing that ACh did not appear to inhibit I_{Ca} in the one
cardiac tissue where it might be expected to have the greatest impact, namely,
the mammalian S-A mode (Noma and Trautwein 1978).

Problems inherent in the tissue studies (simultaneous ACh activation of K
currents and different degrees of sympathetic "tone") were circumvented in
studies on dialyzed myocytes. As a result of the latter studies, a somewhat different picture has now emerged. Iijima et al. (1985) found that high concentrations of ACh had little, if any, effect on basal I_{Ca} in guinea pig atrial
myocytes. Similar results with ACh up to 10^{-4} M were obtained using guinea
pig ventricular myocytes (Hescheler et al. 1986), frog ventricular myocytes
(Fischmeister and Hartzell 1986), and frog atrial myocytes (Hartzell and Simmons 1987). By contrast, Shibata et al. (1986) reported that ACh $> 10^{-6} M$
inhibited basal I_{Ca} by 50% – 100% in frog atrial myocytes *whose K channels
were not blocked*. A similar result using the same type of myocyte was obtained by Hartzell and Simmons (1987). However, they also demonstrated
that when K channels *were blocked*, isolated basal I_{Ca} was unaffected by

ACh. They found a similar situation in experiments on frog ventricular myocytes, and therefore these results supported the conclusion of Iijima et al. (1985), i.e., that the simultaneous activation of a transient, I_{Ca}-overlapping, outward K current by ACh was one reason why others might have concluded that ACh suppresses basal I_{Ca}. In the absence of efficient K channel block, transient outward current activation by ACh could induce action potential shortening (and, thereby, negative inotropy), depression of Ca-dependent slow action potentials, and masking of the inward I_{Ca} transient.

It remains to be seen whether K current activation by ACh can explain all of the results attributed to depression of basal I_{Ca}. An investigation of "sympathetic tone" as a confounding factor in the ACh response of avian ventricular muscle was conducted by Biegon and Pappano (1980). They found that ACh responses persisted in the presence of β-adrenergic antagonists as well as in catecholamine-depleted tissues.

Despite the apparent lack of effect of ACh on basal I_{Ca} (see also Sect. 4.4.2.1), and in keeping with earlier extensive work on the antagonism of adrenergic stimulation by ACh (for reviews see Watanabe et al. 1984; Sperelakis 1988), recent studies on dialyzed myocytes clearly show that muscarinic receptor stimulation can drastically curtail the stimulation of I_{Ca} by β-adrenergic agonists. These studies on myocytes from frog atrium (Breitwieser and Szabo 1985; Hartzell and Simmons 1987), frog ventricle (Fischmeister and Hartzell 1986), and guinea pig ventricle (Hescheler et al. 1986; Trautwein et al. 1986) have yielded complementary results as discussed below.

4.4.2 Muscarinic Inhibition of Up-Regulated Cardiac Calcium Current

It is likely that ACh interaction with muscarinic receptors inhibits catecholamine-stimulated cardiac I_{Ca} by at least two mechanisms: (a) activation of receptor-coupled G_i and resultant inhibition of adenyl cyclase and lowering on elevated cAMP, and (b) activation of guanyl cyclase and elevation of cGMP, with consequent effects of elevated cAMP.

4.4.2.1 Action via the Adenyl Cyclase System

The evidence for the first of these mechanisms is as follows. (a) When G_i was functionally uncoupled from the receptor by PTX pretreatment, and therefore unavailable for activation and inhibition of adenyl cyclase, ACh had no effect on catecholamine-stimulated I_{Ca} (Hescheler et al. 1986). (b) When both G_s and G_i were activated by perfusion of the nonhydrolysable GTP analogue, GMP-PNP, ACh did not reduce the resultant slightly stimulated (30%) I_{Ca} or the subsequent forskolin-stimulated I_{Ca} (ca. 100% stimulation; Hescheler et al. 1986). (c) When forskolin alone was used to stimulate I_{Ca} by activation of adenyl cyclase, ACh was an effective antagonist (Trautwein et al.

1986). (d) When I_{Ca} was stimulated by intracellular perfusion of cAMP, ACh was ineffective (Hescheler et al. 1986; Fischmeister and Hartzell 1986). Finally, (e) when I_{Ca} was stimulated by superfusion of the phosphodiesterase inhibitor IBMX, ACh depressed the I_{Ca} increment by roughly 40% (Trautwein et al. 1986).

The investigators cited above concluded that the ACh attenuation of catecholamine-stimulated I_{Ca} is due to an inhibition of adenyl cyclase activity and reduction of cAMP levels, rather than to inhibition of cAMP-dependent PKA or stimulation of protein phosphatase. The inhibition of adenyl cyclase is mediated by the receptor-coupled G_i protein (Breitwieser and Szabo 1985; Hescheler et al. 1986). It remains unclear why ACh has no effect on nonstimulated basal I_{Ca}. Fischmeister and Hartzell (1986) noted the possibility that ACh might be incapable of lowering basal adenyl cyclase activity, a possibility that is supported by biochemical evidence (see Watanabe and Besch 1975; Keeley et al. 1978). However, the fact that IBMX-enhanced I_{Ca} was inhibited by ACh suggested to Hescheler et al. (1986) that ACh is, in fact, capable of lowering basal adenyl cyclase activity (see also Linden et al. 1985). Their explanation for the lack of effect of ACh on basal I_{Ca} was that even a marked lowering of cAMP below basal levels (≈ 1 μM) by ACh would not affect I_{Ca} because I_{Ca} is insensitive to cAMP concentration below of 1 μM (cf. Kameyama et al. 1985).

4.4.2.2 Action via the Guanyl Cyclase System

Since ACh elevates cGMP levels in the heart (George et al. 1970; Watanabe and Besch 1975; Flitney and Singh 1981), this constitutes another pathway by which muscarinic receptor activation might modulate cardiac Ca channel activity. During the 1970s there was considerable sentiment for the idea that cAMP had stimulatory properties on the heart, whereas cGMP had inhibitory ones (cf. the "Yin-Yan hypothesis": Goldberg et al. 1975). However, the expectations were not always fulfilled, and publications detailing cGMP-related negative inotropism (e.g., Wilkerson et al. 1976; Kohlhardt and Haap 1978; Singh and Flitney 1981) and depression of Ca influx (Nawrath 1977; Trautwein et al. 1982; Bkaily and Sperelakis 1985) were balanced by reports to the contrary (e.g., Diamond et al. 1977; Endoh and Shimizu 1979; Linden and Brooker 1979; Nargeot et al. 1983).

More recently, Hescheler et al. (1986) observed that basal I_{Ca} in guinea pig ventricular myocytes was unaffected by dialysates containing up to $10^{-4}\,M$ cGMP, and a similar result was obtained by Hartzell and Fischmeister (1986) from experiments on dialyzed frog ventricular myocytes. However, Fischmeister and Hartzell (1987) also found that the 6.6-fold increase in I_{Ca} produced by bath application of 2 μM isoproterenol or intracellular perfusion with 5 μM cAMP was subsequently reduced by nearly two-thirds upon intracellular perfusion with 20 μM cGMP (half-maximal effect at 0.6 μM cGMP). However, 8-bromo-cGMP, a potent activator of cGMP-dependent protein kinase (which affects snail neuronal Ca channels: Paupardin-Tritsch et al. 1986), did not affect I_{Ca} stimulation by cAMP. In addition, cAMP antagonism of cAMP-stimulated I_{Ca} was partially suppressed following treatment with the phosphodiesterase inhibitor IBMX. Thus, the authors concluded that cGMP acts by promoting cAMP hydrolysis via stimulation of a cyclic nucleotide phosphodiesterase, a hypothesis originally suggested by Flitney and Singh (1981) to account for observations that cAMP levels decline when cGMP levels rise. Since there is ample evidence that ACh increases cGMP level in the heart (e.g., Lincoln and Keely 1981; Flitney and Singh 1981), cGMP-stimulated phosphodiesterase activity would be a second way in which cholinergic neurotransmitters could affect I_{Ca} up-regulated by cAMP-dependent phosphorylation.

4.4.3 Action via Stimulation of Phospholipase C

The stimulation of phosphoinositide hydrolysis by agonists is an important signal transduction mechanism in many tissues (Berridge and Irvine 1984; Majerus et al. 1986). Receptor occupation, presumably with GTP-binding proteins (G_p) serving as intermediates (Cockcroft and Gomperts 1985; Jones et al. 1988), stimulates phospholipase C and this results in the generation of diacylglycerol (DAG) and inositol trisphosphate (IP_3) from phosphatidylinositol 4,5-bisphosphate (PIP_2). While IP_3 promotes the rapid release of Ca from intracellular stores (Berridge and Irvine 1984), DAG increases the activity of Ca-activated phospholipid-sensitive protein kinase C (PKC; Takai et al. 1982; Wise et al. 1982). PKC normally resides in the cytoplasm, but when activated a major fraction translocates and binds with high affinity to the cell membrane (Kraft and Anderson 1983; Nishizuka 1986; Matthies et al. 1987; Navarro 1987). PKC is thought to phosphorylate a number of endogenous proteins, including cardiac troponin subunits (Katoh et al. 1983), cardiac sarcoplasmic reticulum (Limas 1980; Movsesian et al. 1984), β-adrenergic receptor in erythrocyte membranes (Kelleher et al. 1984; Sibley et al. 1984), and perhaps all receptors coupled to the DAG-PKC system (Nishizuka 1988) as well as ion channels and pumps (Kaczmarek 1987). There are also indications that PKC can directly phosphorylate Ca channel protein. In one study on DHP receptor purified from rabbit skeletal muscle T-tubular membrane, the purified preparation contained three proteins (165, 55, and 33 kDa) of which the two largest contain sites that can be phosphorylated by PKC (Nastainczyk et al. 1987).

A number of tumor-promoting phorbol esters, including 12-0-tetradecanoylphorbol-13-acetate (TPA; Castagna et al. 1982), are known to cause direct persistent activation of PKC, whereas others (e.g., 4-α-phorbol-12,13-didecanoate, αPDD) are without effect. The action on protein kinases may be restricted to PKC since even chronic treatment of neuronal cells for several days with active phorbol esters does not affect the activities of PKA or Ca/calmodulin-dependent protein kinase (Matthies et al. 1987). A variety of DAG analogues are also capable of activating PKC (Nishizuka 1988).

The PKC-activating action of phorbol esters has been invaluable in delineating roles for PKC in the regulation of a large number of cellular events. This is particularly true in neuronal tissue where PKC may regulate the activity of a variety of ionic channels including those selective for Ca (Nishizuka 1986; Kaczmarek 1987). In neuronal, as well as in nonneuronal cells, PKC modulation can be either stimulatory or inhibitory. For example, TPA stimulated I_{Ca} (DeRiemer et al. 1985) and "recruited" covert Ca channels (Strong et al. 1987) in *Aplysia* bag cell neurones (but see Hammond et al. 1987). Phorbol esters also activated Ca channels in neuroblastoma cells (Osugi et al. 1986), but depressed Ca conductance in chick dorsal root ganglion (DRG) cells (Rane and Dunlap 1986) and inhibited voltage-activated DHP-sensitive Ca influx in PC 12 secretory cells (Di Virgilio et al. 1986; Harris et al. 1986; Messing et al. 1986), while paradoxically also *stimulating* Ca-dependent depolarization-evoked ^{3}H-norepinephrine (NE) release (Harris et al. 1986).

Muscarinic agonists have been shown to stimulate phosphoinositide hydrolysis in smooth muscle (Baron et al. 1984; Takuwa et al. 1986), rat ventricular myocytes (Brown et al. 1985), and rat heart tissue (Poggioli et al. 1986). In chick heart cells, muscarinic agonists and GTP-γ-S stimulated phosphoinositide hydrolysis (Jones et al. 1988), suggesting the involvement of a G protein. It does not seem to be the adenyl cyclase inhibitory G_i since, in contrast to carbachol-mediated inhibition of adenyl cyclase, carbachol-mediated phosphoinositide hydrolysis was insensitive to PTX treatment (Tajima et al. 1987).

The foregoing provide good reasons to enquire whether ACh-induced activation of PKC could affect Ca channels and in that manner play a role in cholinergic regulation of the heart. Shibata et al. (1987) used the diglyceride 1,2-diolein to test the possible involvement of PKC in ACh inhibition of isoproterenol-stimulated I_{Ca} in frog atrial myocytes. They reported that 1,2-diolein, but not other inactive diglycerides, consistently inhibited the increment in I_{Ca}. Although Shibata et al. (1987) did not report on the effects of 1,2-diolein on basal I_{Ca}, a possible inhibitory action of PKC on cardiac Ca channels would be one possible explanation for TPA-induced negative inotropy in rat papillary muscle (Uglesity et al. 1987) and rat ventricular myocytes (Capogrossi et al. 1987). The contractile force of spontaneously contracting embryonic chick ventricular cells in culture (Leatherman et al. 1987) was also depressed by an active phorbol ester (4β-phorbol-12-myristate-13-acetate, PMA). The onset of the negative inotropy occurred within 2 min (including bath lag time), maximum effect (10^{-6} M PMA, 50% depression) was reached after 7 min without change in beating rate, and inactive αPDD was without effect. A moderate depression of ^{45}Ca influx with no change in efflux was also measured in these cells (Leatherman et al. 1987), and fura-2 fluorescence indicated that Ca_i fell from 96 to 72 nM over the time course of the negative inotropy. The authors suggested that PKC might inhibit I_{Ca} but not by affecting adenyl cyclase since the contractile response to isoproterenol was unchanged; the latter result also seemed to rule out possible phosphorylation and desensitization of β-adrenergic receptors by PKC (as reported for erythrocyte membranes: Kelleher et al. 1984).

TPA also decreased the amplitude of contraction by nearly 50% in spontaneously contracting or driven neonatal rat ventricular cells in culture (Dösemeci et al. 1988). Unlike the study above on chick cells, TPA simultaneously increased the beating rate by up to 100%. In addition, Dösemeci et al. (1988) reported that phorbol ester increased I_{Ca} through L-type channels by about 30% with little change in kinetics or E_{rev}. TPA also stimulated ^{32}P incorporation into two specific regions of total cell protein (32 and 83 kDa) at concentrations as low as 16 nM. The authors provided two possible explanations for the unusual finding of an increase in I_{Ca} coupled with a reduction in contractile force: reduction of Ca_i (Leathermen et al. 1987; Uglesity et al. 1987) by an unknown effect of PKC and reduced sensitivity of myofilaments to Ca_i after PKC phosphorylation (Katoh et al. 1983).

Here we make a brief detour to discuss PKC effects on noncardiac muscle cells before returning to heart cells once again. In vascular smooth muscle tissue, TPA causes a slowly developing contracture (Rasmussen et al. 1984). This may be related to an increase in Ca influx (Sperti and Colucci 1987) and a 30% increase in Ca channel current (Fish et al. 1988) in aortic A7r5 cells. In agreement with these results on vascular smooth muscle cells, Vivaudou et al. (1988) found that diC_8, a permeable analogue of DAG that activates PKC, increased the amplitude of high-threshold L-like (but not low-threshold T-like) I_{Ca} by about 30% in isolated smooth muscle cells from toad stomach. While this analogue also slowed the rate of I_{Ca} inactivation, other inactive analogues were without any effect. ACh affected the amplitude and the rate of decay of I_{Ca} in a manner that was similar to diC_8 (Vivaudou et al. 1988).

Returning now to heart cells, we focus first on a recent important paper by Lacerda et al. (1988). This group has examined the effects of TPA on Ca channel current and cAMP levels in dialyzed neonatal rat ventricular myocytes. Although they were not successful in delineating effects on whole-cell Ca channel currents, perhaps due to washout of intracellular factors, they found a biphasic effect on Na-carried single Ca channel currents in cell-attached membrane patches. Following a short lag (≈ 2 min), 200 nM TPA markedly increased L-type channel current by increasing the number of openings per trace, without affecting mean open time, voltage dependence, or the number of active channels. The stimulation was also observed with Ba as charge carrier, was blocked by 2 μM nifedipine, and was not elicited by

αPDD. The authors concluded that the TPA-mediated increase in Ca channel activity arises from a modification by PKC of functional preexisting DHP-sensitive Ca channels, rather than from recruitment of covert channels as described in snail neurons by Strong et al. (1987). This modification seemed to be voltage-dependent: when TPA was applied at two holding potentials (-40 and 0 mV) that normally evoked only occasional low-level channel activity, the compound stimulated channel openings with a mean lag time of 104 ms at -40 mV, but just 6 ms at 0 mV. In agreement with other studies (e.g., Shenolikar et al. 1986; Teutsch et al. 1988), 200 nM TPA did not affect myocyte cAMP levels, leading to the conclusion that PKC activation of single Ca channel currents is due to channel phosphorylation that is independent of the cAMP cascade.

Lacerda et al. (1988) also observed an increase in depolarization-evoked, nifedipine-sensitive [45]Ca uptake by neonatal rat heart cells when 200 nM TPA was added to the bath just prior to the 5-s uptake measurement period. However, when cells were pretreated with TPA for 20 min, [45]Ca uptake during the 5-s uptake period was markedly depressed, even when the pretreatment was with TPA as low as 10 nM. This result after longer exposure to TPA agrees with their voltage clamp measurements showing that a stimulatory period lasting for approximately 10 min was followed by strong inhibition that almost abolished Ca channel activity. The secondary inhibition of channel activity was not related to Ca entry since it was observed with either Na or Ba as charge carrier. On the other hand, a specific phorbol ester action on PKC, or on cellular moieties other than PKC (Kreutter et al. 1985), seems to be involved since the DAG analogue, 1,2-dioctanoyl glycerol (DOG), exerted only stimulatory effects on Ca channel current.

The delayed inhibition by TPA observed by Lacerda et al. (1988) may in part help explain earlier, seemingly contradictory findings. For example, the stimulation of [45]Ca influx and Ca channel current in aortic A7r5 cells shortly after TPA administration (Sperti and Colucci 1987; Fish et al. 1988) may not be completely at odds with the inhibition of [45]Ca influx observed in similar cells by Galizzi et al. (1987) since the latter measurement followed a 10-min preincubation period with phorbol ester. Unfortunately, and in contrast to the lack of inhibition observed by Lacerda et al. (1988) with DOG, Galizzi et al. (1987) also reported inhibition of both depolarization-evoked and Bay K8644-stimulated [45]Ca influx after 10-min pretreatment with DAG analogues or DAG-activating peptides (vasopressin, bombesin, oxytocin).

Clear-cut interpretations are also complicated by the fact that phorbol esters can produce seemingly simultaneous and contradictory responses in some cells. One example is the increase in transmitter secretion coincident with depressed [45]Ca influx in PC12 cells (Harris et al. 1986; Messing et al. 1986). Another is the sustained contraction in vascular smooth muscle versus (delayed?) depression of Ca channel activity (Galizzi et al. 1987), and a third is the increase in I_{Ca} accompanied by inotropic depression in neonatal rat ventricular myocytes (Dösemeci et al. 1988). It will not be easy to sort out the various anomalies induced by agonists, DAG analogues, and phorbol esters because the effects of IP_3 on Ca movements (see below) must be taken into account, prolonged exposure to TPA can cause PKC down-regulation (Shenolikar et al. 1986; Matthies et al. 1987), and PKC can feed back and exert inhibitory actions at several points in the pathway leading from the receptor to kinase (e.g., Llano and Marty 1987; Nishizuka 1988). Species differences may also be relevant since in contrast to the stimulatory effects of phorbol esters on I_{Ca} in neonatal rat ventricular cells (Dösemeci et al. 1988; Lacerda et al. 1988), PKC stimulation had no effect on I_{Ca} in guinea pig ventricular myocytes at either 22° or 32°C (Walsh and Kass 1988). Finally, there is the anomaly that TPA applied internally is ineffective whereas TPA applied externally strongly inhibits I_{Ca} in dialyzed neurones (Hockberger et al. 1989) and dialyzed guinea pig ventricular cardiomyocytes (unpublished observation).

Phorbol esters can also affect the binding of organic Ca channel blockers. In PC 12 cells, PMA depressed depolarization-evoked ^{45}Ca influx and the binding of ^{3}H (+) PN 200-110 (Messing et al. 1986). The depressed binding was attributed to a twofold increase in K_D with no change in binding site number. Quite the opposite pattern was obtained by Navarro (1987) in embryonic chick myotubes. When myotubes were treated with PMA, there was a large translocation of activated PKC from cytosol to membrane, and depolarization-evoked, DHP-sensitive ^{45}Ca influx was increased two- to threefold. After 30 min PMA, there was a twofold increase in the specific binding of labeled PN 200-110 to normally polarized cells and this was attributed to an increase in the number of binding sites rather than to a change in K_D. Navarro (1987) also raised the interesting point that depolarization of skeletal muscle produces activation of phospholipase C (cf. Vergara et al. 1985), and therefore speculated that depolarization should also activate PKC which in turn would increase the number of DHP receptors (especially during tetany?). Here, one is reminded of the apparent increase in muscle DHP receptors induced by depolarization (Schwartz et al. 1985), and of the depolarization-induced shortening of the lag time for TPA stimulation of cardiac Ca channels (Lacerda et al. 1988).

The possibility that PKC modulation of T-tubular Ca channels is voltage dependent leads to the consideration of another consequence of phospholipase C activation, the generation of IP_3. In nonmuscle cells, IP_3 is thought to release Ca from the endoplasmic reticulum (Berridge 1987). By analogy, an increase in the second messenger IP_3, elicited by T-tubular depolarization, is postulated to couple excitation to contraction by promoting the opening of Ca channels in the SR (Vergara et al. 1985; Volpe et al. 1985). However, the following points suggested to Vilven and Coronado (1988) that IP_3 might act on T-tubular Ca channels: (a) they were unsuccessful in activating isolated SR-release channels with IP$_3$, (b) Hannon et al. (1988) suggested that IP$_3$ might act on T-tubular Ca channels, and (c) skinned fibers with "depolarized" T-tubules have been shown to be more sensitive to IP$_3$ than fibers with "polarized" tubules (Donaldson et al. 1988). When they investigated the effect of IP$_3$ on Ca channels incorporated into lipid bilayers from vesicular rabbit T-tubular membrane, Vilven and Coronado (1988) observed a nitrendipine-sensitive channel activation by 20 μM IP$_3$ with either Ba or Na as charge carrier. A further interesting result was that the addition of both Bay K8644 and IP$_3$ produced a synergistic rather than a simple additive effect on channel opening.

4.5 Responses to Other Neurohumoral Agents

Space considerations dictate that this section on other neurohumoral agents be fairly brief. Therefore, we have restricted ourselves to short reviews on three such agents, adenosine, atrial natriuretic factor, and angiotensin II.

4.5.1 Adenosine

Adenosine modulates many of the physiological functions of the heart. In so doing, its catalogue of actions is very similar, if not identical, to that of ACh. For example, adenosine depresses pacemaker automaticity, hyperpolarizes S-A nodal cells, and shortens the atrial action potential (Belardinelli and Isenberg 1983a; West and Belardinelli 1985). In guinea pig atrial myocytes, adenosine induces an extra K current whose properties resemble those of a K current activated by ACh (Belardinelli and Isenberg 1983b). Since this adenosine effect is PTX-sensitive and absent after K current enhancement by intracellular GTP-γ-S, Isenberg et al. (1987) concluded that occupancy of the A$_1$ receptor by adenosine is linked to K channel opening via a GTP-binding regulatory (G) protein, presumably G$_i$. Cardiac muscarinic receptors are also coupled to K channels via G$_i$ (Pfaffinger et al. 1985; Codina et al. 1987), and it has been proposed that adenosine and ACh activate an identical population of K channels in cardiac atrial cells (Kurachi et al. 1986).

As expected from its connection with G$_i$ activation, adenosine can also modulate Ca channel activity by affecting the adenyl cyclase-cAMP system. Schrader et al. (1977) attributed the inhibition by adenosine of catecholamine-stimulation of myocardial function to an action on the cAMP system. Direct support for this view was provided by Belardinelli et al. (1982) who demonstrated that adenosine reverses the elevation of cAMP by isoproterenol in embryonic chick ventricle. In the absence of β-adrenergic stimulation, 1 mM adenosine had no effect on I$_{Ca}$ in guinea pig ventricular myocytes

(Belardinelli and Isenberg 1983a; Isenberg et al. 1987). However, lower concentrations of the transmitter markedly attenuated the increase in I_{Ca} induced by isoproterenol, an effect that was much weaker after pretreatment of cells with PTX (Isenberg et al. 1987). I_{Ca} and cAMP levels stimulated by forskolin were also sensitive to inhibition by adenosine and its analogue L-phenylisopropyladenosine (West et al. 1986), but I_{Ca} stimulated by 30 μM cAMP dialysate was unaffected by A_1 receptor stimulation. These results emphasize the similarity between adenosine and ACh in their actions on heart cells, and call attention to activation of similar intracellular regulatory mechanisms.

4.5.2 Atrial Natriuretic Factors

Atrial natriuretic factors (ANF) are secreted by cardiac atrial cells and have potent regulatory actions on body salt balance, water balance, and blood pressure (DeBold 1985; Cantin and Genest 1985). Huang et al. (1986) suggested that cGMP is the second messenger for ANF, and Cramb et al. (1987) have shown that ANF elevates cGMP in rabbit ventricular myocytes. Gisbert and Fischmeister (1988) investigated whether ANF has a profile of action on I_{Ca} that is similar to those of ACh and cGMP in heart cells (see Sects. 4.4.1, 4.4.2). They applied extracellular synthetic ANF to frog ventricular myocytes and found negligible effects on basal I_{Ca}, even at concentrations as high as 200 nM. However, 3 nM ANF depressed catecholamine-stimulated I_{Ca} by 33% within 3–5 min and had a somewhat smaller inhibitory effect on cAMP-stimulated current. The latter antagonism by ANF was not prevented by atropine, but was partially blocked by pretreatment with IBMX. These actions bear a strong resemblance to those of ACh and cGMP under similar circumstances and suggested to Gisbert and Fischmeister that cGMP-stimulated cyclic nucleotide phosphodiesterase (Martins et al. 1982; Harrison et al. 1986) was at least partially responsible for ANF action, although an inhibitory effect by the peptide on cardiac adenyl cyclase (Anand-Srivastava and Cantin 1986) might also be involved.

ANF also antagonizes contractions induced by norepinephrine and angiotensin II in vascular smooth muscle (Fujii et al. 1986; Taylor and Meisheri 1986) but the basis of this inhibition has not yet been resolved.

4.5.3 Angiotensin II

Angiotensin II (Ang II) is an important hormone that in nanomolar concentrations stimulates neurosecretion, vascular constriction (Peach 1977), and myocardial contractile force (in some species; see Koch-Weser 1965; Dempsey et al. 1971; Bonnardeaux and Regoli 1974). The mode of action of this peptide was obscure for a long time but now seems certain to be related to a stimulation of phosphoinositide hydrolysis, as observed in liver cells (Creba et al. 1983), pituitary cells (Canonico and MacLeod 1986), and vascular smooth muscle cells (Alexander et al. 1985).

The first solid indication that Ang II can affect cardiac Ca channels was reported by Kass and Blair (1981). They found that nanomolar concentrations of the peptide increased Ca-dependent net inward current and contractile force during step depolarizations of calf Purkinje fibers. These D 600-sensitive effects were not blocked by 1 μM propranolol, indicating that the effects were not related to β-adrenergic receptor stimulation. There is now good evidence showing that there are specific high-affinity Ang II receptors in myocardial sarcolemmal membranes (Wright et al. 1983; Baker et al. 1984; Rogers 1984), and a recent study by Allen et al. (1988) indicates that Ang II has no effect on basal or isoproterenol-stimulated adenyl cyclase activity or cAMP levels in neonatal rat ventricular myocytes. However, Allen et al. (1988) found that Ang II increased inositol monophosphate (IP) and inositol biphosphate (IP$_2$) within 30 s, and that these elevations were sustained throughout the observation period (10 min) and were unaffected by pretreatment with 10 μM nifedipine. In addition, L-type I_{Ca} was increased by more that 50% with little change in kinetics. The increase in I_{Ca} was accompanied by a 100% increase in the frequency of spontaneous beating and a paradoxical 50% reduction in contraction amplitude that was unrelated to the change in frequency. Thus, the response to Ang II was nearly identical to that elicited by TPA in similar cells (Dösemeci et al. 1988) and strongly suggests that the hormone acts via the phosphoinositide-PKC system.

Baker and Singer (1988) have identified and characterized high-affinity Ang II receptors in guinea atria and ventricle. Since nonhydrolysable analogues of GTP had pronounced effects on the binding of ^{125}I-Ang II, they implicated a G-type protein in the coupling of the agonist stimulation of phosphoinositide hydrolysis. In contrast to the inhibition of rat ventricular contractility noted above, Baker and Singer (1988) reported that Ang II had no effect on guinea pig heart contractility.

Bkaily et al. (1988) showed in single rabbit aortic cells that Ang II (10^{-8} M) increases Ba-carried current through Ca channels and displaces the peak I-V relation to more negative potentials. Simultaneously, the potassium current (I_k) in these cells was blocked. The Ang II antagonist [Leu^8] Ang II at 10^{-8} M rapidly reversed the effects of Ang II on I_{Ba} and I_k. The authors speculated that Ca channel augmentation and K channel block by Ang II may be due to IP$_3$ formation and/or C kinase activation or to direct effects of Ang II, and may explain a part of the vasoconstrictor action of this hormone in vascular smooth muscle.

4.6 Direct Regulation of Calcium Channels by G Proteins

In the foregoing sections we have reviewed the evidence that extracellular signals such as neurotransmitters and hormones can activate receptor-associated G proteins which dictate the levels of intracellular signal molecules such as cAMP, cGMP, DAG, and IP_3. In turn, at least some of these messengers activate protein kinases which modulate Ca channel activity by phosphorylation of Ca channel protein. However, there is growing evidence showing that G proteins may also be able to regulate ionic channels *without* the involvement of cytosolic protein kinases. In neuronal tissue, for example, there are indications that G_i and G_o may exert direct effects on Ca channels (Scott and Dolphin 1987; Hescheler et al. 1988; Miller 1988), and it is not ruled out that a muscarinic-activated, PTX-insensitive G protein (G_p?) can also act directly (Wanke et al. 1987).

In muscle cells, the search for direct regulation of Ca channel activity by G proteins has been led by Brown, Birnbaumer, and colleagues. They began by demonstrating that muscarinic receptor-regulated K channels in cardiac atrial myocytes were under direct control of a PTX-sensitive G protein which they termed G_k (Yatani et al. 1987b; Codina et al. 1987) – G_k is presumably a G_i-like moiety. The direct action, independent of protein kinase phosphorylation, was proven by applying GTP-γ-S-activated G_k protein to the cytoplasmic side of isolated membrane patches while monitoring single channel K currents (Brown and Birnbaumer 1988).

A similar method was employed to investigate the possibility of a direct action by G_s on Ca channels in membrane patches excised from guinea pig ventricular myocytes (Yatani et al. 1987a). Under control conditions, single Ca channel activity was quickly extinguished upon patch excision, confirming earlier observations on excised cardiac membrane patches (Cavalié et al. 1983). Although Ca channel activity was somewhat enhanced and prolonged when isoproterenol was included in the pipette and bath solutions, the further addition of GTP or GTP-γ-S produced an immediate pronounced stimulation. This stimulation could not be duplicated in the absence of guanine nucleotides, even when the adenyl cyclase-cAMP system was maximally stimulated. G_s purified from human erythrocytes and activated by GTP-γ-S reproduced the stimulation, as did the activated subunit α_s^*, but their unactivated counterparts or preactivated G_i-like protein did not. The facilitatory role of isoproterenol in the bath solution was assumable by Bay K8644.

Further investigation by this group confirmed an action of G_s^* and α_s^* on Ca channels from bovine ventricular sarcolemmal vesicles incorporated into planar lipid bilayers (Imoto et al. 1988). In the presence of Bay K8644, G_s^* and α_s^* increased P_o by an average of fivefold during 2–4-min observation periods, despite a background of declining control P_o with time. Latency to first opening on depolarizing pulses was unchanged by G protein, but inactivation was greatly slowed as judged by enhanced reopening during the second half of the pulses. *

G_s can also increase the opening probability of Ca channels from rabbit muscle T-tubular membrane vesicles incorporated into lipid bilayers (Yatani et al. 1988). In this model system, the predominant channel opening had a conductance of 10 pS (100 mM Ba), and there were smaller contributions from 4.6 and 13 pS conductances. The proportion of open time (P_o) for the N channels in the patch (NP_o) stimulated by Bay K8644 was increased another four fold by either GTP-γ-S-activated G_s or GTP-γ-S-activated α_s subunit. The activated test substances were only effective when applied to the cytoplasmic face of the channel. G_s activated by CTX was also effective in the presence of GTP, whereas the β-γ dimers, nonactivated G_s, and activated G_k failed to enhance Ca channel activity. The addition of GTP-γ-S by itself provoked a large increase in NP_o (Yatani et al. 1988). This result led the authors to conclude that the preparation contained G_s that must be endogenous to skeletal muscle T-tubular membrane. A further indication of the presence of endogenous G_s associated with incorporated channels was that isoproterenol and GTP increased NP_o in three of eight trials under ATP-free conditions.

Another important outcome from the Yatani et al. (1988) study was that prior stimulation of Ca channel opening by Bay K 8644 and phosphorylation (ATP-γ-S, PKA) did not preclude additional activation by G_s^*. Conversely, prior stimulation by DHP or phosphorylation was not a prerequisite for G_s^* activation; G_s^* and α_s^* alone increased extremely low basal activity by 25-fold and 16-fold, respectively.

5 ATP, Calcium, Magnesium, Hydrogen and Other Myoplasmic Factors Affecting Calcium Channel Activity

5.1 Overview

It is clear from Sect. 4 that cAMP-dependent channel phosphorylation, and possibly PKC phosphorylation and G_s association, are important regulators of Ca channel activity. It is equally clear that there are many other cellular constituent that can exert powerful actions on Ca channels. The list includes ATP, Ca_i, Mg_i, and H_i.

All of these factors are affected when cell energetic metabolism is compromised (McDonald and McLeod 1973; Carmeliet 1978; Allen et al. 1985; White and Hartzell 1988) and, as discussed below, primarily in a direction that results in restriction of Ca channel opening. This physiological regulation may well be a question of life and death under untoward circumstances.

---

* See p. 206 for text added in proof.

For instance, a temporary shutting down of Ca channels during an episode of myocardial ischemia could have a major bearing on the survival of Ca-overloaded cells.

There are probably other unknown cytoplasmic factors that affect Ca channels and that are at least in part responsible for the phenomenon of "rundown" or "washout" of L-type Ca channels that is a common finding in a wide variety of dialyzed cells (Hagiwara and Byerly 1981; Fenwick et al. 1982; Kostyuk 1984) and is particularly acute in excised membrane patches (Cavalié et al. 1983; Nilius et al. 1985; Yatani et al. 1987a). In this section, the possible regulatory roles of ATP, Ca_i, and Mg_i are reviewed first. Rundown, which is intertwined with ATP and Ca_i, is discussed after that, and we close with a subsection on internal protons.

5.2 Modulation by ATP

Intact ventricular myocytes respond to β-adrenergic stimulation with an increase in cAMP (Powell and Twist 1976). Pursuant to the elevation of cAMP, there is a requirement for ATP to phosphorylate the Ca channel; putative channel proteins incorporate phosphate in the presence of ^{32}P-ATP and the catalytic subunit of cAMP protein kinase (Flockerzi et al. 1986), and Ca current in cultured guinea pig atrial cells cannot be elicited unless cAMP and ATP are included in the cell dialysate (Bechem and Pott 1985).

The dependence of cardiac Ca current on intracellular ATP has been examined in a number of different ways. After guinea pig ventricular myocytes were superfused with glucose-free solution containing cyanide, a procedure expected to reduce ATP concentration by as much as 80% (McDonald and MacLeod 1973; Hayashi et al. 1987), the injection of ATP restored the greatly shortened action potential and increased Ca current (Taniguchi et al. 1983). In similar myocytes, Irisawa and Kokubun (1983) reported that Ca current increased about 2.4-fold when the ATP concentration of the intracellular dialysate was raised from 2 to 9.5 mM; the dialysate contained a constant 5 mM EGTA and 30 μM cAMP. When ATP concentration was varied over the range 0.5–20 mM, Ca current was maximum at about 5 mM and less than 10% maximum at 0.5 mM ATP; in the absence of 10 mM creatine phosphate, higher ATP concentrations were required for the maintenance of I_{Ca} (Noma and Shibasaki 1985). Inclusion of 5 mM ATP in the pipette solution augmented I_{Ca} and slowed inactivation in smooth muscle cells from rabbit portal vein (Ohya et al. 1987).

Intracellular dialysis of guinea pig ventricular myocytes with ATP analogues has also underlined a role for ATP in Ca channel function (Kameyama et al. 1986a; Trautwein et al. 1986). For example, infusion of ATP-γ-S, which on utilization leads to thiophosphorylated protein that is resistant to dephosphorylation, produced a doubling of I_{Ca} within 10 min and, when isoprenaline was applied, a larger than normal increment in peak current. Conversely, adenylimidodiphosphate, an ATP analogue whose γ-phosphate is not available for protein phosphorylation, markedly depressed the Ca current augmentation by isoprenaline.

The foregoing indicates that ATP is important for Ca channel activity. However, there is almost irrefutable evidence that ATP, even in the presence of cAMP-phosphorylating machinery, is either insufficient for the maintenance of normal L-type Ca channel activity or cannot prevent its suppression by other factors. The evidence comes from experiments on excised cardiac membrane patches. L-type channel activity in such patches only lasts for a few minutes when the cytoplasmic membrane face is bathed with simple intracellular-like solutions (Cavalié et al. 1983; Nilius et al. 1985). The addition of ATP and PKA does not prevent the extinction of activity (Kameyama et al. 1987; Yatani et al. 1987a), whereas other "cocktails" are more helpful (see Sect. 5.5).

5.3 Modulation by Intracellular Calcium

5.3.1 Inhibition of Calcium Channel Activity

Most of the material in this subsection deals with the effects of acute changes in Ca_i. To some extent, the effects of acute changes in Ca_i can provide insight into the Ca-induced inactivation process. However, it cannot be expected that they duplicate the scene set by *transient* elevation of Ca_i in and around the Ca channel (see Sect. 2.3.2). Likewise, *chronic* Ca overload may be quite a different situation from acutely high Ca_i, and it is deferred to Sect. 5.5.

In the absence of EGTA in the solution dialyzing guinea pig ventricular myocytes, Irisawa and Kokubun (1983) observed a rapid disappearence of Ca current elicited by pulses to 0 mV. Since millimolar Ca_o and micromolar Ca_i (nominally Ca-free dialysate) provide a strong driving force for inward Ca current, the rapid abolition of current in the absence of EGTA was a strong indication that Ca_i exerted effects independent of changes in driving force. Similar conclusions concerning Ca_i suppression of Ca channel current had been reached from studies on noncardiac cells (Kostyuk and Krishtal 1977; Hagiwara and Byerly 1981; Plant et al. 1983). In their subsequent study on guinea pig ventricular myocytes, Kokubun and Irisawa (1984) found that increasing Ca_i (EGTA-buffered dialysates) from pCa (negative logarithm of the calcium ion concentration) 9 to pCa 7.4 and pCa 6.8 reduced Ca current by 20% and 85%, respectively. The depression of Ca current upon elevating Ca_i was also observed when Sr replaced Ca as the charge carrier. Since kinetics and steady-state inactivation were unaffected by Ca_i, they suggested that high Ca_i reduces calcium conductance (g_{Ca}) by reducing the number of functional channels, the single channel conductance, and/or the single channel opening probability.

High Ca_i also depresses Ca channel current in smooth muscle cells. As with most other cell types, millimolar concentrations of EGTA appear to be required for satisfactory experimental investigation of I_{Ca} (e.g., Ganitkevich et al. 1986; Benham et al. 1987; Ohya et al. 1987; Fish et al. 1988). Ohya et al. (1988) investigated the relation between I_{Ca} and Ca_i in isolated rabbit portal vein cells by dialyzing cells with 10 mM EGTA-buffered Ca solutions. I_{Ca} was reversibly inhibited by high Ca_i dialysates, the K_D being about 0.1 μM (suggesting significant steady-state Ca-induced inactivation under physiological conditions). Inactivation curves were not shifted by 1 μM Ca_i, a concentration which reduced I_{Ca} to 10% maximum. Ba-carried currents were inhibited to the same extent as I_{Ca}, but even 100 μM Ba_i had no effect on I_{Ca}. The latter result agrees with that of Brown et al. (1981) who reported that, in contrast to Ca_i (half-inhibition at 1 μM), Ba_i (up to 1 mM) had no effect on I_{Ca} in *Helix aspersa* neurones.

The mechanism of Ca_i-induced inhibition of Ca channel current is still not known. Plant et al. (1983) have proposed that there is a Ca binding site linked to the inactivation mechanism near the inside mouth of the Ca channel. Indeed, Tanabe et al. (1987) have pointed out that a glutamate-/aspartate-rich region within the amino acid sequence of the putative Ca channel protein is a plausible Ca ion binding site. Since the intracellular application of trypsin increases I_{Ca} and delays its inactivation, Hescheler and Trautwein (1988) proposed that the tryptic action might be on the Ca ion binding site.

A major problem has now arisen with regard to the postulated Ca ion binding site: Ba currents through single cardiac Ca channels in bilayers were depressed by millimolar Ca_i (as expected from Ca-Ba interaction in the pore) but channel activity was clearly present, and intrinsic inactivation was unchanged by Ca_i as high as 10 mM (Rosenberg et al. 1988). Given the overwhelming evidence that Ca_i inhibits Ca channel current in intact cells, Rosenberg et al. (1988) concluded that the absence of Ca_i dependence must have been due to the loss of a cytoplasmic factor or membrane component loosely associated with the Ca channel, perhaps a Ca-dependent phosphatase (cf. Chad and Eckert 1986; see also Sect. 5.5 below).

5.3.2 Facilitation of Calcium Channel Activity

Despite the conclusive evidence that Ca entry and Ca_i inhibit Ca channel activity, there are also experimental results pointing to a facilitation of I_{Ca} by Ca_i, as though there were a bell-shaped relation between Ca_i and Ca channel conductance (see Marban and Tsien 1982; McDonald 1982). It is possible that in some of the studies, including those indicating a postrest positive staircase increase in frog atrial I_{Ca} (Noble and Shimoni 1981; Shimoni 1981) and in rat atrial I_{Ca} (Payet et al. 1981), undetected changes in overlapping K currents or residual I_{Na} produced an impression that I_{Ca} was facilitated. However, there are two recent reports of the phenomenon existing in ventricular myocytes dialyzed with high Cs solution and bathed in Na, K-free solution. Lee (1987) observed a positive I_{Ca} staircase in guinea pig myocytes repetitively pulsed from -90 mV to voltages between -10 and $+30$ mV. Tseng (1988) observed up to a twofold larger I_{Ca} on the second of paired pulses, as well as a positive postrest staircase, in guinea pig and dog ventricular myocytes. The facilitation was absent or inverted when the holding potential was positive to -60 mV (the holding potential usually used in studies on myocytes bathed in Na-containing solutions). Both authors noted that potentiated I_{Ca} had a much slower rate of decay, and one also noted a lengthening in time to peak current (Lee 1987). Facilitation of I_{Ca} has also been observed in molluscan neurones (Heyer and Lux 1976) and in chromaffin cells (Hoshi et al. 1984).

Noble and Shimoni (1981) felt that the potentiation of I_{Ca} was connected to both Ca_i and membrane depolarization. Lee (1987) discounted a role for Ca_i since he also observed facilitation of Ba-carried current. He concluded that repetition of (normal) Ca channel opening enhances transitions into a second long-lived open state. Conversely, Tseng (1988) specifically implicated Ca_i, one reason being that facilitation was almost abolished when Ba replaced Ca_o. Fedida et al. (1988 a, b) have also found that postrest stimulation of Ca channel current in guinea pig ventricular myocytes is enhanced by increasing Ca_o and abolished in external Ba solution. Since they also observed slowed inactivation of facilitated I_{Ca}, a feature of I_{Ca} enhanced by β-adrenergic stimulation (Tsien et al. 1986), they concluded that the entry of Ca might enhance Ca channel phosphorylation. If the Ca/Ba feature is substantiated in future work on cells sensitive to I_{Ca} enhancement by Ca-dependent PKC activation, it would suggest the existence of a physiological positive feedback loop that in muscle cells would not necessarily require the participation of an agonist. Where an agonist was involved, the system would only require a short kick-start (see also remarks in De Riemer et al. 1985). The system could be held in check by feedback down-regulation of the phosphoinositide cascade (see Sect. 4.4.3) and by Ca-induced inactivation.

5.4 Intracellular Magnesium

There is little information on whether Mg_i affects I_{Ca} in muscle cells. One recent paper (White and Hartzell 1988) suggests that the level of Mg_i in heart cells (Hess et al. 1982) may regulate I_{Ca} under particular conditions. When free Mg_i was varied over 0.3 to 3 mM, basal I_{Ca} in frog cardiomyocytes was unaffected. However, I_{Ca} elevated by isoproterenol, cAMP, or the catalytic subunit of PKA, was reduced by 50% when Mg_i was 3 mM. By contrast, I_{Ca} elevated by DHP agonists was affected to a smaller degree. White and Hartzell (1988) concluded that Mg_i might have a direct effect on the phosphory-lated channel, or that it may act by regulating the activity of a protein phos-phatase. In guinea pig cardiomyocytes, Agus et al. (1989) reported that $9.4 \, mM$ Mg_i almost completely blocked I_{Ca} due to enhanced steady-state inactivation. The effect was observed on both basal and phosphorylation-enhanced I_{Ca}.

5.5 The Rundown of Calcium Channels

When cells are dialyzed with an artificial intracellular solution, there is al-most invariably a "rundown" or "washout" of L-type Ca channel activity (N-type but not T-type Ca channels also run down: Nilius et al. 1985; Nowycky et al. 1985; Carbone and Lux 1987a,b, 1988; Fox et al. 1987b). Independent of the charge carrier, the amplitude of whole-cell Ca channel current becomes progressively smaller with time. Complete rundown of L-type current can oc-cur in less than 10 min (Irisawa and Kokubun 1983; Klöckner and Isenberg 1985; Chad and Eckert 1986), and efforts at preventing rundown have fo-cussed on the inclusion of cAMP, PKA, and ATP, as well as on the provision of a large (EGTA) and/or fast (BAPTA, citrate) Ca-buffering capability in the dialysate. A pertinent subjective observation is that L-type Ca channel current is smaller and more susceptible to washout in unhealthy cardiac (Belles et al. 1988a) and smooth muscle (Benham et al. 1987) cells.

The precise reason for the rundown of whole-cell Ca channel current is not known. The inclusion of the agents mentioned above is aimed at counteracting the probable dilution of cellular phosphorylation factors as well as the mainte-nance of low Ca_i. Poor buffering of Ca_i, which can be the case even with $10 \, mM$ EGTA in the pipette solution (Byerly and Moody 1984; Bechem and Pott 1985), may activate Ca-dependent proteases and lead to proteolytic degra-dation of Ca channels (Chad and Eckert 1986). However, there appears to be an absolute requirement for ATP as well (Byerly and Moody 1984; Klöckner and Isenberg 1985; Chad and Eckert 1986; Belles et al. 1988a), whether for channel phosphorylation, Ca_i regulation, or phosphorylation-conferred pro-tection of proteins against enzymatic hydrolysis (Belles et al. 1988a).

Chad and Eckert (1986) provided substance for the possibility that, in neu-rones at least, Ca-dependent proteases may be a major factor in rundown. They demonstrated that leupeptin, a Ca-dependent protease inhibitor, was a very effective antidote against rundown, especially when combined with ATP. A similar hypothesis was explored in heart cells, but here the extralysosomal Ca-dependent proteases calpain I and II were implicated. Calpain is abundant in muscle cells, and is inhibited by calpastatin (Barth and Elce 1981; Mellgren 1987). In guinea pig ventricular myocytes dialyzed with $0.1 \, mM$ EGTA solu-tion, Belles et al. (1988b) found that calpain accelerated rundown in a con-centration-dependent manner, whereas calpastatin retarded the rate of control rundown. The authors concluded that calpain might reduce Ca channel activity during myocardial metabolic dysfunction, and might also be involved in normal degradation and turnover of the channels.

Crude myoplasmic extracts of homogenized rabbit skeletal muscle slowed the rundown of Ca channels in guinea pig ventricular myocytes (Belles et al. 1988a). Since intracellular dialysis with extract and 0.1 mM EGTA was more effective than dialysis with 10 mM EGTA, the effectiveness of the extract may have been due to improved buffering of Ca_i, perhaps by calmodulin (Belles et al. 1988a).

Despite the evidence suggesting that Ca-dependent protease is responsible for the rundown of whole-cell Ca channel current, a different mechanism causes the rundown of isolated Ca channels. As noted above, Ca did not cause the rundown of cardiac Ca channels in bilayers, yet rundown was clearly evident (Rosenberg et al. 1988). Conversely, the lifetime of L-type channels in excised cardiac membrane patches is not lengthened by superfusion of the cytoplasmic face with high EGTA and protease inhibitors (e.g., Nilius et al. 1985; Kaibara and Kameyama 1988). It is also not lengthened (but see Armstrong and Eckert 1987) by application of PKA (Kameyama et al. 1987; Yatani et al. 1987a). These negative results led Kameyama et al. (1988) to look for life-lengthening factors in the supernatant fraction of homogenized guinea pig and bovine ventricle. The rapid rundown of single channel activity in inside-out patches from guinea pig ventricular myocytes was slowed or completely prevented in seven of ten experiments, with particularly good results when 10 µM cAMP and 3 mM MgATP were included in the 0.1 mM EGTA/tissue extract solution. Rundown and complete recovery was fast and repeatable upon exchanging tissue extract and control solutions. The extract was inactivated after heating at 80°C or by trypsin treatment, and experiments with fractionated extracts localized the channel-enhancing activity to a fraction with an apparent molecular weight of 200–300 kDa. These effects seem to have been independent of GTP-binding protein activation, and perhaps also of PKC activation, both of which prolong isolated Ca channel lifetime and/or stimulate low basal activity (Yatani et al. 1987a; Imoto et al. 1988; Yatani et al. 1988).

5.6 Modulation by Intracellular Protons

The final intracellular regulator to be considered here is the H ion. Kurachi (1982) found that the pressure injection of a pH 9.3 solution into guinea pig ventricular myocytes lengthened the action potential and increased I_{Ca}; the opposite effects were obtained when low pH (3.7–4.7) solutions were injected. Quantitative studies on dialyzed guinea pig ventricular cells gave the following results (Sato et al. 1985; Irisawa and Sato 1986). – When the dialysis pipette contained solution at pH 5.5 or higher, the influx of protons into the cell was counterbalanced by proton efflux via the Na-H exchange mechanism. – When the exchanger was blocked with 1 mM amiloride or Na-free external solution, I_{Ca} could be completely suppressed by protons with half-maximal inhibition at pH 6.7. Under these conditions, there were no appreciable shifts in I-V relations or in the inactivation time courses.

Kaibara and Kameyama (1988) examined the effects of intracellular protons on I_{Ba} through single Ca channels in guinea pig ventricular myocytes. The patches were cell-attached, and efficient exchange of intracellular fluid and external test solution was achieved after mechanical disruption of cell segments distal to the centrally located patch. Changes in pH between 8.2 and 5.0 had no effect on the number of functional channels in a patch and only a small effect on the elementary current amplitude. However, open-state probability declined at pH <8, with half-maximum inhibition near pH 6.6. The curve relating p_o to internal pH (pH_i) had a Hill coefficient of one, suggesting that a proton-binding site having a pK of 6.6 might be involved in the H_i-induced reduction of p_o. When pH_i was changed from 6.2 to 7.2−7.4 and back to 6.2, the alkalinization produced a two- to threefold increase in p_o and average current amplitude. Although both the steady-state activation and inactivation curves were shifted by about 10 mV in the negative potential direction, these shifts were not the cause of the acid-induced inhibition. Further, the inhibition was not due to changes in fast kinetics since mean open and closed times only changed by about 10% between pH_i 8.2 and 6.2. The primary effect was rather on the slow gating of the channel as observed by a marked increase in the percentage of blanks. When pH_i was lowered from 8.2 to 6.2, the percentage of blanks in an ensemble climbed from 27% to 60%, and the average lifetime of runs of blank sweeps increased from 3.3 to 5.0 s, whereas the average lifetime of runs of nonblank records declined from 9.2 to 3.3 s.

Although the action of low pH_i on slow gating is reminiscent of the action of organic Ca channel blockers and opposite to that of channel phosphorylation (cf. Trautwein and Pelzer 1985a; Tsien et al. 1986), Kaibara and Kameyama (1988) concluded that the spectrum of low-pH_i effects was inconsistent with binding to a DHP site or with dephosphorylation.

6 The Structure of Dihydropyridine-Sensitive Calcium Channels

6.1 Overview

Molecular biology provides a powerful tool for structural characterization of ionic channels. Cloning of different types of Ca channels in different cells may eventually reveal the structural basis for their distinct electrophysiological properties and patterns of regulation in different tissues (see Sect. 7). Since the low density of binding sites for Ca channel blockers in nonskeletal muscle tissues makes purification from such sources very difficult (Janis et al. 1987; Campbell et al. 1988d; Catterall 1988; Glossmann and Striessnig 1988; Hosey and Lazdunski 1988), skeletal muscle Ca antagonist receptor complexes have been important for molecular biology approaches. In comparison to what is known about the molecular structure of Ca channels in skeletal muscle, the molecular analysis of Ca channels in cardiac and smooth muscle as well as in the brain is still in its early infancy (see Glossmann and Striessnig 1990).

6.2 Cloning and Expression of the Skeletal Muscle Dihydropyridine-Sensitive Calcium Channel

Numa's group recently reported the primary structure of the α_1 subunit ($M_r = 170$ kDa) of the rabbit muscle DHP receptor (DHPR) complex as deduced from its complementary DNA sequence (Tanabe et al. 1987). This polypeptide contains 1873 amino acids, and four distinct internal repeats that exhibit sequence similarity were identified. Based on the hydropathy pattern, it was assumed that these four units span the membrane and form the channel pore. Each unit consists of five hydrophobic (S1, S2, S3, S5, and S6) segments and one hydrophilic (S4) segment, all of which are thought to be membrane-spanning α helices. S4 is thought to form a positively charged spiral (possible voltage sensor). One region in the α_1 subunit (assigned to the cytoplasmic side, residues 740–752) and the sequence of residues 155–164 (assigned mostly to the extracellular side) resembled the EF hand (the calmodulin fold, or EF-hand, consists of an α helix ("E" of the EF-hand), symbolized by the forefinger of a right hand, a loop about the Ca ion (bent mid finger), and a second α helix ("F" of the EF-hand), the symbolic thumb) and the consensus repeat of Ca-dependent membrane binding proteins (Kretzinger et al. 1988), respectively (possible cation binding sites along the path of ion permeation).

According to the Tanabe et al. (1987) model, two out of five potential N-glycosylation sites are present on the extracellular side, and all six potential cAMP-dependent phosphorylation sites are located on the cytoplasmic side of the Ca channel. cAMP-dependent protein kinase rapidly phosphorylates serine-687 which is localized between the transmembrane regions II and III (Röhrkasten et al. 1988a, b). Serine-1617, located on the COOH-terminal domain of the α_1 subunit is also phosphorylated, however, with a slow time course. Thus, it was anticipated that phosphorylation of serine-687 (Röhrkasten et al. 1988a, b) affects the opening probability of the skeletal muscle DHPR complex reconstituted as a functional L-type-like Ca channel (Flockerzi et al. 1986; Trautwein et al. 1987; Pelzer et al. 1988).

Although the primary structure of the α_1 subunit of the DHPR complex from skeletal muscle, deduced by cloning and sequence analysis of DNA complementary to its mRNA (Tanabe et al. 1987), suggested that the α_1 polypeptide ($M_r = 170$ kDa) might be both the voltage sensor for excitation-contraction (EC) coupling as well as a functional Ca channel, this suggestion was at that time difficult to prove. Initially, it was not possible to obtain a functional Ca channel by corresponding cDNA expression studies in *Xenopus* oocytes. However, a collaboration between Numa's group and Beam's group has now been successful (Tanabe et al. 1988). In cultured skeletal muscle cells from mice with muscular dysgenesis, these investigators were able to restore EC coupling and DHP-sensitive Ca channel current by microinjection of an expression plasmid (pCAC6) carrying the DHPR cDNA.

They first had to identify the genetic defect that leads to muscular dysgenesis. Normal skeletal muscle cells acutely isolated from vertebrates or embryonic skeletal muscle cells grown in primary tissue culture have a slowly activated Ca current which is blocked by DHP (I_{slow}) and a DHP-insensitive transient Ca current (I_{fast}) (Cognard et al. 1986; Lamb and Walsh 1987; Beam and Knudson 1988a). Besides DHP sensitivity, I_{fast} and I_{slow} in rodent myotubes have properties similar to those of the T- and L-type Ca channel current, respectively, found in many other cell types (see Sect. 7). In comparison to normal skeletal muscle cells, the autosomal recessive trait of the muscular dysgenesis mutation (Gluecksohn-Waelsch 1963) is characterized by failure of EC coupling (Powell and Fambrough 1973; Klaus et al. 1983) and lack of DHP-sensitive slow Ca current in skeletal muscle cells (Beam et al. 1986; Rieger et al. 1987), but not in heart or brain cells (Beam et al. 1986). Tanabe et al. (1988) then analyzed the genomic DNA of normal mice and of animals that were homozygous or heterozygous for the muscular dysgenesis mutation. In diseased mice, they found a difference in two regions of the structural gene encoding the DHPR, and this was responsible for a large reduction of the corresponding messenger RNA. Microinjection of relevant expression plasmid into nuclei of cultured myotubes from dysgenic mice restored spontaneous and electrically-evoked contractions and DHP-sensitive slow Ca current. The contractions were independent of the flow of I_{slow}, indicating that the DHPR has two functions, that of the voltage sensor involved in EC coupling (see Schneider and Chandler 1973; Rios and Brum 1987) and that of a Ca channel (Flockerzi et al. 1986).

It is not clear whether other proteins which copurify with the 170 kDa polypeptide (α_1; see Sect. 6.3) are present in muscle from dysgenic mice and whether they participate in the functional properties of Ca channels. Perez-Reyes et al. (1989) therefore selected a cell line (murine L cells) which has no endogenous Ca currents or α_2 subunit (see Ellis et al. 1988 for primary structure), and probably no δ subunit, for stable transformation with complementary DNA of the α_1 subunit. The transformed cells expressed DHP-sensitive, voltage-gated Ca channels, indicating that the minimum structure of these channels is at most an $\alpha_1\beta\gamma$ complex and possibly an α_1 subunit alone. In line with the latter suggestion, reconstitution experiments of isolated α_1 subunit in lipid bilayers indicate that the α_1 subunit of the skeletal muscle DHP receptor complex is sufficient to serve as a functional voltage-dependent Ca channel which retains sensitivity to D600, BAY K 8644, and cAMP-dependent phosphorylation (Pelzer et al. 1989b).

6.3 Biochemistry and Immunology of the Skeletal Muscle Dihydropyridine-Sensitive Calcium Channel

At least three additional polypeptides from skeletal muscle T-tubules copurify with the α_1 peptide subunit: α_2 ($M_r \approx 145$ kDa) disulfide linked to δ ($M_r <$ 28 kDa), β ($M_r \approx 50$ kDa) and γ ($M_r \approx 30$ kDa) (see Catterall 1988; Glossmann and Striessnig 1988). They form a 1:1:1:1 complex with $M_r \approx$ 400 kDa (Seagar et al. 1988). The main evidence for the existence of a functional complex with this composition comes from immunological studies with antibodies against each of the four polypeptides (cf. Vaghy et al. 1988; Campbell et al. 1988e). Specific antibodies which selectively recognize α_1 (Leung et al. 1987; Morton and Froehner 1987; Takahashi et al. 1987; Fitzpatrick et al. 1988), α_2 (Lazdunski et al. 1987; Norman et al. 1987; Takahashi and Catterall 1987a, b; Vandaele et al. 1987; Sharp et al. 1988), β (Leung et al. 1988), γ (Sharp et al. 1988), and δ (Lazdunski et al. 1987; Vandaele et al. 1987) subunits immunoprecipitate all subunits of drug-receptor complexes, suggesting that these polypeptides are distinct but closely associated under nondenaturating conditions. Cross-reactivity patterns also suggest that the ≈ 50 K β subunit and the ≈ 30 K γ subunit are not fragments of the two larger components α_1 and α_2/δ. Moreover, antibodies against α_1 (Morton et al. 1988), β, and γ (Vilven et al. 1988) subunits modulate Ca channel function. Anti-β enhances Ca channel currents and prevents the inhibitory action of nitrendipine, whereas anti-γ inhibits Ca channel activity. Taken together, it is possible that α_1 must associate with one or more of the other subunits to acquire "skeletal muscle" properties (see Perez-Reyes et al. 1989). *

There is also evidence for cross-reactivity of antibodies against the heart ADP/ATP carrier of the inner mitochondrial membrane with the cardiac L-type Ca channel (Schultheiss et al. 1988). In rat ventricular myocytes, these antibodies specifically increase I_{Ca} amplitude and slow inactivation. Finally, there is growing evidence that endogenous DHP-displacing peptides of low molecular weight (≤ 1 kDa to about 8 kDa) regulate L-type Ca channel currents in neuronal and cardiac cells either directly or by activation of an as yet undefined intracellular mediator (Hanbauer et al. 1988; Janis et al. 1988).

7 Types of Calcium Channels

7.1 Overview

Ca channels are present in the surface membranes of all known excitable cells and are among the most interesting intrinsic membrane proteins controlling

---

* See p. 207 for text added in proof.

transmembrane ion flow and cellular function. They have frequently been grouped into two major types: potential-operated Ca channels (POCC) responsive to changes in membrane potential elicited by chemical or electrical stimuli, and receptor-operated Ca channels (ROCC) sensitive to chemical stimuli but not to membrane potential (e.g., Bolton 1979; van Breemen et al. 1979; Hagiwara and Byerly 1981; Cauvin et al. 1983, 1985; Reuter 1983, 1985, 1987; Tsien 1983, 1987; Högestatt 1984; Bülbring and Tomita 1987; Janis et al. 1987; Miller 1987 a, b; Tsien et al. 1987, 1988).

There are several tissue-specific subtypes of POCCs as identified by their differing voltage sensitivities, conductances, and activation and inactivation kinetics. In addition, subtypes of POCCs are recognized by differences in their responses to organic Ca channel blockers and activators. In particular, the DHP compounds serve as regulatory ligands for at least one major class of POCCs (L-type). ROCCs are problematic to define since, unlike POCCs, they are classified largely by exclusionary criteria, notably their apparent insensitivity to activation by depolarizing stimuli and to the organic Ca channel blockers. ROCCs may simply be a separate category of Ca channels that lack or fail to express voltage sensitivity. Alternatively, their voltage sensitivity remains to be detected, and they do not contain or are not coupled to Ca-antagonist binding sites. They may represent part of a continuum of channels which, depending upon stimulant, degree of activation, or membrane potential, have varying degrees of sensitivity to the Ca channel blockers (e.g., Cauvin et al. 1983). Conversely, ROCCs could be regarded as an entirely separate category of Ca channels. Whether ROCCs are viewed as specific components of single receptors or as multisensitive components associated with several receptors impacts on the search for channel-directed ligands (e.g., Janis et al. 1987).

7.2 Potential-Dependent Calcium Channels

It was long believed that excitable cells possessed a single class of voltage-sensitive Ca channels. The coexistence of multiple types of potential-dependent, Ca-selective channels within a given type of cell was first described in voltage-clamp studies on starfish eggs (Hagiwara et al. 1975). In vertebrate neurons, multiple Ca conductances were also predicted on the basis of voltage recordings (e.g., Llinás and Sugimori 1980: Purkinje cell somata in mammalian cerebellar slices; Llinás and Yarom 1981: mammalian inferior olivary neurones). Subsequent voltage-clamp and patch-clamp experiments demonstrated the presence of two populations of voltage-sensitive Ca channels in many species and tissues, and a third kind of potential-dependent Ca channel has been identified in neuronal and similar membranes (see below). In this review, we keep the classification T-, N- and L-type Ca channels as proposed by Tsien and colleagues (Nilius et al. 1985; Nowycky et al. 1985).

7.2.1 Cell Types with Multiple Kinds of Voltage-Activated Calcium Channels

Bovine adrenal chromaffin cells apparently possess only L-type Ca channels
(Fenwick et al. 1982; Hoshi 1985), whereas neoplastic B lymphocytes possess
only T-like Ca channels (Fukushima and Hagiwara 1985). T- and L-like Ca
channels have been identified in *Polychaete* eggs (Fox and Krasne 1984), ciliates
(Deitmer 1984), chick DRG neurons (Carbone and Lux 1984a, b; Nowycky et
al. 1984), newborn rat DRG neurons (Fedulova et al. 1985), rat petrose and
nodose neurons (Bossu et al. 1985), rat hippocampal pyramidal CA1 cells
(Docherty and Brown 1986), cultured mammalian hippocampal neurons (Yaari
et al. 1987), neuroblastoma (N1E-115) cells (Fishman and Spector 1981;
Narahashi et al. 1987), GH3 (pituitary) cells (Armstrong and Matteson 1985;
Matteson and Armstrong 1986), rat pituitary cells (De Riemer and Sakmann
1986), nerve terminals (Penner and Dreyer 1986), guinea pig pancreatic α_2
cells (Rorsman 1988), and rat pancreatic B cells (Hiriart and Matteson 1988).
Their coexistence has also been demonstrated in cardiomyocytes from dog
atrium (Bean 1985), frog atrium (Bonvallet 1987), and guinea pig ventricle
(Nilius et al. 1985; Mitra and Morad 1986), as well as in smooth muscle cells
from rabbit artery and vein (Bean et al. 1986; Worley et al. 1986; Benham et
al. 1987), A10 aortal line (Friedman et al. 1986), rat portal vein (Loirand et al.
1986), dog saphenous vein (Yatani et al. 1987c), and dog azygous vein (Sturek
and Hermsmeyer 1986). They are found in frog skeletal muscle (Cota and
Stefani 1986), mouse myoblasts (Beam et al. 1986), rat myoballs (Cognard et
al. 1986), embryonic rat and mouse skeletal myotubes (Beam and Knudson
1988a, b), and neonatal rat FDB (flexor digitorum brevis) fibers (Beam and
Knudson 1988a, b), as well as in 3T3 fibroblasts (Chen et al. 1988a, b) and
astrocytes (Barres et al. 1985).

N- and L-type Ca channels coexist in rat (Wanke et al. 1987; Hirning et al.
1988) and frog (Lipscombe and Tsien 1987; Lipscombe et al. 1988) sympathetic
neurons, and the coexistence of all three types of Ca channels, T, N and L, has
been demonstrated in chick DRG neurons (Nowycky et al. 1985; Fox et al.
1987a, b), rat petrose ganglion neurons (Dupont et al. 1986), mouse DRG
neurons (Kostyuck et al. 1987, 1988), rat hippocampal granules (Gray and
Johnston 1986), and rat hippocampal pyramidal CA3 cells (Madison et al.
1987).

7.2.2 Properties, Distribution and Function of T, N- and L-Type Calcium Channels

7.2.2.1 L-Type Calcium Channels

The L-type Ca channel has a low-voltage threshold, slow voltage- and Ca-
dependent inactivation, large unitary Ba conductance (25 pS in 110 mM Ba),

high sensitivity to DHP and other organic Ca channel blockers, and is modulated by a wide variety of neurotransmitters, enzymes, and drugs (e.g., this review and the accompanying reviews by Glossmann and Striessnig 1990 and by Porzig 1990). L-type Ca channels are highly permeable to a number of monovalent cations in the absence of divalent cations; the monovalent currents are blocked by micromolar Ca in a voltage-dependent manner (e.g., Tsien et al. 1987). However, L-type Ca channels show significant differences in their interactions with divalent cations (e.g., Tsien et al. 1987). Moreover, the large family of DHP-sensitive L-type Ca channels with its tissue-specific subtypes cannot always be clearly dissected by pharmacological means and transmitter-induced modulatory responses (e.g., this review and accompanying reviews by Glossmann and Striessnig 1990 and by Porzig 1990).

DHP-sensitive L-type Ca channels are found in all neurons, gland cells, and muscle cells. Their voltage- and time-dependent kinetics facilitate the conversion of membrane depolarization into an intracellular Ca signal that can trigger cellular responses. Examples of such voltage-response transductions include excitation-contraction coupling in heart and smooth muscle, activation of glycolytic metabolism in skeletal muscle, excitation-secretion coupling in endocrine and exocrine cells, and neurotransmitter release from peripheral neurons (e.g., Godfraind et al. 1986; Perney et al. 1986; Janis et al. 1987; Rane et al. 1987).

7.2.2.2 N-Type Calcium Channels

The N-type Ca channel has an activation threshold that is intermediate between T- and L-type channels, inactivates fairly rapidly, has an inactivation V_h intermediate between that measured for T- and L-type channels, an intermediate unitary slope conductance ($\approx$ 13 pS with 110 mM Ba), sensitivity to ω-CgTX VIA, Cd, and various neurochemicals, and resistance to Ni and DHP (e.g., Nowycky et al. 1985; Fox et al. 1987a, b; McCleskey et al. 1987; Miller 1987a, b; Tsien et al. 1988). Flux measurements together with electrical recordings in brain synaptosomes and neurons from dorsal root and sympathetic ganglia suggest that N-type Ca channels are quite similar to L-type Ca channels with regard to ion permeation (e.g., Tsien et al. 1987).

It is possible that the expression of N-type channels is limited to neuronal or similar membranes (e.g., Fox et al. 1987b; Miller 1987a, b; Tsien et al. 1988). The pharmacological properties of N-type Ca channels, sensitivity to Cd and ω-CgTX VIA, and insensitivity to DHP, are in line with the characteristics of Ca entry pathways underlying transmitter release from sympathetic neurons (Hirning et al. 1988), motor nerve terminals (Kerr and Yoshikami 1984; Quastel et al. 1986), synaptosomes (Nachshen 1985; Reynolds et al. 1986), and brain slices (Middlemiss and Spedding 1985). Selective block of N-type Ca channels has been found in mouse DRG neurons by dynorphin

(Gross and Macdonald 1987), in rat hippocampal neurons by adenosine (Madison et al. 1987), in frog sympathetic neurons by noradrenaline (Lipscombe and Tsien 1987; Lipscombe et al. 1988), and in rat and bullfrog DRG and sympathetic neurons by norepinephrine, somatostatin and dynorphin (Bean 1989). In the latter study, the reduction of current was strongly voltage-dependent with large effects on currents activated by small and moderate depolarizations and much smaller (or no) effects on currents activated by large depolarizations. Inhibition of N-type Ca channels may provide a simple explanation for the ability of the above-mentioned agents to inhibit transmitter release (e.g., Miller 1987 a, b; Hirning et al. 1988; Tsien et al. 1988).

7.2.2.3 T-Type Calcium Channels

The T-type Ca channel can be activated with weak depolarizations, undergoes rapid voltage-dependent inactivation, has a relative insensitivity to DHP and other organic Ca agonists and antagonists, as well as a small unitary conductance (8 pS in 110 mM Ba; e.g., Carbone and Lux 1984 a, b; Bean 1985; Nilius et al. 1985; Nowycky et al. 1985; Mitra and Morad 1986; Carbone and Lux 1987 a, b, 1988; Fox et al. 1987 a, b).

T-type Ca channels resemble L-type Ca channels in that they are permeable to Na in the absence of divalent cations and are blocked by micromolar Ca in a voltage-dependent manner (e.g., Carbone and Lux 1987 b; Tsien et al. 1987; Lux et al. 1988). T-type Ca channels show equal sensitivity to block by Ca ions as L-type Ca channels (e.g., Bean 1985). However, unlike L-type Ca channels, T-type Ca channels have equal unitary conductances for Ca and Ba (e.g., Nilius et al. 1985). They are more resistant to block by Cd ions than L-type channels (e.g., Nilius et al. 1985; Nowycky et al. 1985; Narahashi et al. 1987), and block by ω-CgTX VIA is weak and reversible (e.g., Fox et al. 1987 a, b). In addition, they are largely insensitive to the application of β-adrenergic agonists (Bean 1985; Mitra and Morad 1986; Tytgat et al. 1988) but are preferentially sensitive to block by Ni ions (Hagiwara et al. 1988) and amiloride (Tang and Morad 1988; Tang et al. 1988). T-channel activity is long-lived in detached membrane patches (e.g., Nilius et al. 1985; Carbone and Lux 1987 b), in marked contrast to L-type channel activity (e.g., Fenwick et al. 1982; Cavalié et al. 1983; Nilius et al. 1985).

T-type Ca channels are found along with L-type Ca channels in a wide variety of neuronal, endocrine, exocrine, heart, smooth muscle, and skeletal muscle cells (see above). The cellular distribution and properties (activation at negative membrane potentials) of T-type Ca channels suggest a possible role in pacemaker activity and rebound excitation (e.g., Llinás and Yarom 1981; Hagiwara et al. 1988).

T-type Ca channels are also found in certain rapidly proliferating cells such as neoplastic B lymphocytes (Fukushima and Hagiwara 1985), astrocytes

(Barres et al. 1985), and 3T3 fibroblasts (Chen et al. 1988a, b), but their significance remains unclear.

7.2.3 Relationship Between Various Channel Types

All three types of Ca channels are fairly similar in at least two fundamental respects (Fox et al. 1987b). Open-closed kinetics are characterized by a predominant mode of gating with brief openings that occur in bursts (mode 1 by the definition of Hess et al. 1984), and occasional transitions into modes of gating with much longer openings (mode 2 by the definition of Hess et al. 1984). Furthermore, the selectivity for Ca seems to involve high-affinity Ca binding sites along the path of ion permeation.

Fox and colleagues (1987b) state that no interconversion between channel types has yet been detected in the course of continuous cell-attached recordings. In the light of the significant differences in conductance, kinetics, and pharmacology among the various channel types, they view them as different but closely related proteins. A similar conclusion has been reached from bilayer recordings of multiple Ca channels with different conductances, kinetics, and pharmacology, even when these channels were reconstituted from highly purified skeletal muscle DHP receptor complex preparations (Talvenheimo et al. 1987; Pelzer et al. 1988). However, this hypothesis awaits direct verification from structural analysis.

There are already examples of voltage-gated Ca channels in vertebrate preparations that are difficult to categorize as T-, N-, or L-type (e.g., Fox et al. 1987b; Rosenberg et al. 1988). Thus, a substantial revision of the above classification is to be expected as new information on Ca channel pharmacology, gating, and unitary conductances gradually accumulates.

7.3 Receptor-Operated Calcium Channels

The existence of ROCCs was suggested on the basis of smooth muscle responses which seemed to depend on Ca entry when membrane potential did not change, and/or when POCCs were blocked by "specific Ca channel blockers" (e.g., Bolton 1979; van Breemen et al. 1979). ROCCs have also been invoked to explain Ca entry into cells that are not electrically excitable (e.g., thrombin-stimulated blood platelets). Although ROCCs are an attractive way of explaining phenomena such as voltage-independent and blocker-insensitive Ca entry, it has never been really clear whether these phenomena could be explained by other means (see Janis et al. 1987; Rink 1988). In recent studies on cells that seem to lack voltage-gated Ca entry (e.g., platelets, parotid gland, and endothelial cells), Rink and Hallam (1984) and Merritt and Rink (1987) observed that receptor-mediated elevation of cytosolic Ca is greater and more prolonged in the presence of external Ca than in a Ca-free medium. This suggested, but did not prove (Hallam and Rink 1985), that receptor occupation stimulates Ca entry through some form of ROCC. Additional support for this hypothesis comes from demonstrations of stimulated uptakes of radiolabelled Ca (Massini and Lüscher 1976) and Mn (Benham and Tsien 1988 b). Recordings of Ca currents through single channels that show no voltage-dependent gating and are insensitive to "Ca antagonists" is one way of demonstrating the existence of ROCCs. The ATP channel examined by patch clamp in isolated smooth muscle cells (Benham and Tsien 1988b) appears to meet these criteria since it has a 3:1 selectivity for Ca over Na and is insensitive to voltage. Other examples include receptor- and stretch-activated channels which promote Ca entry into endothelial cells (Lansman et al. 1987). In T lymphocytes (Kuno and Gardner 1987) and in mast cells (Penner et al. 1988), evidence has been presented that intracellular IP_3 can promote Ca channel activity. A role for IP_4 in Ca entry remains controversial (e.g., Penner et al. 1988).

Perhaps the most striking demonstration of a ROCC is that reported by Zschauer and colleagues (1988). Membrane vesicles from resting and thrombin-stimulated human platelets were incorporated into planar lipid bilayers where single channels with *30-fold* more selectivity for Ba over Na were demonstrable after incorporation of thrombin-stimulated (but not resting) preparations. The single-channel conductance with symmetrical 150 mM Ba solution was about 10 pS, and the opening and closing kinetics were not influenced by membrane potential or by 10 µM nisoldipine. However, these channels were blocked by 20 mM Ni, in agreement with earlier studies. Both Cd and Ni have been shown to have a potent inhibitory effect on thrombin-stimulated Ca uptake and aggregation in mammalian platelets (Blache et al. 1985; Hallam and Rink 1985; Jy and Haynes 1987), whereas organic Ca channel blockers, especially 1,4-dihydropyridine analogues, are ineffective (Blache et al. 1985).

The requirement of intact platelets for the activation of channels by thrombin implies that channel activation is more complex than a direct ligand-induced conformational change of a channel molecule. Coupling of the thrombin receptor to channel opening may instead be effected by the generation of a second messenger, as suggested by the 200 ms delay between thrombin application and Ca entry in platelets (Sage and Rink 1987). It has been proposed that IP_3 is the messenger in neutrophils (Kuno and Gardner 1987) and that IP_4 serves this function in sea urchin eggs (Houslay 1987). It will be interesting to see if, in contrast to thrombin which promotes a Ca influx that lasts many minutes, rapidly desensitising ligands such as ADP or platelet-activating factor also induce these channels.

Finally, it should be emphasized that the 3:1 to 30:1 selectivity of ROCCs for divalent cations over monovalent cations is considerably lower than the

$\geqslant 1000 : 1$ selectivity for divalent cations over monovalent cations of POCCs (e.g., Tsien et al. 1987; Pelzer et al. 1989a; see Sect. 3.3). Thus, in marked contrast to POCCs, ROCCs primarily transport Na, or in the best case equal amounts of Na and Ca, under physiological conditions.

8 Future Prospects

Some of the most exciting developments in the study of Ca channels still lie ahead of us. As a matter of fact, more questions can be presently asked about Ca channels than answers can be given. How closely related are various types of muscle Ca channels to each other, to functionally similar Ca channels in other tissues, and to other types of channels? What forms the basis for the distinct electrophysiological properties and patterns of regulation among the various types of Ca channels in different tissues? With efforts to apply biophysical, biochemical, and molecular genetic methods for channel characterization, purification and reconstitution, and cloning and expression, answers to such questions may not be long in coming.

Acknowledgements. We are grateful to Ms. A. Smith and Mrs. H. Leser for competent secretarial help, to Mrs. J. Crozsman, Mr. T. Asai and Dr. J. Terada for invaluable assistance, and to Dr. W. Trautwein for expert comments on the final version of the manuscript. This work was supported by the Deutsche Forschungsgemeinschaft (SFB 246), the Medical Research Council of Canada, and the Nova Scotia Heart Foundation.

References

Aaronson PI, Bolton TB, Lang RJ, MacKenzie I (1988) Calcium currents in single isolated smooth muscle cells from the rabbit ear artery in normal-calcium and high-barium solutions. J Physiol (Lond) 405:57–75

Affolter H, Coronado R (1985) Agonist Bay-K8644 and CGP-28392 open calcium channels reconstituted from skeletal muscle transverse tubules. Biophys J 48:341–347

Agus ZS, Kelepouris E, Dukes I, Morad M (1989) Cytosolic magnesium modulates calcium channel activity in mammalian ventricular cells. Am J Physiol 256:C452–C455

Akaike N, Nishi K, Oyama Y (1983) Characteristics of manganese current and its comparison with currents carried by other divalent cations in snail soma membranes. J Membr Biol 76:289–297

Alexander RW, Brock TA, Gimbrone MA Jr, Rittenhouse SE (1985) Angiotensin increases inositol trisphosphate and calcium in vascular smooth muscle. Hypertension 7:447–451

Allen DG, Morris PG, Orchard CH, Pirolo JS (1985) A nuclear magnetic resonance study of metabolism in the ferret heart during hypoxia and inhibition of glycolysis. J Physiol (Lond) 361:185–204

Allen IS, Cohen NM, Dhallan RS, Gaa ST, Lederer WJ, Rogers TB (1988) Angiotensin II increases spontaneous contractile frequency and stimulates calcium current in cultured neonatal rat heart myocytes: insights into the underlying biochemical mechanisms. Circ Res 62:524–534

Almers W, McCleskey EW (1984) Non-selective conductance in calcium channels of frog muscle: calcium selectivity in a single-file pore. J Physiol (Lond) 353:585–608

Almers W, Palade PT (1981) Slow calcium and potassium currents across frog muscle membrane: measurements with a vaseline-gap technique. J Physiol (Lond) 312:159–176

Almers W, Fink R, Palade PT (1981) Calcium depletion in frog muscle tubules: the decline of calcium current under maintained depolarization. J Physiol (Lond) 312:177–207

Almers W, McCleskey EW, Palade PT (1984) A non-selective cation conductance in frog muscle membrane blocked by micromolar external calcium ions. J Physiol (Lond) 353:565–583

Amédée T, Mironneau C, Mironneau J (1987) The calcium channel current of pregnant rat single myometrial cells in short-term primary culture. J Physiol (Lond) 392:253–272

Anand-Srivastava MB, Cantin M (1986) Atrial natriuretic factor receptors are negatively coupled to adenylate cyclase in cultured atrial and ventricular cardiocytes. Biochem Biophys Res Commun 138:427–436

Anand-Srivastava MB, Franks DJ, Cantin M, Genest J (1984) Atrial natriuretic factor inhibits adenylate cyclase activity. Biochem Biophys Res Commun 121:855–862

Anderson M (1983) Mn ions pass through calcium channels. A possible explanation. J Gen Physiol 81:805–827

Anderson NC, Ramon F, Snyder A (1971) Studies on calcium and sodium in uterine smooth muscle excitation under current-clamp and voltage-clamp conditions. J Gen Physiol 58:322–339

Armstrong CM, Matteson DR (1985) Two distinct populations of calcium channels in a clonal line of pituitary cells. Science 227:65–67

Armstrong D, Eckert R (1987) Voltage-activated calcium channels that must be phosphorylated to respond to membrane depolarization. Proc Natl Acad Sci USA 84:2518–2522

Arreola J, Calvo J, García MC, Sánchez JA (1987) Modulation of calcium channels of twitch skeletal muscle fibres of the frog by adrenaline and cyclic adenosine monophosphate. J Physiol (Lond) 393:307–330

Ashby CD, Walsh DA (1972) Characterization of the interaction of a protein inhibitor with adenosine 3′,5′monophosphate-dependent protein kinase. J Biol Chem 247:6637–6642

Baker KM, Singer HA (1988) Identification and characterization of guinea pig angiotensin II ventricular and atrial receptors: coupling to inositol phosphate production. Circ Res 62:896–904

Baker KM, Campanile CP, Trachte GJ, Peach MJ (1984) Identification and characterization of the rabbit angiotensin II myocardial receptor. Circ Res 54:286–293

Ballermann BJ, Brenner BM (1986) Role of atrial peptides in body fluid homeostasis. Circ Res 58:619–630

Baron CB, Cunningham M, Strauss JF III, Coburn RF (1984) Pharmacomechanical coupling in smooth muscle may involve phosphatidylinositol metabolism. Proc Natl Acad Sci USA 81:6899–6903

Barres BA, Chun LLY, Corey DP (1985) Voltage-dependent ion channels in glial cells. Soc Neurosci Abstr 11:147

Barth R, Elce JS (1981) Immunofluorescent localization of a Ca^{2+}-dependent neutral protease in hamster muscle. Am J Physiol 240:E493–E498

Bassingthwaighte JB, Reuter H (1972) Calcium movements and excitation-contraction coupling in cardiac cells. In: De Mello WC (ed) Electrical phenomena in the heart. Academic, New York, pp 353–395

Beam KG, Knudson CM (1988a) Calcium currents in embryonic and neonatal mammalian skeletal muscle. J Gen Physiol 91:781–798

Beam KG, Knudson CM (1988b) Effect of postnatal development on calcium currents and slow charge movement in mammalian skeletal muscle. J Gen Physiol 91:799–815

Beam KG, Knudson CM, Powell JA (1986) A lethal mutation in mice eliminates the slow calcium current in skeletal muscle cells. Nature 320:168–170

Bean BP (1985) Two kinds of calcium channels in canine atrial cells. Differences in kinetics, selectivity, and pharmacology. J Gen Physiol 86:1–30

Bean BP (1989) Multiple types of calcium channels in heart muscle and neurons: modulation by drugs and neurotransmitters. Ann NY Acad Sci 560:334–345

Bean BP, Nowycky MC, Tsien RW (1984) β-Adrenergic modulation of calcium channels in frog ventricular heart cells. Nature 307:371–375

Bean BP, Sturek M, Puga A, Hermsmeyer K (1986) Calcium channels in muscle cells isolated from rat mesenteric arteries: modulation by dihydropyridine drugs. Circ Res 59:229–235

Beaty GN, Stefani E (1976) Inward calcium current in twitch muscle fibres of the frog. J Physiol (Lond) 260:27P

Beavo JA, Hardmann JG, Sutherland EW (1971) Stimulation of adenosine 3′,5′monophosphate hydrolysis by guanosine 3′,5′monophosphate. J Biol Chem 246:3841–3846

Bechem M, Pott L (1985) Removal of Ca current inactivation in dialysed guinea-pig atrial cardioballs by Ca chelators. Pflügers Arch 404:10–20

Beeler GW Jr, Reuter H (1970) Membrane calcium current in ventricular myocardial fibres. J Physiol (Lond) 207:191–209

Belardinelli L, Isenberg G (1983a) Actions of adenosine and isoproterenol on isolated mammalian ventricular myocytes. Circ Res 53:287–297

Belardinelli L, Isenberg G (1983b) Isolated atrial myocytes: adenosine and acetylcholine increase potassium conductance. Am J Physiol 244:H734–H737

Belardinelli L, Fenton AR, West A, Linden J, Althaus JS, Berne RM (1982) Extracellular action of adenosine and the antagonism by aminophylline on the atrioventricular conduction of isolated perfused guinea pig and rat hearts. Circ Res 51:569–579

Belles B, Malécot CO, Hescheler J, Trautwein W (1988a) "Run-down" of the Ca current during long whole-cell recordings in guinea pig heart cells: role of phosphorylation and intracellular calcium. Pflügers Arch 411:353–360

Belles B, Hescheler J, Trautwein W, Blomgren K, Karlsson JO (1988b) A possible physiological role of the Ca-dependent protease calpain and its inhibitor calpastatin on the Ca current in guinea pig myocytes. Pflügers Arch 412:554–556

Benham CD, Tsien RW (1988a) Noradrenaline modulation of calcium channels in single smooth muscle cells from rabbit ear artery. J Physiol (Lond) 404:767–784

Benham CD, Tsien RW (1988b) A novel receptor-operated Ca^{2+}-permeable channel activated by ATP in smooth muscle. Nature 328:275–278

Benham CD, Hess P, Tsien RW (1987) Two types of calcium channels in single smooth muscle cells from rabbit ear artery studied with whole-cell and single-channel recordings. Circ Res 61 (Suppl I):I-10–I-16

Bernard C, Cardinaux JC, Potreau D (1976) Long-duration responses and slow inward current obtained from isolated skeletal fibres with barium ions. J Physiol (Lond) 256:18P–19P

Berridge MJ (1987) Inositol trisphosphate and diacylglycerol: two interacting second messengers. Annu Rev Biochem 56:159–193

Berridge MJ, Irvine RF (1984) Inositol trisphosphate, a novel second messenger in cellular signal transduction. Nature 312:315–321

Biegon RL, Pappano AJ (1980) Dual mechanism for inhibition of calcium-dependent action potentials by acetylcholine in avian ventricular muscle. Relationship to cyclic AMP. Circ Res 46:353–362

Birnbaumer L, Codina J, Mattera R, Yatani A, Scherer N, Toro M, Brown AM (1987) Signal transduction by G proteins. Kidney Int 32:514–537

Bkaily G, Sperelakis N (1985) Injection of guanosine 5′-cyclic monophosphate into heart cells blocks calcium slow channels. Am J Physiol 248:H745–H749

Bkaily G, Peyrow M, Sculptoreanu A, Jacques D, Chahine M, Regoli D, Sperelakis N (1988) Angiotensin II increases I_{si} and blocks I_K in single aortic cell of rabbit. Pflügers Arch 412:448–450

Blache D, Ciavatti M, Ojeda C (1985) Platelet aggregation may require functional calcium channels. J Physiol (Lond) 358:68P

Bolton TB (1979) Mechanisms of action of transmitters and other substances on smooth muscle. Physiol Rev 59:606–718

Bonnardeaux JL, Regoli D (1974) Action of angiotensin II and analogues on the heart. Can J Physiol Pharmacol 52:50−60

Bonvallet R (1987) A low threshold calcium current recorded at physiological Ca concentrations in single frog atrial cells. Pflügers Arch 408:540−542

Bossu JL, Feltz A, Thomann JM (1985) Depolarization elicits two distinct calcium currents in vertebrate sensory neurons. Pflügers Arch 403:360−368

Brehm P, Eckert R (1978) Calcium entry leads to inactivation of calcium channels in *Paramecium*. Science 202:1203−1206

Breitwieser GE, Szabo G (1985) Uncoupling of cardiac muscarinic and β-adrenergic receptors from ion channels by a guanine nucleotide analogue. Nature 317:538−540

Brown AM, Birnbaumer L (1988) Direct G protein gating of ion channels. Am J Physiol 254:H401−H410

Brown AM, Morimoto K, Tsuda Y, Wilson DL (1981) Calcium current-dependent and voltage-dependent inactivation of calcium channels in *Helix aspersa*. J Physiol (Lond) 320:193−218

Brown HF, Kimura J, Noble D, Noble SJ, Taupignon A (1984) The slow inward current, i_{si}, in the rabbit sino-atrial node investigated by voltage clamp and computer simulation. Proc R Soc Lond [Biol] 222:305−328

Brown JH, Buxton IL, Brunton LL (1985) α-Adrenergic and muscarinic cholinergic stimulation of phosphoinositide hydrolysis in adult rat cardiomyocytes. Circ Res 57:532−537

Brum G, Flockerzi V, Hofmann F, Osterrieder W, Trautwein W (1983) Injection of catalytic subunit of cAMP-dependent protein kinase into isolated cardiac myocytes. Pflügers Arch 398:147−154

Brum G, Osterrieder W, Trautwein W (1984) β-Adrenergic increase in the calcium conductance of cardiac myocytes studied with the patch clamp. Pflügers Arch 401:111−118

Bülbring E, Tomita T (1987) Catecholamine action on smooth muscle. Pharmacol Rev 39:49−96

Byerly L, Moody WJ (1984) Intracellular calcium ions and calcium currents in perfused neurones of the snail, *Lymnaea stagnalis*. J Physiol (Lond) 352:637−652

Byerly L, Chase PB, Stimers JR (1985) Permeation and interaction of divalent cations in calcium channels of snail neurons. J Gen Physiol 85:491−518

Cachelin AB, de Peyer JE, Kokubun S, Reuter H (1983) Ca^{2+} channel modulation by 8-bromo-cyclic AMP in cultured heart cells. Nature 304:462−464

Campbell DL, Giles WR, Shibata EF (1988a) Ion transfer characteristics of the calcium current in bull-frog atrial myocytes. J Physiol (Lond) 403:239−266

Campbell DL, Giles WR, Hume JR, Noble D, Shibata EF (1988b) Reversal potential of the calcium current in bull-frog atrial myocytes. J Physiol (Lond) 403:267−286

Campbell DL, Giles WR, Hume JR, Shibata EF (1988c) Inactivation of calcium current in bull-frog atrial myocytes. J Physiol (Lond) 403:287−315

Campbell KP, Leung AT, Sharp AH (1988d) The biochemistry and molecular biology of the dihydropyridine-sensitive calcium channel. TINS 11:425−430

Campbell KP, Leung AT, Sharp AH, Imagawa T, Kahl SD (1988e) Ca^{2+} channel antibodies: subunit-specific antibodies as probes for structure and function. In: Morad M, Nayler W, Kazda S, Schramm M (eds) The calcium channel: structure, function and implications. Springer, Berlin Heidelberg New York Tokyo, pp 586−600

Canonico PL, MacLeod RM (1986) Angiotensin peptides stimulate phosphoinositide breakdown and prolactin release in anterior pituitary cells in culture. Endocrinology 118:233−238

Cantin M, Genest J (1985) The heart and the atrial natriuretic factor. Endocr Rev 6:107−127

Capogrossi MC, Kaku T, Pelto DJ, Filburn C, Hansford RG, Spurgeon H, Lakatta EG (1987) Phorbol ester translocates protein kinase C and has a negative inotropic effect in rat cardiac myocytes. Biophys J 51:112a

Carbone E, Lux HD (1984a) A low-voltage-activated calcium conductance in embryonic chick sensory neurones. Biophys J 46:413−418

Carbone E, Lux HD (1984b) A low-voltage-activated, fully inactivating Ca channel in vertebrate sensory neurones. Nature 310:501−502

Carbone E, Lux HD (1987a) Kinetics and selectivity of a low-voltage-activated calcium current in chick and rat sensory neurones. J Physiol (Lond) 386:547–570

Carbone E, Lux HD (1987b) Single low-voltage-activated calcium channels in chick and rat sensory neurones. J Physiol (Lond) 386:571–601

Carbone E, Lux HD (1988) Sodium currents through neuronal calcium channels: kinetics and sensitivity to calcium antagonists. In: Morad M, Nayler W, Kazda S, Schramm M (eds) The calcium channel: structure, function and implications. Springer, Berlin Heidelberg New York Tokyo, pp 115–127

Carmeliet E (1978) Cardiac transmembrane potentials and metabolism. Circ Res 42:577–587

Carmeliet E, Ramon J (1980) Effects of acetylcholine on time-dependent currents in sheep cardiac Purkinje fibers. Pflügers Arch 387:217–223

Carmeliet E, Vereecke J (1979) Electrogenesis of the action potential and automaticity. In: Berne RM (ed) Handbook of physiology, sect 2. The cardiovascular system, vol 1. The heart. American Physiological Society, Bethesda, pp 269–334

Castagna M, Takai Y, Kaibuchi A, Sano K, Kikkawa U, Nishizuka Y (1982) Direct activation of calcium-activated phospholipid-dependent protein kinase by tumor promoting phorbol esters. J Biol Chem 257:7847–7851

Catterall WA (1988) Structure and function of voltage-sensitive ion channels. Science 242:50–61

Cauvin C, Loutzenhiser R, Van Breemen C (1983) Mechanisms of calcium antagonist-induced vasodilatation. Annu Rev Pharmacol Toxicol 23:373–396

Cauvin C, Lukeman S, Cameron J, Hwang O, van Breemen C (1985) Differences in norepinephrine activation and diltiazem inhibition of calcium channels in isolated rabbit aorta and mesenteric resistance vessels. Circ Res 56:822–828

Cavalié A, Ochi R, Pelzer D, Trautwein W (1983) Elementary currents through Ca^{2+} channels in guinea pig myocytes. Pflügers Arch 398:284–297

Cavalié A, McDonald TF, Pelzer D, Trautwein W (1985) Temperature-induced transitory and steady-state changes in the calcium current of guinea pig ventricular myocytes. Pflügers Arch 405:294–296

Cavalié A, Pelzer D, Trautwein W (1986) Fast and slow gating behaviour of single calcium channels in cardiac cells. Relation to activation and inactivation of calcium-channel current. Pflügers Arch 406:241–258

Chad JE, Eckert R (1986) An enzymatic mechanism for calcium current inactivation in dialysed *Helix* neurones. J Physiol (Lond) 378:31–51

Chandler WK, Meves H (1965) Voltage clamp experiments on internally perfused giant axons. J Physiol (Lond) 180:788–820

Chen C, Corbley MJ, Roberts TM, Hess P (1988a) Voltage-sensitive calcium channels in normal and transformed 3T3 fibroblasts. Science 239:1024–1026

Chen C, Corbley MJ, Roberts TM, Hess P (1988b) Dihydropyridine-sensitive and -insensitive Ca^{2+} channels in normal and transformed fibroblasts. In: Morad M, Nayler W, Kazda S, Schramm M (eds) The calcium channel: structure, function and implications. Springer, Berlin Heidelberg New York Tokyo, pp 92–102

Chesnais JM, Coraboeuf E, Sauviat MP, Vassas JM (1975) Sensitivity to H, Li and Mg ions of the slow inward sodium current in frog atrial fibers. J Mol Cell Cardiol 7:627–642

Cockcroft S, Gomperts BD (1985) Role of guanine nucleotide binding protein in the activation of polyphosphoinositide phosphodiesterase. Nature 314:534–536

Codina J, Yatani A, Grenet D, Brown AM, Birnbaumer L (1987) The α subunit of the GTP binding protein G_k opens atrial potassium channels. Science 236:442–445

Cognard C, Romey G, Galizzi J-P, Fosset M, Lazdunski M (1986) Dihydropyridine-sensitive Ca^{2+} channels in mammalian skeletal muscle cells in culture: electrophysiological properties and interactions with Ca^{2+} channel activator (Bay K8644) and inhibitor (PN 200-110). Proc Natl Acad Sci USA 83:1518–1522

Cohen NM, Lederer WJ (1987) Calcium current in isolated neonatal rat ventricular myocytes. J Physiol (Lond) 391:169–191

Cohen P (1982) The role of protein phosphorylation in neural and hormonal control of cellular activity. Nature 296:613–620

Coraboeuf E (1980) Voltage clamp studies of the slow inward current. In: Zipes DP, Bailey JC, Elharrar V (eds) The slow inward current and cardiac arrhythmias. Nijhoff, The Hague, pp 25–95

Coronado R (1987) Planar bilayer reconstittion of calcium channels: lipid effects on single channel kinetics. Circ Res 61 (Suppl II):II-46–II-52

Coronado R, Affolter H (1986) Insulation of the conduction pathway of muscle transverse tubule calcium channels from the surface charge of bilayer phospholipid. J Gen Physiol 87:933–953

Coronado R, Smith JS (1987) Monovalent ion current through single calcium channels of skeletal muscle transverse tubules. Biophys J 51:497–502

Cota G, Stefani E (1984) Saturation of calcium channels and surface charge effects in skeletal muscle fibres of the frog. J Physiol (Lond) 351:135–154

Cota G, Stefani E (1986) A fast-activated inward calcium current in twitch muscle fibres of the frog (Rana montezume). J Physiol (Lond) 370:151–163

Cota G, Nicola Siri L, Stefani E (1984) Calcium channel inactivation in frog (Rana pipiens and Rana moctezuma) skeletal muscle fibres. J Physiol (Lond) 353:99–108

Cramb G, Banks R, Rugg EL, Aiton JF (1987) Actions of atrial natriuretic peptide (ANP) on cyclic nucleotide concentrations and phosphatidylinositol turnover in ventricular myocytes. Biochem Biophys Res Commun 148:962–970

Creba JA, Downes CP, Hawkins PT, Brewster G, Mitchell RH, Kirk CJ (1983) Rapid breakdown of phosphatidylinositol 4-phosphate and phosphatidylinositol 4,5-biphosphate in rat hepatocytes stimulated by vasopressin and other Ca^{2+}-mobilizing hormones. Biochem J 212:733–747

Daly J (1984) Forskolin, adenylate cyclase, and cell physiology: an overview. Adv Cyclic Nucl Res 17:81–89

DeBold AJ (1985) Atrial natriuretic factor: a hormone produced by the heart. Science 230:767–770

Deitmer JW (1984) Evidence for two voltage-dependent calcium currents in the membrane of the ciliate Stylonychia. J Physiol (Lond) 355:137–159

Dempsey JP, McCallum ZT, Kent KM, Cooper T (1971) Direct myocardial effects of angiotensin II. Am J Physiol 220:477–481

DeRiemer SA, Sakmann B (1986) Two calcium currents in normal rat anterior pituitary cells identified by a plaque assay. Exp Brain Res 14:139–154

DeRiemer SA, Strong JA, Albert KA, Greengard P, Kaczmarek LK (1985) Enhancement of calcium current in Aplysia neurons by phorbol ester and protein kinase C. Nature 313:313–316

Diamond J, Ten Eick RE, Trapani AJ (1977) Are increases in cyclic GMP levels responsible for the negative inotropic effects of acetylcholine in the heart? Biochem Biophys Res Commun 79:912–917

Di Virgilio F, Pozzan T, Wollheim CB, Vicentini LM, Meldolesi J (1986) Tumor promoter phorbol myristate acetate inhibits Ca^{2+} influx through voltage-gated Ca^{2+} channels in two secretory cell lines, PC12 and RINm5F. J Biol Chem 261:32–35

Docherty RJ, Brown DA (1986) Interaction of 1,4-dihydropyridines with somatic Ca currents in hippocampal CA1 neurones of the guinea pig in vitro. Neurosci Lett 70:110–115

Donaldson PL, Beam KG (1983) Calcium currents in a fast-twitch skeletal muscle of the rat. J Gen Physiol 82:449–468

Donaldson SK, Goldberg ND, Walseth TF, Huetteman DA (1988) Voltage dependence of inositol 1,4,5-trisphosphate-induced Ca^{2+} release in peeled skeletal muscle fibers. Proc Natl Acad Sci USA 85:5749–5753

Dösemeci A, Dhallan RS, Cohen NM, Lederer WJ, Rogers TB (1988) Phorbol ester increases calcium current and stimulates the effect of angiotensin II on cultured neonatal rat heart myocytes. Circ Res 62:347–357

Droogmans G, Callewaert G (1986) Ca^{2+}-channel current and its modification by the dihydropyridine agonist Bay K 8644 in isolated smooth muscle cells. Pflügers Arch 406:259–265

Droogmans G, Declerck I, Casteels R (1987) Effect of adrenergic agonists on Ca^{2+}-channel currents in single vascular smooth muscle cells. Pflügers Arch 409:7–12

Dupont JL, Bossu JL, Feltz A (1986) Effect of internal calcium concentration on calcium currents in rat sensory neurones. Pflügers Arch 406:433–435

Eckert R, Chad JE (1984) Inactivation of Ca channels. Prog Biophys Mol Biol 44:215–267

Eisenman G, Horn R (1983) Ionic selectivity revisited: the role of kinetic and equilibrium processes in ion permeation through channels. J Membr Biol 76:197–225

Ellis SB, Williams ME, Ways NR, Brenner R, Sharp AH, Leung AT, Campbell KP, McKenna E, Koch WJ, Hui A, Schwartz A, Harpold MM (1988) Sequence and expression of mRNAs encoding the α_1 and α_2 subunits of a DHP-sensitive calcium channel. Science 241:1661–1664

Endoh M, Shimizu T (1979) Failure of dibutyryl and 8-bromo-cyclic GMP to mimic the antagonistic action of carbachol on the positive inotropic effects of sympathomimetic amines in the canine isolated ventricular myocardium. Jpn J Pharmacol 29:423–433

Fabiato A (1985) Time and calcium dependence of activation and inactivation of calcium-induced release of calcium from the sarcoplasmic reticulum of a skinned canine cardiac Purkinje cell. J Gen Physiol 85:247–289

Fatt P, Ginsborg BL (1958) The ionic requirements for the production of action potentials in crustacean muscle fibres. J Physiol (Lond) 142:516–543

Fedida D, Noble D, Shimoni Y, Spindler AJ (1987) Inward current related to contraction in guinea-pig ventricular myocytes. J Physiol (Lond) 385:565–589

Fedida D, Noble D, Spindler AJ (1988a) Use-dependent reduction and facilitation of Ca^{2+} current in guinea-pig myocytes. J Physiol (Lond) 405:439–460

Fedida D, Noble D, Spindler AJ (1988b) Mechanism of the use dependence of Ca^{2+} current in guinea-pig myocytes. J Physiol (Lond) 405:461–475

Fedulova SA, Kostyuk PG, Veselovsky NS (1985) Two types of calcium channels in the somatic membrane of new-born rat dorsal root ganglion neurones. J Physiol (Lond) 359:431–446

Fenwick EM, Marty A, Neher E (1982) Sodium and calcium channels in bovine chromaffin cells. J Physiol (Lond) 331:599–635

Fischmeister R, Hartzell HC (1986) Mechanism of action of acetylcholine on calcium current in single cells from frog ventricle. J Physiol (Lond) 376:183–202

Fischmeister R, Hartzell HC (1987) Cyclic guanosine 3′, 5′-monophosphate regulates the calcium current in single cells from frog ventricle. J Physiol (Lond) 387:453–472

Fischmeister R, Brocas-Randolph M, Lechene P, Argibay JA, Vassort G (1986) A dual effect of cardiac glycosides on Ca current in single cells of frog heart. Pflügers Arch 406:340–342

Fischmeister R, Argibay JA, Hartzell HC (1987) Modifications of cardiac calcium current in cells with inherent differences in current density. Biophys J 51:29a

Fish RD, Sperti G, Colucci WS, Clapham DE (1988) Phorbol ester increases the dihydropyridine-sensitive calcium conductance in a vascular smooth muscle cell line. Circ Res 62:1049–1054

Fishman MC, Spector I (1981) Potassium current suppression by quinidine reveals additional calcium currents in neuroblastoma cells. Proc Natl Acad Sci USA 78:5245–5249

Fitzpatrick LA, Chin H, Nirenberg M, Aurbach GD (1988) Antibodies to an α subunit of skeletal muscle calcium channels regulate parathyroid cell secretion. Proc Natl Acad Sci USA 85:2115–2119

Fleming JW, Strawbridge RA, Watanabe AM (1987) Muscarinic receptor regulation of cardiac adenylate cyclase activity. J Mol Cell Cardiol 19:47–61

Flitney FW, Singh J (1981) Evidence that cyclic GMP may regulate cyclic AMP metabolism in the isolated frog ventricle. J Mol Cell Cardiol 13:963–979

Flockerzi V, Oeken HJ, Hofmann F, Pelzer D, Cavalié A, Trautwein W (1986) Purified dihydropyridine-binding site from skeletal muscle t-tubules is a functional calcium channel. Nature 323:66–68

Fox AP, Krasne S (1984) Two calcium currents in *Neanthes arenaceodentatus* egg cell membranes. J Physiol (Lond) 356:491–505

Fox AP, Hess P, Lansman JB, Nowycky MC, Tsien RW (1984) Slow variations in the gating properties of single calcium channels in guinea-pig heart cells, chick neurones and neuroblastoma cells. J Physiol (Lond) 353:75P

Fox AP, Nowycky MC, Tsien RW (1987a) Kinetic and pharmacological properties distinguishing three types of calcium currents in chick sensory neurones. J Physiol (Lond) 394:149−172

Fox AP, Nowycky MC, Tsien RW (1987b) Single-channel recordings of three types of calcium channels in chick sensory neurones. J Physiol (Lond) 394:173−200

Friedman ME, Suarez-Kurtz G, Kaczorowski GJ, Katz GM, Reuben JP (1986) Two calcium currents in a smooth muscle cell line. Am J Physiol 250:H699−H703

Fujii K, Ishimatsu T, Kuriyama H (1986) Mechanism of vasodilation induced by α-human atrial natriuretic polypeptide in rabbit and guinea-pig renal arteries. J Physiol (Lond) 377:315−332

Fukuda J, Kawa K (1977) Permeation of manganese, cadmium, zinc, and beryllium through calcium channels of an insect muscle membrane. Science 196:309−311

Fukushima Y, Hagiwara S (1985) Currents carried by monovalent cations through calcium channels in mouse neoplastic B lymphocytes. J Physiol (Lond) 358:255−284

Galizzi JP, Qar J, Fosset M, Van Renterghem C, Lazdunski M (1987) Regulation of calcium channels in aortic muscle cells by protein kinase C activators (diacylglycerol and phorbol esters) and by peptides (vasopressin or bombesin) that stimulate phosphoinositide breakdown. J Biol Chem 262:6947−6950

Ganitkevich VYa, Shuba MF, Smirnov SV (1986) Potential-dependent calcium inward current in a single isolated smooth muscle cell of the guinea-pig taenia caeci. J Physiol (Lond) 380:1−16

Ganitkevich VYa, Shuba MF, Smirnov SV (1987) Calcium-dependent inactivation of potential-dependent calcium inward current in an isolated guinea-pig smooth muscle cell. J Physiol (Lond) 392:431−449

Ganitkevich VYa, Shuba MF, Smirnov SV (1988) Saturation of calcium channels in single isolated smooth muscle cells of guinea-pig taenia caeci. J Physiol (Lond) 399:419−436

Garnier D, Rougier O, Gargouil YM, Coraboeuf E (1969) Electrophysiological analysis of myocard membrane properties during the plateau of the action potential, existence of a slow inward current in solutions without divalent ions. Pflügers Arch 313:321−342

Garnier D, Nargeot J, Ojeda C, Rougier O (1978) The action of acetylcholine on background conductance in frog atrial trabeculae. J Physiol (Lond) 274:381−396

George WJ, Polson JB, O'Toole AG, Golberg N (1970) Elevation of 3′,5′-cyclic phosphate in rat heart after perfusion with acetylcholine. Proc Natl Acad Sci USA 66:398−403

Giles W, Noble SJ (1976) Changes in membrane currents in bullfrog atrium produced by acetylcholine. J Physiol (Lond) 261:103−123

Gisbert MP, Fischmeister R (1988) Atrial natriuretic factor regulates the calcium current in frog isolated cardiac cells. Circ Res 62:660−667

Glossmann H, Striessnig J (1988) Calcium channels. Vitam Hormon 44:155−328

Glossmann H, Striessnig J (1990) Molecular properties of calcium channels. In: Blaustein MP et al. (eds) Reviews of Physiology, Biochemistry and Pharmacology, Vol 114. Springer Berlin Heidelberg New York, pp 1−106

Gluecksohn-Waelsch S (1963) Lethal genes and analysis of differentiation. Science 142:1269−1276

Godfraind T, Miller R, Wibo M (1986) Calcium antagonists and calcium entry blockade. Pharmacol Rev 38:321−416

Goldberg ND, Haddox MK (1977) Cyclic GMP metabolism and involvement in biologic regulation. Annu Rev Biochem 46:823−896

Goldberg ND, Haddox MK, Nichol SE, Glass DB, Sanford CH, Kuehl FA, Estensen R (1975) Biologic regulation through opposing influences of cyclic GMP and cyclic AMP: the Yin-Yang hypothesis. Adv Cyclic Nucl Res 5:307−330

Gray R, Johnston D (1986) Multiple types of calcium channels in acutely exposed neurons from the adult guinea pig hippocampus. J Gen Physiol 88:25a

Graziano MP, Gilman AG (1987) Guanine nucleotide-binding regulatory proteins: mediators of transmembrane signaling. Trends Pharmacol Sci 8:478−481

Gross RA, Macdonald RL (1987) Dynorphin A selectively reduces a large transient (N-type) calcium current of mouse dorsal root ganglion neurons in cell culture. Proc Natl Acad Sci USA 84:5469–5473

Hadley RW, Hume JR (1987) An intrinsic potential-dependent inactivation mechanism associated with calcium channels in guinea-pig myocytes. J Physiol (Lond) 389:205–222

Hadley RW, Hume JR (1988) Calcium channel antagonist properties of Bay K8644 in single guinea pig ventricular cells. Circ Res 62:97–104

Hagiwara N, Irisawa H, Kameyama M (1988) Contribution of two types of calcium currents to the pacemaker potentials of rabbit sino-atrial node cells. J Physiol (Lond) 395:233–253

Hagiwara S, Byerly L (1981) Calcium channel. Annu Rev Neurosci 4:69–125

Hagiwara S, Fukuda J, Eaton DC (1974) Membrane currents carried by Ca, Sr, and Ba in barnacle muscle fiber during voltage clamp. J Gen Physiol 63:564–578

Hagiwara S, Ozawa S, Sand O (1975) Voltage clamp analysis of two inward current mechanisms in the egg cell membrane of a starfish. J Gen Physiol 65:617–644

Hallam TJ, Rink TJ (1985) Agonists stimulate divalent cation channels in the plasma membrane of human platelets. FEBS Lett 186:175–179

Hamill OP, Marty A, Neher E, Sakmann B, Sigworth FJ (1981) Improved patch-clamp techniques for high-resolution current recording from cells and cell-free membrane patches. Pflügers Arch 391:85–100

Hammond C, Paupardin-Tritsch D, Nairn AC, Greengard P, Gerschenfeld HM (1987) Cholecystokinin induces a decrease in Ca^{2+} current in snail neurons that appears to be mediated by protein kinase C. Nature 325:809–811

Hanbauer I, Sanna E, Callewaert G, Morad M (1988) An endogenous purified peptide modulates Ca^{2+} channels in neurons and cardiac myocytes. In: Morad M, Nayler W, Kazda S, Schramm M (eds) The calcium channel: structure, function and implications. Springer, Berlin Heidelberg New York Tokyo, pp 611–618

Hannon JD, Lee NKM, Blinks JR (1988) Calcium release by inositol trisphosphate in amphibian and mammalian skeletal muscle is an artifact of cell disruption, and probably results from depolarization of sealed-off T-tubules. Biophys J 53:607a

Harris KM, Kongsamut S, Miller RJ (1986) Protein kinase C mediated regulation of calcium channels in PC-12 pheochromocytoma cells. Biochem Biophys Res Commun 134:1298–1305

Harrison SA, Reifsnyder DH, Gallis B, Cadd GG, Beavo JA (1986) Isolation and characterization of bovine cardiac muscle cGMP-inhibited phosphodiesterase: a receptor for new cardiotonic drugs. Mol Pharmacol 29:506–514

Hartzell HC, Fischmeister R (1986) Cyclic GMP and cyclic AMP produce opposite effects on Ca current in single heart cells. Nature 323:273–275

Hartzell HC, Simmons MA (1987) Comparison of effects of acetylcholine on calcium and potassium currents in frog atrium and ventricle. J Physiol (Lond) 389:411–422

Hayashi H, Watanabe T, McDonald TF (1987) Action potential duration in ventricular muscle during selective metabolic block. Am J Physiol 253:H373–H379

Hescheler J, Trautwein W (1988) Modification of L-type calcium current by intracellulary applied trypsin in guinea-pig ventricular myocytes. J Physiol (Lond) 404:259–274

Hescheler J, Kameyama M, Trautwein W (1986) On the mechanism of muscarinic inhibition of the cardiac Ca current. Pflügers Arch 407:182–189

Hescheler J, Kameyama M, Trautwein W, Mieskes G, Söling HD (1987) Regulation of the cardiac calcium channel by protein phosphatases. Eur J Biochem 165:261–266

Hescheler J, Rosenthal W, Hinsch KD, Wulfern M, Trautwein W, Schultz G (1988) Angiotensin II-induced stimulation of voltage-dependent Ca^{2+} currents in an adrenal cortical cell line. EMBO J 7:619–624

Hess P, Tsien RW (1984) Mechanism of ion permeation through calcium channels. Nature 309:453–456

Hess P, Metzger P, Weingart R (1982) Free magnesium in sheep, ferret and frog striated muscle at rest measured with ion-selective microelectrodes. J Physiol (Lond) 333:173–188

Hess P, Lansman JB, Tsien RW (1984) Different modes of Ca channel gating behaviour favoured by dihydropyridine Ca agonists and antagonists. Nature 311:538–544

Hess P, Lansman JB, Tsien RW (1986) Calcium channel selectivity for divalent and monovalent cations. Voltage and concentration dependence of single channel current in ventricular heart cells. J Gen Physiol 88:293–319

Heyer CB, Lux HD (1976) Properties of a facilitating calcium current in pace-maker neurones of the snail, *Helix pomatia*. J Physiol (Lond) 262:319–348

Hille B (1984) Ionic channels of excitable membranes. Sinauer, Sunderland

Hille B, Schwarz W (1978) Potassium channels as multi-ion single-file pores. J Gen Physiol 72:409–442

Hino N, Ochi R (1980) Effect of acetylcholine on membrane currents in guinea-pig papillary muscle. J Physiol (Lond) 307:183–197

Hiraoka M, Sano T (1978) Role of slow inward current on premature excitation in ventricular muscle. In: Kobayashi T, Sano T, Dhalla NS (eds) Recent advances in studies on cardiac structure and metabolism, vol 11. University Park Press, Baltimore, pp 31–36

Hiriart M, Matteson DR (1988) Na channels and two types of Ca channels in rat pancreatic B cells identified with the reverse hemolytic plaque assay. J Gen Physiol 91:617–639

Hirning LD, Fox AP, McCleskey EW, Olivera BM, Thayer SA, Miller RJ, Tsien RW (1988) Dominant role of N-type Ca^{2+} channels in evoked release of norepinephrine from sympathetic neurons. Science 239:57–61

Hockberger P, Toselli M, Swandulla D, Lux HD (1989) A diacylglycerol analogue reduces neuronal calcium currents independently of protein kinase C activation. Nature 338:340–342

Hodgkin AL, Huxley AF (1952) The dual effect of membrane potential on sodium conductance in the giant axon of *Loligo*. J Physiol (Lond) 116:497–506

Högestätt ED (1984) Characterization of two different calcium entry pathways in small mesenteric arteries from rat. Acta Physiol Scand 122:483–495

Hosey MM, Lazdunski M (1988) Calcium channels: molecular pharmacology, structure and regulation. J Membr Biol 104:81–106

Hoshi T (1985) Gating of voltage-dependent calcium channels in adrenal chromaffin cells. PhD Thesis, Yale University

Hoshi T, Rothlein J, Smith SJ (1984) Facilitation of Ca^{2+}-channel currents in bovine adrenal chromaffin cells. Proc Natl Acad Sci USA 81:5871–5875

Houslay MD (1987) Ion channels controlled by guanine nucleotide regulatory proteins. Trends Biol Sci 12:167–168

Huang CL, Ives HE, Cogan MG (1986) In vivo evidence that cGMP is the second messenger for atrial natriuretic factor. Proc Natl Acad Sci USA 83:8015–8018

Hume JR, Giles W (1983) Ionic currents in single isolated bullfrog atrial cells. J Gen Physiol 81:153–194

Iijima T, Irisawa H, Kameyama M (1985) Membrane currents and their modification by acetylcholine in isolated single atrial cells of the guinea-pig. J Physiol (Lond) 359:485–501

Ikemoto Y, Goto M (1977) Effects of Ach on slow inward current and tension components of the bullfrog atrium. J Mol Cell Cardiol 9:313–326

Imoto Y, Ehara T, Goto M (1985) Calcium channel currents in isolated guinea-pig ventricular cells superfused with Ca-free EGTA solution. Jpn J Physiol 35:917–932

Imoto Y, Yatani A, Reeves JP, Codina J, Birnbaumer L, Brown AM (1988) α-Subunit of G_s directly activates cardiac calcium channels in lipid bilayers. Am J Physiol (Lond) 255:H722–H728

Ingebritsen TS, Cohen P (1983) Protein phosphatases: properties and role in cellular regulation. Science 221:331–338

Inomata H, Kao CY (1976) Ionic currents in the guinea-pig taenia coli. J Physiol (Lond) 255:347–378

Inoue D, Hachisu M, Pappano AJ (1983) Acetylcholine increases resting membrane potassium conductance in atrial but not ventricular muscle during muscarinic inhibition of Ca^{2+}-dependent action potentials in chick heart. Circ Res 53:158–167

Inui J, Imamura H (1977) Effects of acetylcholine on calcium-dependent electrical and mechanical responses of the guinea-pig papillary muscle partially depolarized by potassium. Naunyn-Schmiedebergs Arch Pharmacol 299:1−7

Irisawa H (1984) Electrophysiology of single cardiac cells. Jpn J Physiol 34:375−388

Irisawa H, Kokubun S (1983) Modulation by intracellular ATP and cyclic AMP of the slow inward current in isolated single ventricular cells of the guinea-pig. J Physiol (Lond) 338:321−337

Irisawa H, Sato R (1986) Intra- and extracellular actions of proton on the calcium current of isolated guinea pig ventricular cells. Circ Res 59:348−355

Isenberg G (1977) Cardiac Purkinje fibres. The slow inward current component under the influence of modified $[Ca^{2+}]_i$. Pflügers Arch 371:61−69

Isenberg G, Klöckner U (1980) Glycocalyx is not required for slow inward calcium current in isolated rat heart myocytes. Nature 284:358−360

Isenberg G, Klöckner U (1982) Calcium currents of isolated bovine ventricular myocytes are fast and of large amplitude. Pflügers Arch 395:30−41

Isenberg G, Klöckner U (1985 a) The electrophysiological properties of the isolated adult heart cell: an overview. Basic Res Cardiol 80 (Suppl 2):51−54

Isenberg G, Klöckner U (1985 b) Calcium currents of smooth muscle cells isolated from the urinary bladder of the guinea-pig: inactivation, conductance and selectivity is controlled by micromolar amounts of $[Ca]_o$. J Physiol (Lond) 358:60P

Isenberg G, Klöckner U (1985 c) Elementary currents through single Ca channels in smooth muscle cells isolated from bovine coronary arteries. Effects of nifedipine and Bay K 8644. Pflügers Arch 403:R23

Isenberg G, Cerbai E, Klöckner U (1987) Ionic channels and adenosine in isolated heart cells. In: Gerlach E, Becker BF (eds) Topics and perspectives in adenosine research. Springer, Berlin Heidelberg New York Tokyo, pp 323−335

Janis RA, Silver PJ, Triggle DJ (1987) Drug action and cellular calcium regulation. Adv Drug Res 16:309−589

Janis RA, Johnson DE, Shrikhande AV, McCarthy RT, Howard AD, Greguski R, Scriabine A (1988) Endogenous 1,4-dihydropyridine-displacing substances acting on L-type Ca^{2+} channels: isolation and characterization of fractions from brain and stomach. In: Morad M, Nayler W, Kazda S, Schramm M (eds) The calcium channel: structure, function and implications. Springer, Berlin Heidelberg New York Tokyo, pp 564−574

Jmari K, Mironneau C, Mironneau J (1986) Inactivation of calcium channel current in rat uterine smooth muscle: evidence for calcium- and voltage-mediated mechanisms. J Physiol (Lond) 380:111−126

Jmari K, Mironneau C, Mironneau J (1987) Selectivity of calcium channels in rat uterine smooth muscle: interactions between sodium, calcium and barium ions. J Physiol (Lond) 384:247−261

Jones LG, Goldstein D, Brown JH (1988) Guanine nucleotide-dependent inositol trisphosphate formation in chick heart cells. Circ Res 62:299−305

Josephson I, Sperelakis N (1982) On the ionic mechanism underlying adrenergic-cholinergic antagonism in ventricular muscle. J Gen Physiol 79:69−86

Josephson IR, Sanchez-Chapula J, Brown AM (1984) A comparison of calcium currents in rat and guinea pig single ventricular cells. Circ Res 54:144−156

Jy W, Haynes DH (1987) Thrombin-induced calcium movements in platelet activation. Biochim Biophys Acta 929:88−102

Kaczmarek LK (1987) The role of protein kinase C in the regulation of ion channels and neurotransmitter release. TINS 10:30−34

Kaibara M, Kameyama M (1988) Inhibition of the calcium channel by intracellular protons in single ventricular myocytes of the guinea-pig. J Physiol (Lond) 403:621−640

Kameyama A, Nakayama T (1988) Calcium efflux through cardiac calcium channels reconstituted into liposomes − flux measurement with fura-2. Biochem Biophys Res Commun 154:1067−1074

Kameyama M, Hofmann F, Trautwein W (1985) On the mechanism of β-adrenergic regulation of the Ca channel in the guinea-pig heart. Pflügers Arch 405:285–293

Kameyama M, Hescheler J, Hofmann F, Trautwein W (1986a) Modulation of Ca current during the phosphorylation cycle in the guinea pig heart. Pflügers Arch 407:123–128

Kameyama M, Hescheler J, Mieskes G, Trautwein W (1986b) The protein-specific phosphatase 1 antagonizes the β-adrenergic increase of the cardiac Ca current. Pflügers Arch 407:461–463

Kameyama M, Kameyama A, Kaibara M, Nakayama T (1987) Involvement of intracellular factor(s) in "run down" of the cardiac L-type Ca channel. J Physiol Soc Jpn 49:501

Kameyama M, Kameyama A, Nakayama T, Kaibara M (1988) Tissue extract recovers cardiac calcium channels from "run-down". Pflügers Arch 412:328–330

Kass RS, Blair ML (1981) Effects of angiotensin II on membrane current in cardiac Purkinje fibers. J Mol Cell Cardiol 13:797–809

Kass RS, Krafte DS (1987) Negative surface charge density near heart calcium channels. Relevance to block by dihydropyridines. J Gen Physiol 89:629–644

Kass RS, Sanguinetti MC (1984) Inactivation of calcium channel current in the calf cardiac Purkinje fiber. Evidence for voltage- and calcium-mediated mechanisms. J Gen Physiol 84:705–726

Kass RS, Scheuer T (1982) Slow inactivation of calcium channels in the cardiac Purkinje fiber. J Mol Cell Cardiol 14:615–618

Kass RS, Tsien RW (1975) Multiple effects of calcium antagonists on plateau currents in cardiac Purkinje fibers. J Gen Physiol 66:169–192

Katoh N, Wise BC, Kuo JF (1983) Phosphorylation of cardiac troponin inhibitory subunit (troponin I) and tropomyosin binding subunit (troponin T) by cardiac phospholipid-sensitive Ca^{2+}-dependent protein kinase. Biochem J 209:189–195

Katzung BG, Reuter H, Porzig H (1973) Lanthanum inhibits Ca inward current but not Na-Ca exchange in cardiac muscle. Experientia 29:1073–1075

Kawashima Y, Ochi R (1988) Voltage-dependent decrease in the availability of single calcium channels by nitrendipine in guinea-pig ventricular cells. J Physiol (Lond) 402:219–235

Keeley SL Jr, Lincoln TM, Corbin JD (1978) Interaction of acetylcholine and epinephrine on heart cyclic AMP-dependent protein kinase. Am J Physiol 234:H432–H438

Kelleher DJ, Pessin JE, Ruoho AE, Johnson GL (1984) Phorbol ester induces desensitization of adenylate cyclase and phosphorylation of the β-adrenergic receptor in turkey erythrocytes. Proc Natl Acad Sci USA 81:4316–4320

Kerr LM, Yoshikami D (1984) A venom peptide with novel presynaptic blocking action. Nature 308:282–284

Klaus MM, Scordilis SP, Rapalus JM, Briggs RT, Powell JA (1983) Evidence for dysfunction in the regulation of cytosolic Ca^{2+} in excitation-contraction uncoupled dysgenic muscle. Dev Biol 99:152–165

Klöckner U, Isenberg G (1985) Calcium currents of cesium loaded isolated smooth muscle cells (urinary bladder of the guinea pig). Pflügers Arch 405:340–348

Koch-Weser J (1965) Nature of the inotropic action of angiotensin on ventricular myocardium. Circ Res 16:230–237

Kohlhardt M, Haap K (1978) 8-Bromo-guanosine 3′,5′-monophosphate mimics the effect of acetylcholine on slow response action potential and contractile force in mammalian atrial myocardium. J Mol Cell Cardiol 10:573–586

Kohlhardt M, Herdey A, Kübler M (1973) Interchangeability of Ca ions and Sr ions as charge carriers of the slow inward current in mammalian myocardial fibres. Pflügers Arch 344:149–158

Kohlhardt M, Krause H, Kübler M, Herdey A (1975) Kinetics of inactivation and recovery of the slow inward current in the mammalian ventricular myocardium. Pflügers Arch 355:1–17

Kohlhardt M, Haap K, Figulla HR (1976) Influence of low extracellular pH upon the Ca inward current and isometric contractile force in mammalian ventricular myocardium. Pflügers Arch 366:31–38

Kokubun S, Irisawa H (1984) Effects of various intracellular Ca ion concentrations on the calcium current of guinea-pig single ventricular cells. Jpn J Physiol 34:599–611

Kokubun S, Nishimura M, Noma A, Irisawa H (1982) Membrane currents in the rabbit atrioventricular node cell. Pflügers Arch 393:15–22

Kostyuk PG (1984) Intracellular perfusion of nerve cells and its effects on membrane currents. Physiol Rev 64:435–454

Kostyuk PG, Krishtal OA (1977) Effects of calcium and calcium-chelating agents on the inward and outward current in the membrane of mollusc neurones. J Physiol (Lond) 270:569–580

Kostyuk PG, Mironov SL, Shuba YaM (1983) Two ion-selecting filters in the calcium channel of the somatic membrane of mollusc neurons. J Membr Biol 76:83–93

Kostyuk PG, Shuba YaM, Savchenko AN (1987) Three types of calcium channels in the membrane of mouse sensory neurons. Biol Membrany 4:366–373

Kostyuk PG, Shuba YaM, Savchenko AN, Teslenko VI (1988) Kinetic characteristics of different calcium channels in the neuronal membrane. In: Morad M, Nayler W, Kazda S, Schramm M (eds) The calcium channel: structure, function and implications. Springer, Berlin Heidelberg New York Tokyo, pp 442–464

Kraft AS, Anderson WB (1983) Phorbol esters increase the amount of Ca^{2+}, phospholipid-dependent protein kinase associated with plasma membrane. Nature 301:621–623

Kretsinger RH, Moncrief ND, Goodman M, Czelusniak J (1988) Homology of calcium-modulated proteins: their evolutionary and functional relationships. In: Morad M, Nayler W, Kazda S, Schramm M (eds) The calcium channel: structure, function and implications. Springer, Berlin Heidelberg New York Tokyo, pp 16–34

Kreutter D, Caldwell AB, Morin MJ (1985) Dissociation of protein kinase C activation from phorbol ester-induced maturation of HL-60 leukemia cells. J Biol Chem 260;5979–5984

Kuno M, Gardner P (1987) Ion channels activated by inositol 1,4,5'-trisphosphate in plasma membrane of human T-lymphocytes. Nature 326:301–304

Kuo JF, Lee TP, Reyes PL, Walton KG, Donnelly TE, Greengard P (1972) Cyclic nucleotide-dependent protein kinases. X. An assay method for the measurement of guanosine 3',5'monophosphate in various biological materials and a study of agents regulating its level in heart and brain. J Biol Chem 247:16–22

Kurachi Y (1982) The effects of intracellular protons on the electrical activity of single ventricular cells. Pflügers Arch 394:264–270

Kurachi Y, Nakajima T, Sugimoto T (1986) On the mechanism of activation of muscarinic K^+ channels by adenosine in isolated atrial cells: involvement of GTP-binding proteins. Pflügers Arch 407:264–274

Lacerda AE, Rampe D, Brown AM (1988) Effects of protein kinase C activators on cardiac Ca^{2+} channels. Nature 335:249–251

Lamb GD, Walsh T (1987) Calcium currents, charge movement and dihydropyridine binding in fast- and slow-twitch muscles of rat and rabbit. J Physiol (Lond) 393:595–617

Lansman JB, Hess P, Tsien RW (1986) Blockade of current through single calcium channels by Cd^{2+}, Mg^{2+}, and Ca^{2+}. Voltage and concentration dependence of calcium entry into the pore. J Gen Physiol 88:321–347

Lansman JB, Hallam TJ, Rink TJ (1987) Single stretch-activated ion channels in vascular endothelial cells as mechanotransducers. Nature 325:811–813

Lazdunski M, Schmid A, Romey G, Renaud JF, Galizzi JP, Fosset M, Borsotto M, Barhanin J (1987) Dihydropyridine-sensitive Ca^{2+} channels: molecular properties of interaction with Ca^{2+} channel blockers, purification, subunit structure, and differentiation. J Cardiovasc Pharmacol 9:S10–S15

Leatherman GF, Kim D, Smith TW (1987) Effect of phorbol esters on contractile state and calcium flux in cultured chick heart cells. Am J Physiol 253:H205–H209

Lee KS (1987) Potentiation of the calcium-channel currents of internally perfused mammalian heart cells by repetitive depolarization. Proc Natl Acad Sci USA 84:3941–3945

Lee KS, Tsien RW (1982) Reversal of current through calcium channels in dialysed single heart cells. Nature 297:498–501

Lee KS, Tsien RW (1983) Mechanism of calcium channel blockade by verapamil, D600, diltiazem and nitrendipine in single dialysed heart cells. Nature 302:790–794

Lee KS, Tsien RW (1984) High selectivity of calcium channels in single dialysed heart cells of the guinea-pig. J Physiol (Lond) 354:253–272

Lee KS, Akaike N, Brown AM (1980) The suction pipette method for internal perfusion and voltage clamp of small excitable cells. J Neurosci Methods 2:51–78

Lee KS, Marban E, Tsien RW (1985) Inactivation of calcium channels in mammalian heart cells: joint dependence on membrane potential and intracellular calcium. J Physiol (Lond) 364:395–411

Leung AT, Imagawa T, Campbell KP (1987) Structural characterization of the 1,4-dihydropyridine receptor of the voltage-dependent calcium channel from rabbit skeletal muscle. J Biol Chem 262:7943–7946

Leung AT, Imagawa T, Block B, Franzini-Armstrong C, Campbell KP (1988) Biochemical and ultrastructural characterization of the 1,4-dihydropyridine receptor from rabbit skeletal muscle. Evidence for a 52000 Da subunit. J Biol Chem 263:994–1001

Levi R, DeFelice LJ (1986) Sodium-conducting channels in cardiac membranes in low calcium. Biophys J 50:5–9

Levitan IB (1988) Modulation of ion channels in neurons and other cells. Annu Rev Neurosci 11:119–136

Limas CJ (1980) Phosphorylation of cardiac sarcoplasmic reticulum by a calcium-activated phospholipid-dependent protein kinase. Biochem Biophys Res Commun 96:1378–1383

Lincoln TM, Keely SL (1981) Regulation of cardiac cyclic GMP-dependent protein kinase. Biochim Biophys Acta 676:230–244

Linden J, Brooker G (1979) The questionable role of cyclic guanosine 3',5'monophosphate in heart. Biochem Pharmacol 28:3351–3360

Linden J, Hollen C, Patel A (1985) The mechanism by which adenosine and cholinergic agents reduce contractility in rat myocardium. Correlation with cyclic AMP and receptor densities. Circ Res 56:728–735

Lipscombe D, Tsien RW (1987) Noradrenaline inhibits N-type Ca channels in isolated frog sympathetic neurones. J Physiol (Lond) 390:84P

Lipscombe D, Madison DV, Poenie M, Reuter H, Tsien RY, Tsien RW (1988) Spatial distribution of calcium channels and cytosolic calcium transients in growth cones and cell bodies of sympathetic neurons. Proc Natl Acad Sci USA 85:2398–2402

Llano I, Marty A (1987) Protein kinase C activators inhibit the inositol trisphosphate-mediated muscarinic current responses in rat lacrimal cells. J Physiol (Lond) 394:239–248

Llinás R, Sugimori M (1980) Electrophysiological properties of in vitro Purkinje cell somata in mammalian cerebellar slices. J Physiol (Lond) 305:171–195

Llinás R, Yarom Y (1981) Properties and distribution of ionic conductances generating electroresponsiveness of mammalian inferior olivary neurones in vitro. J Physiol (Lond) 315:569–584

Loirand G, Pacaud P, Mironneau C, Mironneau J (1986) Evidence for two distinct calcium channels in rat vascular smooth muscle cells in short-term primary culture. Pflügers Arch 407:566–568

Lux HD, Carbone E, Zucker H (1988) Block of sodium currents through a neuronal calcium channel by external calcium and magnesium ions. In: Morad M, Nayler W, Kazda S, Schramm M (eds) The calcium channel: structure, function and implications. Springer, Berlin Heidelberg New York Tokyo, pp 128–137

Madison D, Fox AP, Tsien RW (1987) Adenosine reduces an inactivating component of calcium current in hippocampal CA3 neurons. Biophys J 51:30a

Majerus PW, Connolly TM, Deckman H, Ishii H, Bansal VS, Wilson DB (1986) The metabolism of phosphoinositide-derived messenger molecules. Science 234:1519–1526

Marban E, Tsien RW (1982) Enhancement of calcium current during digitalis inotropy in mammalian heart: positive feed-back regulation by intracellular calcium? J Physiol (Lond) 329:589–614

Marban E, Wier WG (1985) Ryanodine as a tool to determine the contributions of calcium entry and calcium release to the calcium transient and contraction of cardiac Purkinje fibers. Circ Res 56:133–138

Martins TJ, Mumby MC, Beavo JA (1982) Purification and characterization of a cyclic GMP-stimulated cyclic nucleotide phosphodiesterase from bovine tissues. J Biol Chem 25:1973–1979

Massini P, Lüscher EF (1976) On the significance of the influx of calcium ions into stimulated human blood platelets. Biochim Biophys Acta 436:652–663

Matsuda H (1986) Sodium conductance in calcium channels of guinea-pig ventricular cells induced by removal of external calcium ions. Pflügers Arch 407:465–475

Matsuda H, Noma A (1984) Isolation of calcium current and its sensitivity to monovalent cations in dialysed ventricular cells of guinea-pig. J Physiol (Lond) 357:553–573

Matteson DR, Armstrong CM (1986) Properties of two types of calcium channels in clonal pituitary cells. J Gen Physiol 87:161–182

Matthies HJG, Palfrey HC, Hirning LD, Miller RJ (1987) Down regulation of protein kinase C in neuronal cells: effects on neurotransmitter release. J Neurosci 7:1198–1206

McCleskey EW, Almers W (1985) The Ca channel in skeletal muscle is a large pore. Proc Natl Acad Sci USA 82:7149–7153

McCleskey EW, Hess P, Tsien RW (1985) Interaction of organic cations with the cardiac Ca channel. J Gen Physiol 86:22a

McCleskey EW, Fox AP, Feldman D, Cruz LJ, Olivera BM, Tsien RW, Yoshikami D (1987) ω-Conotoxin: direct and persistent blockade of specific types of calcium channels in neurones but not muscle. Proc Natl Acad Sci USA 84:4327–4331

McDonald TF (1982) The slow inward calcium current in the heart. Annu Rev Physiol 44:425–434

McDonald TF, MacLeod DP (1973) Metabolism and the electrical activity of anoxic ventricular muscle. J Physiol (Lond) 229:559–582

McDonald TF, Pelzer D, Trautwein W (1981) Does the calcium current modulate the contraction of the accompanying beat? A study of E-C coupling in mammalian ventricular muscle using cobalt ions. Circ Res 49:576–583

McDonald TF, Pelzer D, Trautwein W (1984) Cat ventricular muscle treated with D600: characteristics of calcium channel block and unblock. J Physiol (Lond) 352:217–241

McDonald TF, Cavalié A, Trautwein W, Pelzer D (1986) Voltage-dependent properties of macroscopic and elementary calcium channel currents in guinea pig ventricular myocytes. Pflügers Arch 406:437–448

McDonald TF, Pelzer D, Trautwein W (1989) Dual action (stimulation, inhibition) of D600 on contractility and calcium channels in guinea pig and cat heart cells. J Physiol (Lond) 414:569–586

Mellgren R (1987) Calcium-dependent proteases: an enzyme system active at cellular membranes? FASEB J 1:110–115

Mentrard D, Vassort G, Fischmeister R (1984) Calcium-mediated inactivation of the calcium conductance in cesium-loaded frog heart cells. J Gen Physiol 83:105–131

Merritt JE, Rink TJ (1987) Regulation of cytosolic free calcium in fura-2-loaded rat parotid acinar cells. J Biol Chem 262:17362–17369

Messing RO, Carpenter CL, Greenberg DA (1986) Inhibition of calcium flux and calcium channel antagonist binding in the PC12 neural cell line by phorbol esters and protein kinase C. Biochem Biophys Res Commun 136:1049–1056

Meves H, Vogel W (1973) Calcium inward currents in internally perfused giant axons. J Physiol (Lond) 235:225–265

Middlemiss DN, Spedding M (1985) A functional correlate for the dihydropyridine binding site in rat brain. Nature 314:94–96

Miller RJ (1987a) Multiple calcium channels and neuronal function. Science 235:46–52

Miller RJ (1987b) Calcium channels in neurons. Receptor Biochem Methodol 9:161–264

Miller RJ (1988) G proteins flex their muscles. Trends in Neurol Sci 11:3–6

Mironneau J (1974) Voltage clamp analysis of the ionic currents in uterine smooth muscle using the double sucrose gap method. Pflügers Arch 352:197–210

Mironneau J, Eugene D, Mironneau C (1982) Sodium action potentials induced by calcium chelation in rat uterine smooth muscle. Pflügers Arch 395:232–238

Mitchell MR, Powell T, Terrar DA, Twist VW (1983) Characteristics of the second inward current in cells isolated from rat ventricular muscle. Proc R Soc Lond [Biol] 219:447–469

Mitchell MR, Powell T, Terrar DA, Twist VW (1985) Influence of a change in stimulation rate on action potentials, currents and contractions in rat ventricular cells. J Physiol (Lond) 364:113–130

Mitra R, Morad M (1986) Two types of calcium channels in guinea pig ventricular myocytes. Proc Natl Acad Sci USA 83:5340–5344

Morton ME, Froehner SC (1987) Monoclonal antibody identifies a 200-kDa subunit of the dihydropyridine-sensitive calcium channel. J Biol Chem 262:11904–11907

Morton ME, Caffrey JM, Brown AM, Froehner SC (1988) Monoclonal antibody to the α_1-subunit of the dihydropyridine-binding complex inhibits calcium currents in BC 3H1 myocytes. J Biol Chem 263:613–616

Movsesian MA, Nishikawa M, Adelstein RS (1984) Phosphorylation of phospholamban by calcium-activated, phospholipid-dependent protein kinase: stimulation of cardiac sarcoplasmic reticulum calcium uptake. J Biol Chem 259:8029–8032

Murad F, Chi YM, Rall TW, Sutherland E (1962) Adenylcyclase III. The effect of catecholamines and choline esters on the formation of adenosine 3',5'-phosphate by preparations from cardiac muscle and liver. J Biol Chem 237:1233–1238

Nachshen DA (1985) The early time course of potassium-stimulated calcium uptake in presynaptic nerve terminals isolated from rat brain. J Physiol (Lond) 361:251–268

Nakazawa K, Matsuki N, Shigenobu K, Kasuya Y (1987) Contractile response and electrophysiological properties in enzymatically dispersed smooth muscle cells of rat vas deferens. Pflügers Arch 408:112–119

Nakazawa K, Saito H, Matsuki N (1988) Fast and slowly inactivating components of Ca-channel current and their sensitivities to nicardipine in isolated smooth muscle cells from rat vas deferens. Pflügers Arch 411:289–295

Narahashi T, Tsunoo A, Yoshii M (1987) Characterization of two types of calcium channels in mouse neuroblastoma cells. J Physiol (Lond) 383:231–249

Nargeot J, Garnier D (1982) The action of muscarinic agonists and antagonists on frog atrial fibers. Electrophysiological studies. J Pharmacol 13:431–445

Nargeot J, Nerbonne JM, Engels J, Lester HA (1983) Time course of the increase in the myocardial slow inward current after a photochemically generated concentration jump of intracellular cAMP. Proc Natl Acad Sci USA 80:2395–2399

Nastainczyk W, Röhrkasten A, Sieber M, Rudolph C, Schächtele C, Marmé D, Hofmann F (1987) Phosphorylation of the purified receptor for calcium channel blockers by cAMP kinase and protein kinase C. Eur J Biochem 169:137–142

Nathan RD, Kanai K, Clark RB, Giles W (1988) Selective block of calcium current by lanthanum in single bullfrog atrial cells. J Gen Physiol 91:549–572

Navarro J (1987) Modulation of [^{3}H] dihydropyridine receptors by activation of protein kinase C in chick muscle cells. J Biol Chem 262:4649–4652

Nawrath H (1977) Does cyclic GMP mediate the negative inotropic effect of acetylcholine in the heart? Nature 267:72–74

Nelson MT, Standen NB, Brayden JE, Worley JF III (1988) Noradrenaline contracts arteries by activating voltage-dependent calcium channels. Nature 336:382–385

Nilius B, Benndorf K (1986) Joint voltage- and calcium-dependent inactivation of Ca channels in frog atrial myocardium. Biomed Biochim Acta 45:795–811

Nilius B, Hess P, Lansman JB, Tsien RW (1985) A novel type of cardiac calcium channel in ventricular cells. Nature 316:443–446

Nishizuka Y (1986) Studies and perspectives of protein kinase C. Science 233:305–312

Nishizuka Y (1988) The molecular heterogeneity of protein kinase C and its implications for cellular regulation. Nature 334:661–665

Noble D (1984) The surprising heart: a review of recent progress in cardiac electrophysiology. J Physiol (Lond) 353:1–50

Noble S, Shimoni Y (1981) The calcium and frequency dependence of the slow inward current "staircase" in frog atrium. J Physiol (Lond) 310:57–75

Noma A, Shibasaki T (1985) Membrane current through adenosine-triphosphate-regulated potassium channels in guinea-pig ventricular cells. J Physiol (Lond) 363:463−480

Noma A, Trautwein W (1978) Relaxation of the Ach-induced potassium current in the rabbit sinoatrial node cell. Pflügers Arch 377:193−200

Noma A, Kotake H, Irisawa H (1980) Slow inward current and its role mediating the chronotropic effect of epinephrine in the rabbit sinoatrial node. Pflügers Arch 388:1−9

Norman RI, Burgess AJ, Allen E, Harrison TM (1987) Monoclonal antibodies against the 1,4-dihydropyridine receptor associated with voltage-sensitive Ca^{2+} channels detect similar polypeptides from a variety of tissues and species. FEBS Lett 212:127−132

Nowycky MC, Fox AP, Tsien RW (1984) Two components of calcium channel current in chick dorsal root ganglion cells. Biophys J 45:36a

Nowycky MC, Fox AP, Tsien RW (1985) Three types of neuronal calcium channel with different calcium agonist sensitivity. Nature 316:440−443

Ochi R (1970) The slow inward current and the action of manganese ions in guinea-pig's myocardium. Pflügers Arch 316:81−94

Ochi R (1975) Manganese action potentials in mammalian cardiac muscle. Experientia 31:1048−1049

Ochi R, Hino N, Okuyama H (1986) β-Adrenergic modulation of the slow gating process of cardiac calcium channels. Jpn Heart J 27 (Suppl):51−55

Ohmori H, Yoshii M (1977) Surface potential reflected in both gating and permeation mechanisms of sodium and calcium channels of the tunicate egg cell membrane. J Physiol (Lond) 267:429−463

Ohya Y, Kitamura K, Kuriyama H (1987) Modulation of ionic currents in smooth muscle balls of the rabbit intestine by intracellularly perfused ATP and cyclic AMP. Pflügers Arch 408:465−473

Ohya Y, Kitamura K, Kuriyama H (1988) Regulation of calcium current by intracellular calcium in smooth muscle cells of rabbit portal vein. Circ Res 62:375−383

Osterrieder W, Brum G, Hescheler J, Trautwein W, Flockerzi V, Hofmann F (1982) Injection of subunits of cyclic AMP-dependent protein kinase into cardiac myocytes modulates Ca^{2+} current. Nature 298:576−578

Osugi T, Imaizumi T, Mizushima A, Uchida S, Yoshida H (1986) 1-Oleoyl-2-acetyl-glycerol and phorbol diester stimulate Ca^{2+} influx through Ca^{2+} channels in neuroblastoma × glioma hybrid NG108-15 cells. Eur J Pharmacol 126:47−51

Pacaud P, Loirand G, Mironneau C, Mironneau J (1987) Opposing effects of noradrenaline on the two classes of voltage-dependent calcium channels of single vascular smooth muscle cells in short-term primary culture. Pflügers Arch 410:557−559

Palade PT, Almers W (1985) Slow calcium and potassium currents in frog skeletal muscle: their relationship and pharmacologic properties. Pflügers Arch 405:91−101

Paupardin-Tritsch D, Hammond C, Gerschenfeld HM, Nairn AC, Greengard P (1986) cGMP-dependent protein kinase enhances Ca^{2+} current and potentiates the serotonin-induced Ca^{2+} current increase in snail neurones. Nature 323:812−814

Payett MD, Schanne OF, Ruiz-Ceretti E (1981) Frequency dependence of the ionic currents determining the action potential repolarization in rat ventricular muscle. J Mol Cell Cardiol 13:207−215

Peach MJ (1977) Renin-angiotensin system: biochemistry and mechanism of action. Physiol Rev 57:313−370

Pelzer D, Trautwein W, McDonald TF (1982) Calcium channel block and recovery from block in mammalian ventricular muscle treated with organic channel inhibitors. Pflügers Arch 394:97−105

Pelzer D, Cavalié A, Trautwein W (1985) Guinea-pig ventricular myocytes treated with D600: mechanism of calcium-channel blockade at the level of single channels. In: Lichtlen PR (ed) Recent aspects in calcium antagonism. Schattauer, Stuttgart, pp 3−26

Pelzer D, Cavalié A, Trautwein W (1986a) Activation and inactivation of single calcium channels in cardiac cells. Exp Brain Res 14:17−34

Pelzer D, Cavalié A, McDonald TF, Trautwein W (1986b) Macroscopic and elementary currents through cardiac calcium channels. Prog Zool 33:83–98

Pelzer D, Cavalié A, Flockerzi V, Hofmann F, Trautwein W (1988) Reconstitution of solubilized and purified dihydropyridine receptor from skeletal muscle microsomes as two single calcium channel conductances with different functional properties. In: Morad M, Nayler W, Kazda S, Schramm M (eds) The calcium channel: structure, function and implications. Springer, Berlin Heidelberg New York Tokyo, pp 217–230

Pelzer D, Cavalié A, McDonald TF, Trautwein W (1989a) Calcium channels in single heart cells. In: Piper HM, Isenberg G (eds) Isolated adult cardiomyocytes II. CRC Press, Boca Raton, pp 29–73

Pelzer D, Grant AO, Cavalié A, Pelzer S, Sieber M, Hofmann F, Trautwein W (1989b) Calcium channels reconstituted from the skeletal muscle dihydropyridine receptor protein complex and its α_1 peptide subunit in lipid bilayers. Ann NY Acad Sci 560:138–154

Penner R, Dreyer F (1986) Two different presynaptic calcium currents in mouse motor nerve terminals. Pflügers Arch 406:190–197

Penner R, Matthews G, Neher E (1988) Regulation of calcium influx by second messengers in rat mast cells. Nature 334:499–504

Perez-Reyez E, Kim HS, Lacerda AE, Horne W, Wei X, Rampe D, Campbell KP, Brown AM, Birnbaumer L (1989) Induction of calcium currents by the expression of the α_1-subunit of the dihydropyridine receptor from skeletal muscle. Nature 340:233–236

Perney TM, Hirning LD, Leeman SE, Miller RJ (1986) Multiple calcium channels mediate neurotransmitter release from peripheral neurons. Proc Natl Acad Sci USA 83:6656–6659

Pfaffinger PJ, Martin JM, Hunter DD, Nathanson NM, Hille B (1985) GTP-binding proteins couple cardiac muscarinic receptors to a K channel. Nature 317:536–538

Pietrobon D, Prod'hom B, Hess P (1988) Conformational changes associated with ion permeation in L-type calcium channels. Nature 333:373–376

Plant TD, Standen NB, Ward TA (1983) The effects of injection of calcium ions and calcium chelators on calcium channel inactivation in *Helix* neurones. J Physiol (Lond) 334:189–212

Poggioli J, Sulpice JC, Vassort G (1986) Inositol phosphate production following α_1-adrenergic, muscarinic or electrical stimulation in isolated rat heart. FEBS Lett 206:292–298

Porzig H (1990) Pharmacological modulation of voltage-dependent Ca channels in intact cells. In: Blaustein MP et al. (eds) Reviews of Physiology, Biochemistry and Pharmacology, Vol 114. Springer, Berlin Heidelberg New York, pp 209–262

Potreau D, Raymond G (1982) Existence of a sodium-induced calcium release mechanism on frog skeletal muscle fibres. J Physiol (Lond) 333:463–480

Powell JA, Fambrough DM (1973) Electrical properties of normal and dysgenic mouse skeletal muscle in culture. J Cell Physiol 82:21–38

Powell T, Twist VW (1976) Isoprenaline stimulation of cyclic AMP production by isolated cells from adult rat myocardium. Biochem Biophys Res Commun 72:1218–1225

Prod'hom B, Pietrobon D, Hess P (1987) Direct measurement of proton transfer rates to a group controlling the dihydropyridine-sensitive Ca^{2+} channel. Nature 329:243–246

Prosser CL, Kreulen DL, Weigel RJ, Yau W (1977) Prolonged potentials in gastrointestinal muscles induced by calcium chelation. Am J Physiol 233:C19–C24

Quastel DM, Saint DA, Guan YY (1986) Does the motor nerve terminal have only one transmitter release system and only one species of Ca^{2+} channel? Soc Neurosci Abstr 12:28

Rane SG, Dunlap K (1986) Kinase C activator 1,2-oleoylacetylglycerol attenuates voltage-dependent calcium current in sensory neurons. Proc Natl Acad Sci USA 83:184–188

Rane SG, Holz IV GG, Dunlap K (1987) Dihydropyridine inhibition of neuronal calcium current and substance P release. Pflügers Arch 409:361–366

Rapoport RM, Ginsburg R, Waldman SA, Murad F (1986) Effects of atriopeptins on relaxation and cyclic GMP levels in human coronary artery in vitro. Eur J Pharmacol 124:193–196

Rasmussen H, Forder J, Kojima K, Scriabine A (1984) TPA-induced contraction of isolated rabbit vascular smooth muscle. Biochem Biophys Res Commun 122:776–784

Reuter H (1965) Über die Wirkung von Adrenalin auf den zellulären Ca-Umsatz des Meerschweinchenvorhofs. Naunyn-Schmiedebergs Arch Pharmacol 251:401–412

Reuter H (1966) Strom-Spannungsbeziehungen von Purkinje-Fasern bei verschiedenen extrazellulären Kalzium-Konzentrationen und unter Adrenalineinwirkung. Pflügers Arch 287:357–367

Reuter H (1967) The dependence of slow inward current in Purkinje fibres on the extracellular calcium-concentration. J Physiol (Lond) 192:479–492

Reuter H (1973) Divalent cations as charge carriers in excitable membranes. Prog Biophys Mol Biol 26:3–43

Reuter H (1974) Localization of β-adrenergic receptors, and effects of noradrenaline and cyclic nucleotides on action potentials, ionic currents and tension in mammalian cardiac muscle. J Physiol (Lond) 242:429–451

Reuter H (1979) Properties of two inward membrane currents in the heart. Annu Rev Physiol 41:413–424

Reuter H (1983) Calcium channel modulation by neurotransmitters, enzymes and drugs. Nature 301:569–574

Reuter H (1985) A variety of calcium channels. Nature 316:391

Reuter H (1987) Modulation of ion channels by phosphorylation and second messengers. News Physiol Sci 2:168–171

Reuter H, Scholz H (1977a) A study of the ion selectivity and the kinetic properties of the calcium dependent slow inward current in mammalian cardiac muscle. J Physiol (Lond) 264:17–47

Reuter H, Scholz H (1977b) The regulation of the calcium conductance of cardiac muscle by adrenaline. J Physiol (Lond) 264:49–62

Reuter H, Stevens CF, Tsien RW, Yellen G (1982) Properties of single calcium channels in cardiac cell culture. Nature 297:501–504

Reuter H, Cachelin AB, de Peyer JE, Kokubun S (1983) Modulation of calcium channels in cultured cardiac cells by isoproterenol and 8-bromo-cAMP. Cold Spring Harbor Symp Quant Biol 48:193–200

Reynolds IJ, Wagner JA, Snyder SH, Thayer SA, Olivera BM, Miller RJ (1986) Brain voltage-sensitive calcium channel subtypes of differentiated by ω-conotoxin fraction GVIA. Proc Natl Acad Sci USA 83:8804–8807

Rieger F, Bournaud R, Shimahara T, Garcia L, Pincon-Raymond M, Romey G, Lazdunski M (1987) Restoration of dysgenic muscle contraction and calcium channel function by co-culture with normal spinal cord neurons. Nature 330:563–566

Rink TJ (1988) A real receptor-operated calcium channel? Nature 334:649–650

Rink TJ, Hallam TJ (1984) What turns platelets on? Trends Biol Sci 9:215–219

Rios E, Brum G (1987) Involvement of dihydropyridine receptors in excitation-contraction coupling in skeletal muscle. Nature 325:717–720

Rogers TB (1984) High affinity angiotensin II receptors in myocardial sarcolemmal membranes. J Biol Chem 259:8106–8114

Röhrkasten A, Meyer HE, Nastainczyk W, Sieber M, Hofmann F (1988a) cAMP-dependent protein kinase rapidly phosphorylates serine-687 of the skeletal muscle receptor for calcium channel blockers. J Biol Chem 263:15325–15329

Röhrkasten A, Meyer HE, Schneider T, Nastainczyk W, Sieber M, Jahn H, Regulla S, Ruth P, Flockerzi V, Hofmann F (1988b) Site-specific phosphorylation of the skeletal muscle receptor for calcium-channel blockers by cAMP-dependent protein kinase. In: Morad M, Nayler W, Kazda S, Schramm M (eds) The calcium channel: structure, function and implications. Springer, Berlin Heidelberg New York Tokyo, pp 193–199

Rorsman P (1988) Two types of Ca^{2+} currents with different sensitivities to organic Ca^{2+} channel antagonists in guinea pig pancreatic α_2 cells. J Gen Physiol 91:243–254

Rosenberg RL, Hess P, Reeves J, Smilowitz H, Tsien RW (1986) Calcium channels in planar lipid bilayers: new insights into the mechanisms of permeation and gating. Science 231:1564–1566

Rosenberg RL, Hess P, Tsien RW (1988) Cardiac calcium channels in planar lipid bilayers. L-type channels and calcium-permeable channels open at negative membrane potentials. J Gen Physiol 92:27–54

Rougier O, Vassort G, Garnier D, Gargouil YM, Coraboeuf E (1969) Existence and role of a slow inward current during the frog atrial action potential. Pflügers Arch 308:91–110

Sage SO, Rink TJ (1987) The kinetics of changes in intracellular calcium concentration in fura-2-loaded human platelets. J Biol Chem 262:16364–16369

Sakmann B, Neher E (eds) (1983) Single-channel recording. Plenum, New York

Sakmann B, Neher E (1984) Patch clamp techniques for studying ionic channels in excitable membranes. Annu Rev Physiol 46:455–472

Sánchez JA, Stefani E (1978) Inward calcium current in twitch muscle fibres of the frog. J Physiol (Lond) 283:197–209

Sánchez JA, Stefani E (1983) Kinetic properties of calcium channels of twitch muscle fibres of the frog. J Physiol (Lond) 337:1–17

Sato R, Noma A, Kurachi Y, Irisawa H (1985) Effects of intracellular acidification on membrane currents in ventricular cells of the guinea-pig. Circ Res 57:553–561

Satoh H, Hashimoto K, Seyama I (1982) Effects of changes in extracellular pH on the membrane current of rabbit atrial node cells. Jpn Heart J 23:57–59

Schmid A, Renaud JF, Lazdunski M (1985) Short term and long term effects of β-adrenergic effectors and cyclic AMP on nitrendipine-sensitive voltage-dependent Ca^{2+} channels of skeletal muscle. J Biol Chem 260:13041–13046

Schneider MF, Chandler WK (1973) Voltage-dependent charge movement in skeletal muscle: a possible step in excitation-contraction coupling. Nature 242:244–246

Schrader J, Baumann G, Gerlach E (1977) Adenosine as inhibitor of myocardial effects of catecholamines. Pflügers Arch 372:29–35

Schultheiss HP, Janda I, Kühl U, Ulrich G, Morad M (1988) Antibodies against the ADP/ATP carrier interact with the calcium channel and induce cytotoxicity by enhancement of calcium permeability. In: Morad W, Nayler W, Kazda S, Schramm M (eds) The calcium channel: structure, function and implications. Springer, Berlin Heidelberg New York Tokyo, pp 619–631

Schwartz LM, McCleskey EM, Almers W (1985) Dihydropyridine receptors in muscle are voltage-dependent but most are not functional calcium channels. Nature 314:747–750

Scott RH, Dolphin AC (1987) Activation of a G protein promotes agonist responses to calcium channel ligands. Nature 330:760–762

Seager MJ, Takahashi M, Catterall WA (1988) Molecular properties of dihydropyridine-sensitive calcium channels from skeletal muscle. In: Morad M, Nayler W, Kazda S, Schramm M (eds) The calcium channel: Structure, function and implications. Springer, Berlin Heidelberg New York Tokyo, pp 200–210

Seamon K, Daly J (1983) Forskolin, cyclic AMP and cellular physiology. TIPS 4:120–123

Seamon K, Wetzel B (1984) Interaction of forskolin with dually regulated adenylate cyclase. Adv Cyclic Nucl Res 17:91–99

Sharp AH, Gaver M, Kahl SD, Campbell KP (1988) Structural characterization of the 32 kDa subunit of the skeletal muscle 1,4-dihydropyridine receptor. Biophys J 53:231a

Shenolikar S, Karbon EW, Enna SJ (1986) Phorbol esters down-regulate protein kinase C in rat brain cerebral cortical slices. Biochem Biophys Res Commun 139:251–258

Shibata EF, Giles WR (1985) Ionic currents that generate the spontaneous diastolic depolarization in individual cardiac pacemaker cells. Proc Natl Acad Sci USA 82:7796–7800

Shibata EF, Northup JK, Momose Y, Giles W (1986) Muscarinic receptor and guanine nucleotides mediated inhibition of I_{Ca} in single cells from bullfrog atrium. Biophys J 49:349a

Shibata EF, Northup JK, Momose Y, Giles W (1987) Is diacylglycerol a second messenger for the muscarinic guanine nucleotide mediated inhibition of calcium current in bullfrog atrium? Biophys J 51:413a

Shimoni Y (1981) Parameters affecting the slow inward channel repriming process in frog atrium. J Physiol (Lond) 320:269–291

Shimoni Y, Raz S, Gotsman M (1984) Two potentially arrhythmogenic mechanisms of adrenaline action in cardiac muscle. J Mol Cell Cardiol 16:471–476

Shimoni Y, Spindler AJ, Noble D (1987) The control of calcium current reactivation by catecholamines and acetylcholine in single guinea-pig ventricular myocytes. Proc R Soc Lond [Biol] 230:267–278

Sibley DR, Nambi P, Peters JR, Lefkowitz RJ (1984) Phorbol diesters promote β-adrenergic receptor phosphorylation and adenylate cyclase desensitization in duck erythrocytes. Biochem Biophys Res Commun 121:973–979

Singh J, Flitney FW (1981) Inotropic responses of the frog ventricle to dibutyryl cyclic AMP and 8-bromo-cyclic GMP and related changes in endogenous cyclic nucleotide levels. Biochem Pharmacol 30:1475–1481

Sperelakis N (1988) Regulation of calcium slow channels of cardiac muscle by cyclic nucleotides and phosphorylation. J Mol Cell Cardiol 20 (Suppl II):75–105

Sperelakis N, Schneider J (1976) A metabolic control mechanism for calcium ion influx that may protect the ventricular myocardial cell. Am J Cardiol 37:1079–1085

Sperti G, Colucci WS (1987) Phorbol ester-stimulated bidirectional transmembrane calcium flux in A7r5 vascular smooth muscle cells. Mol Pharmacol 32:37–42

Stanfield PR (1977) A calcium-dependent inward current in frog skeletal muscle fibres. Pflügers Arch 368:267–270

Stefani E, Toro L, García J (1987) α- and β-adrenergic stimulation of fast and slow Ca^{2+} channels in frog skeletal muscle. Biophys J 51:425a

Strong JA, Fox AP, Tsien RW, Kaczmarek LK (1987) Stimulation of protein kinase C recruits covert calcium channels in Aplysia bag cell neurons. Nature 325:714–717

Sturek M, Hermsmeyer K (1986) Calcium and sodium channels in spontaneously contracting vascular muscle cells. Science 233:475–478

Tajima T, Tsuji Y, Brown JH, Pappano AJ (1987) Pertussis toxin-insensitive phosphoinositide hydrolysis, membrane depolarization, and positive inotropic effect of carbachol in chick atria. Circ Res 61:436–445

Takahashi M, Catterall WA (1987a) Identification of an α subunit of dihydropyridine-sensitive brain calcium channels. Science 236:88–91

Takahashi M, Catterall WA (1987b) Dihydropyridine-sensitive calcium channels in cardiac and skeletal membranes: studies with antibodies against the α subunit. Biochemistry 26:5518–5526

Takahashi M, Seager MJ, Jones JF, Reber BFX, Catterall WA (1987) Subunit structure of dihydropyridine-sensitive calcium channels from skeletal muscle. Proc Natl Acad Sci USA 84:5478–5482

Takai Y, Minakuchi R, Kikkawa U, Sano K, Kaibuchi K, Yu B, Matsubara T, Nishizuka Y (1982) Membrane phospholipid turnover, receptor function, and protein phosphorylation. Prog Brain Res 56:287–301

Takuwa Y, Takuwa N, Rasmussen H (1986) Carbachol induces a rapid and sustained hydrolysis of polyphosphoinositide in bovine tracheal smooth muscle measurements of the mass of polyphosphoinositides, 1,2-diacylglycerol, and phosphatidic acid. J Biol Chem 261:14670–14675

Talvenheimo JA, Worley JF III, Nelson MT (1987) Heterogeneity of calcium channels from a purified dihydropyridine receptor preparation. Biophys J 52:891–899

Tanabe T, Takeshima H, Mikami A, Flockerzi V, Takahashi H, Kangawa K, Kojima M, Matsuo H, Hirose T, Numa S (1987) Primary structure of the receptor for calcium channel blockers from skeletal muscle. Nature 328:313–318

Tanabe T, Beam KG, Powell JA, Numa S (1988) Restoration of excitation-contraction coupling and slow calcium current in dysgenic muscle by dihydropyridine receptor complementary DNA. Nature 336:134–139

Tang CM, Morad M (1988) Amiloride selectively blocks the low threshold (T) calcium channel. Biophys J 53:22a

Tang CM, Presser F, Morad M (1988) Amiloride selectively blocks the low threshold (T) calcium channel. Science 240:213–215

Taniguchi J, Noma A, Irisawa H (1983) Modification of the cardiac action potential by intracellular injection of adenosine triphosphate and related substances in guinea pig single ventricular cells. Circ Res 53:131–139

Taylor CJ, Meisheri KD (1986) Inhibitory effects of a synthetic atrial peptide on contractions and ^{45}Ca fluxes in vascular smooth muscle. J Pharmacol Exp Ther 237:803–808

Ten Eick R, Nawrath H, McDonald TF, Trautwein W (1976) On the mechanism of the negative inotropic effect of acetylcholine. Pflügers Arch 361:207–213

Teutsch T, Beer W, Frisch G, Weible A, Ruoff HJ (1988) Interaction between phorbol dibutyrate and forskolin on cAMP accumulation and force in isolated left guinea pig atria. Naunyn-Schmiedebergs Arch Pharmacol 337:R61

Toro L, Stefani E (1987) Ca^{2+} and K$^+$ current in cultured vascular smooth muscle cells from rat aorta. Pflügers Arch 408:417–419

Trautwein W, Pelzer D (1985a) Voltage-dependent gating of single calcium channels in the cardiac cell membrane and its modulation by drugs. In: Marmé D (ed) Calcium and cell physiology. Springer, Berlin Heidelberg New York Tokyo, pp 53–93

Trautwein W, Pelzer D (1985b) Gating of single calcium channels in the membrane of enzymatically isolated ventricular myocytes from adult mammalian hearts. In: Zipes DP, Jalife J (eds) Cardiac electrophysiology and arrhythmias. Grune and Stratton, Orlando, pp 31–42

Trautwein W, Pelzer D (1986) Single calcium channels in isolated cardiac cells. In: Bader H, Gietzen K, Rosenthal J, Rüdel R, Wolf HU (eds) Intracellular calcium regulation. Manchester University Press, Manchester, pp 15–34

Trautwein W, Pelzer D (1988) Kinetics and β-adrenergic modulation of cardiac Ca^{2+} channels. In: Morad M, Nayler W, Kazda S, Schramm M (eds) The calcium channel: structure, function and implications. Springer, Berlin Heidelberg New York Tokyo, pp 39–53

Trautwein W, McDonald TF, Tripathi O (1975) Calcium conductance and tension in mammalian ventricular muscle. Pflügers Arch 354:55–74

Trautwein W, Taniguchi J, Noma A (1982) The effect of intracellular cyclic nucleotides and calcium on the action potential and acetylcholine response of isolated cardiac cells. Pflügers Arch 392:307–314

Trautwein W, Kameyama M, Hescheler J, Hofmann F (1986) Cardiac calcium channels and their transmitter modulation. Prog Zool 33:163–182

Trautwein W, Cavalié A, Flockerzi V, Hofmann F, Pelzer D (1987) Modulation of calcium channel function by phosphorylation in guinea pig ventricular cells and phospholipid bilayer membranes. Circ Res 61 (Suppl I):I-17–I-23

Triggle DJ, Janis RA (1987) Calcium channel ligands. Annu Rev Pharmacol Toxicol 27:347–370

Trube G (1983) Enzymatic dispersion of heart and other tissues. In: Sakmann B, Neher E (eds) Single-channel recording. Plenum, New York, pp 69–76

Tseng GN (1988) Calcium current restitution in mammalian ventricular myocytes is modulated by intracellular calcium. Circ Res 63:468–482

Tseng GN, Robinson RB, Hoffman BF (1987) Passive properties and membrane currents of canine ventricular myocytes. J Gen Physiol 90:671–701

Tsien RW (1973) Adrenaline-like effects of intracellular iontophoresis of cyclic AMP in cardiac Purkinje fibers. Nature 245:120–122

Tsien RW (1983) Calcium channels in excitable cell membranes. Annu Rev Physiol 45:341–358

Tsien RW (1987) Calcium currents in heart cells and neurons. In: Kaczmarek LK, Levitan IB (eds) Neuromodulation. The biochemical control of neuronal excitability. Oxford University Press, Oxford, pp 206–242

Tsien RW, Giles WR, Greengard P (1972) Cyclic AMP mediates the effects of adrenaline on cardiac Purkinje fibers. Nature 240:181–183

Tsien RW, Bean BP, Hess P, Lansman JB, Nilius B, Nowycky MC (1986) Mechanisms of calcium channel modulation by β-adrenergic agents and dihydropyridine calcium agonists. J Mol Cell Cardiol 18:691–710

Tsien RW, Hess P, McCleskey EW, Rosenberg RL (1987) Calcium channels: mechanisms of selectivity, permeation, and block. Annu Rev Biophys Biophys Chem 16:265–290

Tsien RW, Lipscombe D, Madison DV, Bley RK, Fox AP (1988) Multiple types of neuronal calcium channels and their selective modulation. TINS 11:431–438

Tsuji Y, Inoue D, Pappano AJ (1985) β-Adrenoceptor agonist accelerates recovery from inactivation of calcium-dependent action potentials. J Mol Cell Cardiol 17:517–521

Tytgat J, Nilius B, Vereecke J, Carmeliet E (1988) The T-type Ca channel in guinea-pig ventricular myocytes is insensitive to isoproterenol. Pflügers Arch 411:704–706

Uehara A, Hume JR (1985) Interactions of organic calcium channel antagonists with calcium channels in single frog atrial cells. J Gen Physiol 85:621–647

Uglesity A, Sharma VK, Sheu SS (1987) Effect of protein kinase C activation on the inotropic response induced by α-adrenoceptor stimulation in rat myocardium. Biophys J 51:264a

Ui M (1984) Islet-activating protein, pertussis toxin: a probe for functions of the inhibitory guanin nucleotide regulatory component of adenylate cyclase. TIPS 5:277–279

Vaghy PL, McKenna E, Schwartz A (1988) Molecular characterization of the 1,4-dihydropyridine receptor in skeletal muscle. In: Morad M, Nayler W, Kazda S, Schramm M (eds) The calcium channel: structure, function and implications. Springer, Berlin Heidelberg, New York Tokyo, pp 211–216

van Breemen C, Aaronson P, Loutzenhiser R (1979) Sodium-calcium interactions in mammalian smooth muscle. Pharmacol Rev 30:167–208

Vandaele S, Fosset M, Galizzi JP, Lazdunski M (1987) Monoclonal antibodies that coimmunoprecipitate the 1,4-dihydropyridine and phenylalkylamine receptors and reveal the Ca^{2+} channel structure. Biochemistry 26:5–9

Vassort G, Rougier O, Garnier D, Sauviat MP, Coraboeuf E, Gargouil YM (1969) Effects of adrenaline on membrane inward currents during the cardiac action potential. Pflügers Arch 309:70–81

Vergara J, Tsien RY, Delay M (1985) Inositol 1,4,5-trisphosphate: a possible chemical link in excitation-contraction coupling in muscle. Proc Natl Acad Sci USA 82:6352–6356

Vilven J, Coronado R (1988) Opening of dihydropyridine calcium channels in skeletal muscle membranes by inositol trisphosphate. Nature 336:587–589

Vilven J, Leung AT, Imagawa T, Sharp AH, Campbell KP, Coronado R (1988) Interaction of calcium channels of skeletal muscle with monoclonal antibodies specific for its dihydropyridine receptor. Biophys J 53:556a

Vivaudou MB, Clapp LH, Walsh JV Jr, Singer JJ (1988) Regulation of one type of Ca^{2+} current in smooth muscle cells by diacylglycerol and acetylcholine. FASEB J 2:2497–2504

Volpe P, Salviati G, Di Virgilio F, Pozzan T (1985) Inositol 1,4,5-trisphosphate induces calcium release from sarcoplasmic reticulum of skeletal muscle. Nature 316:347–349

Walsh JV Jr, Singer JJ (1981) Voltage clamp of single freshly dissociated smooth muscle cells: current-voltage relationships for three currents. Pflügers Arch 390:207–210

Walsh JV Jr, Singer JJ (1987) Identification and characterization of major ionic currents in isolated smooth muscle cells using the voltage-clamp technique. Pflügers Arch 408:83–97

Walsh KB, Kass RS (1988) Regulation of a heart potassium channel by protein kinase A and C. Science 242:67–69

Wanke E, Ferroni A, Malgaroli A, Ambrosini A, Pozzan T, Meldolesi J (1987) Activation of a muscarinic receptor selectively inhibits a rapidly inactivated Ca^{2+} current in rat sympathetic neurons. Proc Natl Acad Sci USA 84:4313–4317

Watanabe AM, Besch HR (1975) Interaction between cyclic adenosine monophosphate and cyclic guanosine monophosphate in guinea pig ventricular myocardium. Circ Res 37:309–317

Watanabe AM, Lindemann JP, Fleming JW (1984) Mechanisms of muscarinic modulation of protein phosphorylation in intact ventricles. Fed Proc 43:2618–2623

Weingart R, Kass RS, Tsien RW (1978) Is digitalis inotropy associated with enhanced slow inward calcium current? Nature 273:389–391

Wendt-Gallitelli MF, Isenberg G (1985) Extra- and intracellular lanthanum: modified calcium distribution, inward currents and contractility in guinea pig ventricular preparations. Pflügers Arch 405:310–322

West GA, Belardinelli L (1985) Correlation of sinus slowing and hyperpolarization caused by adenosine in sinus node. Pflügers Arch 403:75–81

West GA, Isenberg G, Belardinelli L (1986) Antagonism of forskolin effects by adenosine in iso-
    lated hearts and ventricular myocytes. Am J Physiol 250:H769–H777
White RE, Hartzell HC (1988) Effects of intracellular free magnesium on calcium current in
    isolated cardiac myocytes. Science 239:778–780
Wilkerson RD, Paddock RJ, George WJ (1976) Effects of derivatives of cyclic AMP and cyclic
    GMP on contraction force of cat papillary muscles. Eur J Pharmacol 36:247–251
Wilson DL, Morimoto K, Tsuda Y, Brown AM (1983) Interaction between calcium ions and sur-
    face charge as it relates to calcium currents. J Membr Biol 72:117–130
Wise BC, Raynor RL, Kuo JF (1982) Phospholipid-sensitive Ca^{2+}-dependent protein kinase
    from heart. I. Purification and general properties. J Biol Chem 257:8481–8488
Wolff J, Hope-Cook G, Goldhammer A, Londoz C, Hewlett E (1984) *Bordetella pertussis*: mul-
    tiple attacks on host cell cyclic AMP regulation. Adv Cyclic Nucl Res 17:161–172
Worley JF III, Deitmer JW, Nelson MT (1986) Single nisoldipine-sensitive calcium channels in
    smooth muscle cells isolated from rabbit mesenteric artery. Proc Natl Acad Sci USA
    83:5746–5750
Wright GB, Alexander RW, Ekstein LS, Gimbrone MA Jr (1983) Characterization of the rabbit
    ventricular myocardial receptor for angiotensin II. Mol Pharmacol 24:213–221
Yaari Y, Hamon B, Lux HD (1987) Development of two types of calcium channels in cultured
    mammalian hippocampal neurons. Science 235:680–682
Yatani A, Goto M (1983) The effect of extracellular low pH on the plateau current in isolated,
    single rat ventricular cells – a voltage clamp study. Jpn J Physiol 33:403–415
Yatani A, Codina J, Imoto Y, Reeves JP, Birnbaumer L, Brown AM (1987a) A G protein directly
    regulates mammalian cardiac calcium channels. Science 238:1288–1292
Yatani A, Codina J, Brown AM, Birnbaumer L (1987b) Direct activation of mammalian atrial
    muscarinic potassium channels by GTP regulatory protein G_k. Science 235:207–211
Yatani A, Seidel CL, Allen J, Brown AM (1987c) Whole-cell and single-channel calcium cur-
    rents of isolated smooth muscle cells from saphenous vein. Circ Res 60:523–533
Yatani A, Imoto Y, Codina J, Hamilton SL, Brown AM, Birnbaumer L (1988) The stimulatory
    G protein of adenylyl cyclase, G_s, also stimulates dihydropyridine-sensitive Ca^{2+} channels.
    Evidence for direct regulation independent of phosphorylation by cAMP-dependent protein
    kinase or stimulation by a dihydropyridine agonist. J Biol Chem 263:9887–9895
Yoshino M, Yabu H (1985) Single Ca channel currents in mammalian visceral smooth muscle
    cells. Pflügers Arch 404:285–286
Zschauer A, van Breemen C, Bühler FR, Nelson MT (1988) Calcium channels in thrombin-acti-
    vated human platelet membrane. Nature 334:703–705

Text Added in Proof

Direct Regulation of Calcium Channels by G Proteins in Intact Cardiac Cells

In guinea pig ventricular myocytes dialyzed with a substrate-free minimal intracellular solution
containing phosphorylation inhibitory agents, stimulation of basal I_{Ca} by 0.1 μM isoproterenol
(about 35% at +10 mV) could still be evoked after ineffective trials with forskolin (Heßlinger
et al. 1989; Trautwein et al. 1989) suggesting that β-adrenoreceptor occupation can stimulate
I_{Ca} even if the cAMP-dependent phosphorylation pathway is blocked. Nonhydrolyzable GTP
analogues (GTP-γ-S, GMP-PNP) also enhanced I_{Ca} amplitude by 20%–50% (+10 mV), slowed I_{Ca} inactivation and shifted the voltage eliciting maximum I_{Ca} by 5–10 mV in the negative
direction; dialysates containing GTP or GDP-β-S were ineffective, and predialysis with GDP-β-S
blocked stimulation by GTP-γ-S (Heßlinger et al. 1989). The inactivation of G_i by pretreatment
with PTX did not block enhancement, and a G_p-activating regimen was without effect. Finally,
the augmentation of I_{Ca} amplitude and the slowing of I_{Ca} inactivation by GTP analogues was
reproduced by cell dialysis with G_s^* after blockade of cAMP-dependent cytoplasmic signalling
pathways (Trautwein et al. 1989). This direct membrane-delimited G_s pathway between the

β-adrenoreceptor and the Ca channel is fast (time constant of about 150 ms) and may account for the ability of cardiac sympathetic nerves to change heart rate within a single beat (Yatani and Brown 1989). A further attractive possibility is that direct action by activated G_s may prime cardiac Ca channels for up-regulation by cAMP-dependent phosphorylation (Trautwein et al. 1989) or other modulators (McDonald et al. 1989).

Cloning and Expression of the Cardiac Dihydropyridine-Sensitive Calcium Channel

The Ca antagonist receptors in cardiac muscle are contained in a ≥ 185-kDa peptide (α_1) that is significantly larger than, and structurally and immunologically different from, its skeletal muscle counterpart (Schneider and Hofmann 1988; Hosey et al. 1989). The primary structure of this protein as predicted from the cloned cDNA was deduced using the open reading frame corresponding to the amino acid sequence of the skeletal muscle DHP receptor (Mikami et al. 1989). The rabbit cardiac DHP receptor is composed of 2171 amino acids and, like its skeletal muscle counterpart (Tanabe et al. 1987), contains four repeated units of homology. Each repeat has five hydrophobic segments (S1, S2, S3, S5, and S6) and one positively charged segment (S4). The degree of amino acid sequence homology between the cardiac and the skeletal muscle DHP receptor is 66%. The regions corresponding to the four internal repeats are highly conserved, whereas the remaining regions, all of which are assigned to the cytoplasmic side of the membrane, are less well conserved (except for the short segment between repeats III and IV). Both the amino-terminal and the carboxy-terminal region of the cardiac DHP receptor are larger than those of the skeletal muscle DHP receptor. The structural similarity observed suggests that the cardiac DHP receptor has the same transmembrane topology as proposed for its skeletal muscle counterpart (Tanabe et al. 1987). This model is consistent with four potential N-glycosylation sites (Asn residues 183, 358, 1418, and 1469) being located on the extracellular side and with six potential cAMP-dependent phosphorylation sites (Ser residues 124, 1575, 1627, 1700, 1848, and 1928) on the cytoplasmic side; residues 183 (with a shift by one residue) and 358 and residues 1627 and 1700 are conserved in the skeletal muscle DHP receptor. Messenger RNA derived from the cardiac DHP receptor (d_1) cDNA was sufficient to direct the formation of a functional DHP-sensitive Ca channel in *Xenopus* oocytes (Mikami et al. 1989). Co-injection of the skeletal muscle α_2 subunit-specific mRNA together with the mRNA specific for the cardiac α_1 subunit (DHP receptor) induced substantially larger Ca channel currents without affecting their sensitivity to BAY K8644 and nifedipine, or altering the peak current-voltage relationship.

References

Heßlinger B, McDonald TF, Pelzer D, Shuba YM, Trautwein W (1989) Whole-cell calcium current in guinea-pig ventricular myocytes dialysed with guanine nucleotides. J Physiol (Lond) (in press)

Hosey MM, Chang FC, O'Callahan CM, Ptasienski J (1989) L-type calcium channels in cardiac and skeletal muscle: purification and phosphorylation. Ann NY Acad Sci 560:27–38

Mikami A, Imoto K, Tanabe T, Niidome T, Mori Y, Takeshima H, Narumiya S, Numa S (1989) Primary structure and functional expression of the cardiac dihydropyridine-sensitive calcium channel. Nature 340:230–233

Schneider T, Hofmann H (1988) The bovine cardiac receptor for calcium channel blockers is a 195-kDa protein. Eur J Biochem 174:369–375

Trautwein W, Cavalié A, Allen TJA, Shuba YM, Pelzer S, Pelzer D (1989) Direct and indirect regulation of cardiac L-type calcium channels by β-adrenoreceptor agonists. Adv Second Messenger Phosphoprotein Res (in press)

Yatani A, Brown AM (1989) Rapid β-adrenergic modulation of cardiac calcium channel currents by a fast G protein pathway. Science 245:71–74

Rev. Physiol. Biochem. Pharmacol., Vol. 114
© Springer-Verlag 1990

Pharmacological Modulation of Voltage-Dependent Calcium Channels in Intact Cells

HARTMUT PORZIG[1]

Contents

---

[1] Pharmakologisches Institut, Universität Bern, Friedbühlstrasse 49, CH-3010 Bern, Switzerland

1 Introduction

Pharmacology has played a pivotal part in defining different types of voltage-dependent calcium channels and in analyzing their biochemical and functional properties (see the accompanying reviews by Glossmann and Striessnig and by Pelzer). Motivated by the great therapeutic potential, current pharmaceutical research strives to find ever more specific compounds capable of modulating Ca channel properties. Moreover, it has been increasingly recognized that an interaction with Ca channels is an important side effect of a surprisingly large number of established drugs.

These new developments all require a detailed analysis of the mechanism by which drugs and neurotransmitters interact with Ca channels in intact cells. In many instances living cells have to be used because the physiological function and the regulation of Ca channels depend on the membrane potential and on biochemical interactions which require an intact cell metabolism. Potential and metabolism are both lost in purified membrane preparations. In this review I shall briefly summarize what has been learned from intact cell studies about the interactions between drugs and Ca channels. In particular, I shall concentrate on the attempts that have been made to correlate an observed functional change with some defined direct interaction of an extracellularly administered drug with one of the various types of voltage-dependent Ca channels. A detailed discussion of work where the modulation of a Ca channel has been indirectly inferred as a possible mechanism of drug action is beyond the scope of this article.

A number of excellent reviews have recently dealt with various aspects of Ca channel biochemistry and function (Godfraind et al. 1986; Tsien et al. 1987, 1988; Triggle and Janis 1987; Miller 1987a, b; Janis et al. 1987; Schramm and Towart 1988; Narahashi 1988; Glossmann and Striessnig 1988; Hosey and Lazdunski 1988; Campbell et al. 1988; Catterall 1988; Bean 1989). To keep the body of references manageable, I shall frequently refer the reader to these reviews and limit the citation of original work to the more recent contributions.

2 Classification of Calcium Channel Types

In the past few years a variety of different voltage-dependent Ca channels have been described. A generally accepted classification is not yet available. However, to simplify the following discussion a preliminary classification based on a mixture of electrophysiological and pharmacological criteria will be adopted according to the suggestions of Fox et al. (1987a, b) and Cruz et al. (1987). These authors discriminate three main categories of voltage-dependent channels called L, T and N that pass Ca ions with high selectivity. L-channels (*l*ong-lasting) are activated by strong depolarizations from the resting potential, show little inactivation during a 100-ms clamp pulse with Ba^{2+} as charge carrier and are sensitive to dihydropyridines and to various other so-called organic Ca antagonists. L-channels are subdivided into L_m, L_h and L_n types. L_m- and L_h channels occur in skeletal muscle (L_m) and in heart and smooth muscle (L_h), respectively. They are sensitive to dihydropyridines and the other organic Ca channel ligands but not to ω-conotoxin GVIA (ω-CgTX), the peptide toxin from the marine snail *Conus geographus*.

L_n-channels occur in neuronal tissue and are sensitive to block by dihydropyridines and possibly by ω-CgTX. T-channels (*t*ransient) are activated by a small depolarization from the resting potential, show rapid inactivation and are completely inactivated at holding potentials close to $-50\,mV$. They are widely distributed in excitable tissue, occurring in muscular and neuronal tissue alike. These channels are essentially resistant to dihydropyridines and to ω-CgTX. Finally, N-type channels (*n*either T nor L or *N*euronal) are activated by strong depolarizations from hyperpolarized holding potentials, show marked voltage-dependent inactivation and are blocked by ω-CgTX but not by dihydropyridines. These channels, which have been observed only in neuronal tissue, are the most controversial category because it proved difficult to identify unequivocally a macroscopic current component that could be assigned to it (Swandulla and Armstrong 1988). Therefore, some authors prefer to subsume L- and N-channels into a common group called "high-threshold" or "high-voltage-activated" channels (for review see Tsien et al. 1988). In any case, the reader should be aware of the rather preliminary nature of this classification when using the information on channel types in Tables 1 – 3. Depending on the tissue, different channel types seem to be involved in the various cellular functions that depend on voltage-regulated changes in intracellular free Ca concentrations. For example, Ca current through L_m- and L_h-channels delivers Ca required for muscular contraction, current through T-type channels depolarizes sinoatrial pacemaker cells and N-type Ca currents contribute to the increase in intracellular Ca required for neurotransmitter release. Yet, these functional associations are not exclusive and seem to vary according to different cell types. In particular, neurotransmitter release may be supported by Ca flowing through any of the three channel classes (for review see Miller 1987a, b; Tsien 1987; Tsien et al. 1988). Some recent aspects of assigning channel types to specific functions are discussed in the last section of this paper.

Table 1. Drugs with (presumably) direct actions on voltage-gated Ca channels

Compound	Channel type	Sources	Effects observed	Techniques	Suggested mechanism	References
"Classical channel blockers": dihydropyridines, phenylalkylamines, benzothiazepines	L	Contractile tissues, glandular tissues, neuronal tissues	Block of Ca current, reduction of single-channel openings, block of Ca fluxes, inhibition of transmitter release, inhibition of contractility	Whole-cell and single-channel patch clamp, ^{45}Ca fluxes, measurement of various Ca-associated functions, direct binding studies	Block of channel by interaction with specific binding sites on channel protein	Reviews by Godfraind et al. 1986 Janis et al. 1987 Glossmann and Striessnig 1988
Bepridil (0.5 – 30 μM)	L	Rabbit skeletal muscle, guinea pig cardiac muscle	Voltage-dependent block of Ca channels, binding competition	Whole-cell patch clamp, binding studies	Block of channel by interaction with specific binding site	Galizzi et al. 1986a Yatani et al. 1986a
Phenytoin (3 – 100 μM)	L	PC12 cells, guinea pig heart neuroblastoma cells	Inhibition of K-stimulated Ca uptake, binding competition, inhibition of Ca current	Whole-cell patch clamp, ^{45}Ca fluxes, binding studies	Block of channel by interaction with specific site allosterically coupled to dihydropyridine site	Messing et al. 1985 Yatani et al. 1986b Bodewei et al. 1985
Phenytoin (3 – 100 μM)	T	Neuroblastoma cells (N1E-115)	Inhibition of Ca current	Whole-cell patch clamp	Voltage- and frequency-dependent block of channels	Twombly et al. 1988
Diphenylbutylpiperidine neuroleptics (pimozide, fluspirilene)	L	GH$_4$C$_1$ pituitary cells, skeletal muscle myoballs, vascular smooth muscle (rabbit), brain synaptosomes	Inhibition of Ca flux and Ca current, inhibition of prolactin secretion, binding competition	Whole-cell patch clamp, ^{45}Ca fluxes, binding studies	Block of channel by interaction with specific binding site (verapamil site?)	Gould et al. 1983 Flaim et al. 1985 Galizzi et al. 1986b Enyeart et al. 1987 Qar et al. 1987
Diclofurime isomers (0.001 – 10 μM)	L	Guinea pig smooth muscle, brain membranes (rat)	Inhibition of Ca-induced contraction, binding competition	Mechanogram, binding studies	Block of channel by interaction with diltiazem site	Spedding et al. 1987 Mir and Spedding 1987

Compound	Channel	Cell/tissue	Effect	Method	Mechanism	Reference
Reserpine (9–30 μM)	L (?)	Smooth muscle (rabbit), GH_3 pituitary cells, rat seminal vesicles	Inhibition of contraction, binding competition, inhibition of Ca uptake, increase in DHP-binding sites	^{45}Ca fluxes, mechanogram, binding studies	Block of channel by interaction with specific site? Indirect effects of catecholamine depletion?	Casteels and Login 1983; Login et al. 1985; Powers and Colucci 1985
Tetrandrine (0.01–10 μM)	L	GH_3 pituitary cells	Block of Ca currents, binding competition	Whole-cell patch clamp, binding studies	Block of channel by interaction with diltiazem site	King et al. 1988
Calmidazolium	L	GH_4C_1 pituitary cells	Inhibition of prolactin secretion, inhibition of Ca fluxes	^{45}Ca flux	?	Enyeart et al. 1987
Phenylisopropyladenosine (PIA)	L	Myocardial cells (guinea pig)	Competitive antagonism with Bay K 8644	Mechanogram	Competitive interaction with Bay K 8644 binding site	Caparotta et al. 1987
Flunarizine Cinnarizine (5–10 μM)	T, L	Myocardial cells (guinea pig)	Block of Ca current	Whole-cell patch clamp	Block of channel by interaction with specific site	Tytgat et al. 1988; van Skiver et al. 1988
Tetramethrin	T (L)	Neuroblastoma cells	Block of T-type currents	Voltage clamp	?	Tsunoo et al. 1985; Narahashi 1988
Retinoic acid (50 μM)	T	Mouse hybridoma (MHY 206)	Inhibition of Ca current, inhibition of proliferation	Whole-cell patch clamp	Direct block of channel (?)	Bosma and Sidell 1988
Amiloride (30 μM)	T	Frog heart	Block of low-threshold Ca currents	Whole-cell patch clamp	Direct block of channel? Indirect block via intracellular H^+ increase?	Tang et al. 1988
Aminoglycoside antibiotics (IC_{50} 5–160 μM)	N (?) L_n (?)	Cerebral cortex membranes (guinea pig)	Binding competition with ω-conotoxin	Binding studies	Block of channel by direct interaction with specific site	Knaus et al. 1987

Table 2. Drugs acting on voltage-gated Ca channels via secondary (receptor-) mechanisms

Compound	Channel type	Sources	Effects observed	Techniques	Suggested mechanism	References
β-Adrenoceptor agonists	L	Mammalian and frog heart, vascular smooth muscle (rabbit)	Positive inotropic effect, increased Ca influx, increased Ca current	Whole-cell and single-channel patch clamp, ^{45}Ca fluxes, mechanogram	β-Receptor-mediated stimulation of protein kinase A, phosphorylation of Ca channels	Nelson et al. 1988 Reviews: Reuter 1984, 1987; Tsien et al. 1988; Levitan 1988
Benzodiazepines (1–600 μM)	L_n (?)	GH$_3$ pituitary cells, Leech neurons, rat brain nerve terminals	Inhibition of depolarization-induced Ca influx, inhibition of Ca-dependent action potentials	^{45}Ca fluxes (quin-2), slow action potential measurement	Block of channel mediated by interaction with low-affinity benzodiazepine-receptor which may be associated with GABA$_B$-receptor	Taft and De Lorenzo 1984; Johansen et al. 1985; Gershengorn et al. 1988; Rampe and Triggle 1986
Pk11195 (3 μM) (benzodiazepine analogue)	L	Guinea pig heart	Inhibition of the stimulating effect of Bay K 8644	Measurement of slow action potentials	Interaction with low-affinity benzodiazepine receptors? Direct interaction at DHP site?	Mestre et al. 1986a,b
Gamma-aminobutyric acid (GABA) (1 μM)	L_n	Chick dorsal root ganglion, chick embryonic sensory neurons	Decrease in Ca (Ba) current, AP duration decreased	Voltage clamp	G-protein (G$_i$ or G$_o$)-mediated direct or indirect effect on Ca channel. G-protein activated by interaction with GABA$_B$ receptor	Dunlap 1981; Dunlap and Fischbach 1981; Deisz and Lux 1985; Holz et al. 1986
Baclofen (1–100 μM)	L_n	Rat sensory neuron	Inhibition of Ca (Ba) currents	Voltage clamp	G-protein-mediated effect on channel. Activation of G-protein via GABA$_B$ receptor	Dolphin and Scott 1986, 1987; Dunlap 1981; Désarmenien et al. 1984
Ethanol (50–600 μM)	L (?)	PC12 cells	Short-term: decrease in Ca uptake; long-term: increase in Ca uptake	^{45}Ca flux, binding studies	Short-term effect: unknown; long-term effect: increase in DHP-binding sites	Messing et al. 1986b

Compound	Channel type	Cells	Effect	Methods	Mechanism	References
6-Hydroxy-dopamine	L	Rat heart, chick heart	Increase in DHP binding	Binding studies	Unknown (effect on membrane protein synthesis?)	Skattebøl and Triggle 1986; Renaud et al. 1984
Enkephalins (DADLE 1 µM)	L_n	Neuroblastoma cells (NG 108-15), dorsal root ganglion, frog neuromuscular junction	Inhibition of Ca(Ba) current, reduction of neurotransmitter release	Voltage clamp	Inhibitory interaction of G-protein (G_o) with channel, G_o activation via δ-opiate receptor	Hescheler et al. 1987; Bixby and Spitzer 1983; Mudge et al. 1979; Tsunoo et al. 1986; Narahashi 1988
Morphine (10 µM)	L	Neuroblastoma cells (NG 108-CC15; NIE-115)	Increased Ca influx upon depolarization	^{45}Ca flux, membrane potential recording	Activation of channels via µ-opiate receptors	Lorentz et al. 1988
Dynorphin A	N	Mouse dorsal root ganglion	Decrease of Ca current. Decrease of Ca dependent AP-duration	AP measurement, voltage clamp	Inhibition of channel via interaction with κ-opiate receptors	Gross and Macdonald 1987; Werz and Macdonald 1985; North 1986
Angiotensin II (1 nM – 1 µM)	L (T?)	Adrenal glomerulosa cells (bovine), adrenal cortical cell line (Y1)	Stimulation of Ca currents	Whole-cell patch clamp	Coupling of angiotensin II receptors to Ca channel via G_i-type G protein	Hescheler et al. 1988; Kojima et al. 1986; Cohen et al. 1988
Phorbol esters, diacylglycerol	L	Chick dorsal root ganglion, smooth muscle cell line (A7r5), PC12 cells, RINm5F cells	Inhibition of Ca current, inhibition of DHP-inhibitable, depolarization-induced Ca flux, loss of DHP-binding sites	Voltage clamp, ^{45}Ca flux, binding studies	Activation of protein kinase C modulates channel properties	Holz et al. 1986; Rane and Dunlap 1986; Galizzi et al. 1987; Messing et al. 1986a; Di Virgilio et al. 1986
Phorbol esters	L?	Neuroblastoma cells (NG 108-15), rabbit ear artery, A7r5 cells	Stimulation of Ca influx, increase in perfusion pressure, increased contraction, stimulation of Ca uptake	Quin-2, perfusion pressure monitoring, mechanogram, ^{45}Ca flux, whole-cell patch clamp	Activation of protein kinase C? Direct effect on channel?	Fish et al. 1988; Forder et al. 1985; Chin et al. 1988; Osugi et al. 1986

Table 2 (continued)

Compound	Channel type	Sources	Effects observed	Techniques	Suggested mechanism	References
Phorbol esters	N? (non L)	*Aplysia* neurons	Increase in Ca current, enhancement of evoked AP	Voltage clamp	Recruitment of channels via protein kinase C-mediated phosphorylation	Strong et al. 1987; DeRiemer et al. 1985
Peptides: Bombesin (1 µM); Vasopressin (0.5 µM); Oxytocin (1 µM)	L	A7r5 smooth muscle cell line	Inhibition of depolarization-induced Ca influx	^{45}Ca flux, whole-cell patch clamp	Stimulation of PI metabolism and protein kinase activity via V_1 receptors (vasopressin + oxytocin)	Galizzi et al. 1987
LHRH (0.1 µM) (luteotropic hormone-releasing hormone)	L?	GH$_3$ pituitary cells	Stimulation of Ca current	Whole-cell patch clamp	G-protein (G_i)-mediated effect on channels via LHRH receptor	Rosenthal et al. 1988b
Somatostatin (0.1 µM)	L? N?	Pituitary cell lines (AtT-20/D16-16), GH$_3$, neuroblastoma cells (NG 108-15)	Inhibition of Ca current	Whole-cell patch clamp	G-protein (G_o?)-mediated effect on channels via somatostatin receptors	Lewis et al. 1986; Tsunoo et al. 1986; Rosenthal et al. 1988a,b
Neuropeptide Y	N, T (L?)	Rat dorsal root ganglion, myenteric plexus	Inhibition of Ca current	Whole-cell patch clamp, Ca flux (fura-2)	G-protein (G_o)-mediated effect on channels via NPY receptors	Ewald et al. 1988a,b; Walker et al. 1988
2-Chloroadenosine (Adenosine A1 agonist) (0.5 µM)	N?	Rat dorsal root ganglion	Inhibition of Ca current, shortening of AP	AP monitoring, whole-cell patch clamp	G-protein-mediated effect on channels via adenosine A_1 receptors	Dolphin et al. 1986; Dolphin and Scott 1987

Drug	Channel type	Cell type	Effect	Method	Mechanism	References
Acetylcholine (0.4–15 µM)	N	Rat sympathetic ganglion	Inhibition of Ca current	Whole-cell patch clamp, fura-2 monitoring of [Ca]$_i$	G-protein-mediated effect on channels via M$_1$ muscarinic receptors	Wanke et al. 1987
Dopamine (1–10 µM)	T, N?	Chick dorsal root ganglion and sympathetic neurons	Decrease of Ca currents (decrease in activation rates)	Whole-cell and single-channel patch clamp	D2(?)-receptor-mediated effect on channels	Marchetti et al. 1986
Tamoxifen (0.1 µM)	T, N?	Pituitary cells	Inhibition of Ca current	Whole-cell patch clamp	Unclear, inhibition of protein kinase C?	Sartor et al. 1988
Noradrenaline (1–50 µM)	N, T	Chick dorsal root ganglion, frog sympathetic ganglion, rat sympathetic ganglion	Inhibition of Ca current	Voltage clamp, whole-cell and single-channel patch clamp	G-protein-mediated effect on channel via α-adrenoceptors, protein kinase C-mediated effect on channel availability?	Lipscombe and Tsien 1987; Forscher and Oxford 1985; Dunlap and Fischbach 1981; Holz et al. 1986; Marchetti et al. 1986; Galvan and Adams 1982

Table 3. Modulation of voltage-gated Ca channels by toxins and inorganic ions

Compound	Channel type	Sources	Effects observed	Techniques	Suggested mechanism	References
Atrotoxin	L	Guinea pig ventricular cells	Potential-independent activation of Ca (Ba) currents, inhibition of DHP binding	Whole-cell patch clamp, DHP binding, single-channel patch clamp	Calcium channel activation by interaction with binding site allosterically coupled to DHP site	Hamilton et al. 1985; Lacerda and Brown 1986
β-Leptinotarsin	L	Rat brain synaptosomes	Depolarization, increased transmitter release, increased Ca uptake	Transmitter release, ^{45}Ca, flux	Opening of presynaptic Ca channels	Crosland et al. 1984
Goniopora-toxin ($1-5\ \mu M$)	L	Guinea pig smooth muscle, rabbit skeletal muscle, chick cardiac cells	Positive inotropic effect, increased contraction, increased Ca influx	Mechanogram, ^{45}Ca fluxes, [^{3}H]isradipine binding	Calcium channel activation by binding to receptor allosterically coupled to DHP site	Qar et al. 1986
Taicatoxin (nanomolar)	L	Guinea pig ventricular cells, cultured neonatal rat heart cells	Potential-dependent inhibition of Ca current	Single-channel patch clamp, DHP binding	Reduced opening probability of Ca channels, interaction with DHP site	Brown et al. 1987
Maitotoxin ($0.5-400$ ng/ml)	L?	Heart, smooth muscle, PC12 cells, NG 108-15 cells, BC$_3$H$_1$ cells	Positive inotropic effect, increased contraction, increased transmitter release, increase in Ca fluxes	Whole-cell patch clamp, ^{45}Ca fluxes, mechanograms, noradrenaline and dopamine release	Permanent activation of channels? Induction of pore with pharmacological properties of L-channel?	Kobayashi et al. 1985, 1986; Hamilton and Perez 1987; Wu and Narahashi 1988; Sladeczek et al. 1988
Apamin (0.1 nM)	L	Embryonic chick heart cultures	Block of slow action potential	Action potential recording by intracellular electrode	Block of Ca channels	Bkaily et al. 1985

Compound	Channel type	Cell/tissue	Effect	Method	Mechanism	References
ω-Conotoxin GVIA (0.5 nM – 1 μM)	N, L_n?	Chick dorsal root ganglion, frog sympathetic ganglion, rat hippocampus, brain synaptosomes, electric organ nerve endings	Inhibition of Ca current, inhibition of transmitter release, inhibition of Ca fluxes	Whole-cell and single-channel patch clamp, ATP-release, ^{45}Ca fluxes, binding studies	Quasi-irreversible block of Ca channels via direct binding to channel	Kerr and Yoshikami 1984; Olivera et al. 1985; McClesky et al. 1987; Fox et al. 1987a,b; Cruz et al. 1987; Gray et al. 1988; Ahmad and Miljanich 1988
"Lipophilic Na channel toxins" (veratridine batrachotoxin, cevadin, α-dihydrograyanotoxin II)	L_n, N?	N1E-115 neuroblastoma cells, GH$_4$C pituitary cells	Inhibition of Ca currents, inhibition of prolactin secretion, inhibition of Ca fluxes	Whole-cell patch clamp, ^{45}Ca fluxes	Direct block of channels by interaction with binding site on channel protein	Romey and Lazdunski 1982; Kongsamut et al. 1985b; Enyeart et al. 1987
Cadmium (20 – 50 μM)	N, L_n	Chick dorsal root ganglion	Inhibition of Ca (Ba) currents	Whole-cell patch clamp	Direct block of channels	Fox et al. 1987a,b
Cobalt (2 mM)	T, L_n	Canine atria	Inhibition of Ca (Ba) current	Whole-cell patch clamp	Direct block of channels	Bean 1985
Nickel (40 – 200 μM)	T	Rabbit sinoatrial node, chick dorsal root ganglion	Inhibition of Ca current	Whole-cell patch clamp	Direct block of channels	Hagiwara et al. 1988; Fox et al. 1987a,b; Carbone et al. 1987
Gadolinium (0.5 – 20 μM)	N	NG 108-15 neuroblastoma cells	Partial block of Ca currents	Whole-cell patch clamp	Direct block of channels	Docherty 1988

3 L-Type Calcium Channels as Targets for Drug Action

Elementary Ca channels, later classified as L-type, were first identified in mammalian cardiac cells on the basis of their electrophysiological properties (Reuter et al. 1982; Cavalié et al. 1983) and their sensitivity towards a group of organic compounds (Lee and Tsien 1983; Kokubun and Reuter 1984) called "Ca antagonists" (Fleckenstein 1977). These drugs consist mainly of three structurally different chemical groups, the 1,4-dihydropyridines, phenylalkyl-amines, and benzothiazepines (see Glossmann and Striessnig 1989). They were known from earlier pharmacological and electrophysiological work to interfere with transmembrane Ca currents and to inhibit some of the effects of extracellular Ca on cardiac and smooth muscle contractility (for review see Godfraind et al. 1986). Up to now a host of additional drugs, toxins and neu-rotransmitters have been discovered to interact, directly or indirectly, with L-type Ca channels (see Tables 1 – 3). Mainly because of this rich pharmacology, much more is known about the functional role of L-channels and their inter-action with drugs than about any of the other channel types.

3.1 Interaction of Channel Blockers with Functional Calcium Channels

3.1.1 Dihydropyridines

The discovery of high-affinity binding sites for 1,4-dihydropyridines in mem-brane homogenates from excitable tissues (Bellemann et al. 1981; Glossmann et al. 1982; Murphy and Snyder 1982; Bolger et al. 1982; Fosset et al. 1982) initially met with considerable scepticism regarding their functional signifi-cance (Miller and Freedman 1984). The doubts arose because in some tissues, heart and brain in particular, a large discrepancy existed between the apparent (nanomolar) affinity of the binding sites and the (micromolar) drug concen-trations required for a pharmacological effect on Ca channel function (Janis et al. 1984a).

Electrophysiological evidence suggested that at least part of this discrepan-cy might be explained by the voltage-dependent conformational changes of the Ca channel protein. During the course of an action potential the channel cycles through open and closed states (or groups of states) each of which may interact differently with a given drug. The evidence is compatible with the view that channel blockers preferentially bind to and stabilize the inactivated closed conformation of the channel (Bean 1984; Sanguinetti and Kass 1984a; Gurney et al. 1985; Cognard et al. 1986; Kawashima and Ochi 1988). In some cases, they may also bind with high affinity to open channels (Cohen and Mc-Carthy 1987). This interpretation was supported by two main findings: (1) All of the classical Ca channel blockers (dihydropyridines, verapamil, diltiazem)

showed use-dependent effects, albeit to variable degrees. That is, their blocking potency increased when the stimulation frequency was raised and, hence,
the channel cycled more often through the open and inactivated states (Lee
and Tsien 1983; Kanaya et al. 1983; Sanguinetti and Kass 1984a). (2) Holding
the membranes at potentials less negative than the normal resting potential
enhanced the blocking effect of dihydropyridines dramatically (Bean 1984;
Sanguinetti and Kass 1984a). From these experiments it could be calculated
that the apparent binding affinity of resting and inactivated channels for the
dihydropyridine derivative nitrendipine differed by a factor of more than 1000
(Bean 1984). Therefore, binding constants measured in membrane fragments,
where the potential has collapsed and all channels are presumably inactivated,
cannot be compared with the apparent affinity of the drug under in vivo conditions. On the other hand, these observations also explained the very good
correlations between binding affinities of Ca channel blockers in brain,
smooth muscle or heart homogenate and their relaxing effect on smooth muscle contractures in high potassium solution (Bellemann et al. 1983; Bolger et
al. 1983; Janis et al. 1984a; 1987; Sarmiento et al. 1984; for reviews see Godfraind et al. 1986). Under these conditions, binding and effect are both measured in depolarized preparations. Yet, a direct demonstration of the pre-
sumed effect of membrane voltage on the apparent K_D values of channel-blocking drugs in intact cells proved to be very difficult. The first studies
attempting such measurements in dissociated cardiac cells (Green et al. 1985)
or small bundles of cells from skeletal muscle (Schwartz et al. 1985) all
showed an increase in total binding capacity, rather than in binding affinity,
when the cells were depolarized. However, several groups have reported recently that, indeed, an increase in dihydropyridine-binding affinity is associated with membrane depolarization of intact cells originating from heart,
smooth muscle or a neuronal cell line (Reuter et al. 1985; Kokubun et al. 1986;
Greenberg et al. 1986; Morel and Godfraind 1987; Sumimoto et al. 1988) or
even with a depolarization of cardiac sarcolemma vesicles (Schilling and
Drewe 1986). The earlier failure to detect this change in K_D values seems to
have resulted indirectly from this large shift in affinity. At the normal resting
potential of cardiac or striated muscle cells (about -80 mV) the affinity for
most dihydropyridines is too low to allow detection of specific binding with
the usual techniques. In dissociated cell preparations, a variable proportion
of the cells will always have low resting potentials because some unavoidable
damage occurs during the preparation procedure. Such heterogeneous resting
potential levels have been demonstrated directly in apparently viable isolated
ventricular myocytes (Masuda et al. 1987). The small fraction of permanently
depolarized cells will bind the radioligands with high affinity. Upon depolarization *all* cells will exhibit high-affinity binding, thus creating the impression
of an increase in binding capacity (Kokubun et al. 1986). Nevertheless it
should be noted that a conflicting study by Lee et al. (1987) has reported an

increase in dihydropyridine (isradipine) binding upon depolarization in intact chick cardiac cell cultures. The reason for the discrepant binding data in chick is not clear.

Are all the high-affinity binding sites, identified with dihydropyridine binding studies in intact cells, functional Ca channels? A single cell from the myocardium of newborn rats has about 50000 high-affinity binding sites (Kokubun et al. 1986). Maximal Ca currents in whole-cell patch clamp recordings, however, seem to reflect the opening of only 1000–10000 Ca channels (Reuter 1983). This may be due to the fact that the probability of a channel being in the open state is much less than one (e.g. Reuter et al. 1986). Ca currents can be modulated by changes in single-channel availabilities and open state probabilities (for review see Tsien 1987). Thus the discrepancy between receptor site density and the number of functional channels is difficult to quantify because it depends very much on the probability of each channel entering an open state when the total membrane current is maximal. Schwartz et al. (1985) have suggested that in skeletal muscle t-tubules less than 10% of the binding sites may function as Ca channels. However, it could well be more than 30% if a different probability estimate is used (Lamb and Walsh 1987). Recent evidence suggests that the dihydropyridine-binding protein may serve a dual function as a Ca channel and as a voltage sensor for excitation-contraction coupling (Rios and Brum 1987). If the two modes of functioning required different conformations of the receptor protein, the number of conducting channels would be limited according to the equilibrium distribution ratio between the two states. Non-muscular tissues such as neurons and secretory cells usually contain much lower densities of dihydropyridine-binding sites. For example, in PC12 cells 1200–6000 specific sites/cell have been observed (Greenberg et al. 1986; Messing et al. 1986a; H. Porzig and C. Becker, unpublished). These values are in good agreement with the number of functional L-type channels (midrange value 2500 channels/cell) calculated from peak Ca currents in the same cell line by Kunze et al. (1987).

3.1.2 Phenylalkylamines and Benzothiazepines

The two other main classes of organic Ca channel blockers also seem to interact preferentially with inactivated channels (Kanaya et al. 1983; Uehara and Hume 1985). However, compared with dihydropyridines their channel-blocking effect is much more dependent on the stimulation frequency (i.e., "use-dependent") and much less dependent on the steady-state membrane potential (Lee and Tsien 1983; Sanguinetti and Kass 1984a). The more pronounced use-dependence of phenylalkylamines and benzothiazepines is perhaps related to their high degree of ionization at physiological pH values (the pK values of verapamil and of *d-cis*-diltiazem are 8.7 and 7.7, respectively). The ionized compounds seem to need channel opening in order to reach their target site within the channel (see below).

These results could account for the observation that verapamil and *d-cis*-diltiazem, unlike dihydropyridines, have comparable potencies in inhibiting cardiac and smooth muscle contractility. The selectivity ratio for verapamil in favour of smooth muscle was 7 (canine papillary muscle versus coronary artery, Motomura et al. 1987) or 1 (rat papillary muscle versus portal vein, Ljung et al. 1987). In the latter system a ratio of 8.9 was reported for *d-cis*-diltiazem. Consequently, the correlation between negative inotropic potency and binding affinity was better for verapamil and diltiazem than for dihydropyridines (Goll et al. 1986).

3.1.3 Cooperative Interactions Between Calcium Channel Blockers

Binding studies using tissue homogenates from heart, smooth muscle, skeletal muscle and brain have clearly established that dihydropyridines, phenylalkylamines and benzothiazepines occupy distinct but allosterically coupled binding sites on the channel protein (for review see Godfraind et al. 1986; Janis et al. 1987; Glossmann and Striessnig 1988; Hosey and Lazdunski 1988). In intact cardiac cells we have recently shown that allosteric interactions between different classes of channel blocking compounds and also among different groups of dihydropyridines are voltage dependent (Kokubun et al. 1986; Porzig and Becker 1988). *d-cis*-Diltiazem significantly enhanced the affinity of the dihydropyridine derivative isradipine (= PN 200−110) in polarized cardiac cells but was almost ineffective in depolarized cells. On the other hand, (±)-verapamil, which also increased the affinity of isradipine in polarized cells, strongly decreased its affinity in depolarized cells. A number of studies suggest that positive or negative cooperativity among different Ca channel ligands, as defined in binding experiments on intact cells, is also reflected in functional assays. The stimulation of smooth muscle contractile responses with the dihydropyridine Bay K 8644 was potentiated by two other dihydropyridines, nitrendipine and nimodipine, and by diltiazem (Dubé et al. 1985a, b). On the other hand, nimodipine has been found to potentiate the negative inotropic effect of diltiazem on cardiac contractility (De Pover et al. 1983; Garcia et al. 1986).

3.2 Interactions of Channel Activators with Functional Calcium Channels

Except for a number of toxins (see Table 3), drugs with a stimulating effect on L-type Ca channel activity have only been discovered among the dihydropyridine derivatives (for reviews see Godfraind et al. 1986; Janis et al. 1987; Schramm and Towart 1988; Bechem et al. 1988). Pharmacological data are available for five different compounds with activating properties. All of them have an asymmetrical carbon atom and exist in two enantiomeric forms. In those cases where the enantiomers have been separated and tested individually, the channel-activating property was always associated with only one of the two enantiomers (Bay K 8644: Franckowiak et al. 1985; 202-791: Hof et al. 1985; H 160/51: Gjörstrup et al. 1986). The other enantiomer acted as an inhibitor. Binding studies with Bay K 8644 in intact cardiac cells (Bellemann 1984) and in fragmented cardiac membranes (Janis et al. 1984b) showed a good correlation of the apparent K_D values with the concentration required for a half maximal positive inotropic effect in heart or stimulation of contractile activity in smooth muscle (Schramm et al. 1983; Hof et al. 1985; Vaghy et al. 1984a, b). On the other hand, binding competition studies in fragmented heart or brain membranes suggested that dihydropyridine Ca channel blockers and activators competed with high affinity for the same site on inactivated channels (Bellemann 1984; Vaghy et al. 1984a, b; Janis et al. 1984a; Hamilton et al. 1987). In electrophysiological studies, Ca channel activators increased Ca inward current mainly due to a marked prolongation of the single-channel open state (Kokubun and Reuter 1984; Hess et al. 1984; Brown et al. 1984). The effect of the activators, like that of the blockers, was voltage dependent, but in a distinctly different way. Ca currents were only stimulated when elicited from fairly negative membrane-holding potentials. At more positive holding potentials (e.g. above −20 mV in rat cardiac myocytes) inhibition of Ca currents rather than activation was observed (Sanguinetti and Kass 1984b; Kokubun et al. 1986, Sanguinetti et al. 1986; Kass 1987). The strength of this blocking action may be tissue dependent. Thus in guinea pig ventricular cells no significant block was seen although the stimulating effect of Bay K 8644 was also lost at holding potentials less negative than −30 mV (Hamilton et al. 1987). The inhibitory action of channel "activators" arises because these compounds, like blockers, shift the steady-state current inactivation curve towards more negative potentials (Kokubun et al. 1986; Hadley and Hume 1988). In ^{45}Ca flux studies, channel activators enhanced Ca uptake upon potassium-induced depolarization in cardiac, smooth muscle, secretory and neuronal cells (Freedman and Miller 1984; Loutzenhiser et al. 1984; Laurent et al. 1985; Kongsamut et al. 1985). However, it is important to note that a depolarizing step is always required before an increase in ^{45}Ca flux or in current flow can be observed. Apparently, the activators cannot per se open Ca channels by some voltage-independent mechanism.

In summary, functional studies as well as binding experiments clearly indicate that channel activators have high affinities for both open and inactivated channels. Binding to channels stabilizes the open state, thus leading to increased current flow. Binding to inactivated channels induces a shift of the steady-state inactivation curve towards more negative membrane potentials.

3.3 Analysis of Mutual Interactions of Activators and Blockers with Calcium Channels In Situ

The discovery of dihydropyridine Ca channel activators initiated a debate on whether these compounds shared a single common binding site with the blockers or, rather, channel activation required interaction with a second site (Thomas et al. 1984; Janis and Triggle 1984; Dubé et al. 1985a, b; Glossmann et al. 1985a; Maan and Hosey 1987). Although a second low-affinity dihydropyridine binding site has been observed repeatedly (Vaghy et al. 1984b; Janis et al. 1984b; Rogart et al. 1986; Lee et al. 1987), most authors agree that activators and blockers share a common high-affinity binding site in membrane preparations (Schwartz et al. 1984; Vaghy et al. 1984a; Janis et al. 1984a; Williams et al. 1985; Hamilton et al. 1987). Similarly, binding experiments in depolarized, intact cardiac cells showed purely competitive interactions between a channel activator (S-202-791) and a blocker (S-isradipine) (Kokubun et al. 1986). It is clear from these studies that only a single binding site is available for activating and blocking dihydropyridines when the Ca channels are inactivated. However, in a study on intact polarized cells the activating enantiomer of 202-791 showed positive cooperative interactions with the radio-labelled blocker isradipine (Kokubun et al. 1986). The affinity of isradipine was significantly enhanced by small concentrations of S-202-791. In parallel functional experiments, the channel activating potency of S-202-791 was markedly increased by small concentrations of the blocking enantiomer R-202-791. By definition, allosteric interactions between two ligands imply simultaneous occupation of two binding sites on the receptor protein (Monod et al. 1965). Therefore, channel activation seems to be associated with a high-affinity interaction of the activator drug at a channel site which is not accessible, or has a low affinity for blockers. Theoretical calculations of the charge distribution on the surface of activator and blocker molecules by Höltje and Marrer (1987) yielded characteristic differences between the two antagonistically acting types of drugs. Such differences could explain their interaction with non-identical sites in open Ca channels.

The observation of cooperative interactions between activators and blockers of Ca channels in polarized cells also shed new light on the molecular mechanism of channel block by dihydropyridines. In an earlier study it has been assumed that open channel block could be one of the mechanisms by which Ca current is inhibited (Lee and Tsien 1983). However, such a mechanism is incompatible with the observation that the activator-induced increase in mean open time and open time probabilities can be further cooperatively enhanced by a blocker (Kokubun et al. 1986). Moreover, a specific antibody against the beta subunit of the dihydropyridine receptor complex from muscle has been found to stimulate Ca current through L-type channels (Vilven et al. 1988; Campbell et al. 1988). Dihydropyridines should be able to antagonize this effect if acting via open channel block. Such inhibition was not observed.

3.4 Effect of Calcium on Channel-Drug Interactions in Living Cells

Previous studies in membrane homogenates had suggested a strong influence of divalent cations on the binding of Ca channel blockers (for reviews see Glossmann et al. 1985a; Godfraind et al. 1986; Janis et al. 1987; Glossmann and Striessnig 1988). Specific binding of dihydropyridines to brain, heart and smooth muscle preparations (but *not* to skeletal muscle) was found to be markedly reduced or completely abolished by EDTA or other Ca chelators. This was due to a decrease of B_{max} values and could be reversed by readdition of divalent cations. By contrast, an increase in the concentration of divalent cations usually inhibited the binding of phenylalkylamines and benzothiazepines (Reynolds et al. 1983; Galizzi et al. 1984, 1985; Glossmann et al. 1985b). Therefore, it has been postulated that a high-affinity Ca-binding site allosterically coupled to the drug-binding sites is located on the channel protein (Glossmann et al. 1985a; Glossmann and Striessnig 1988). Although these observations do not find direct correlates in functional experiments or binding studies in intact cells, important modulatory effects of Ca and other divalent cations on channel properties have been reported. Part of the differences may be due to the fact that Ca modulates dihydropyridine binding by interaction with an intracellular site that is accessible in membranes but not in intact cells (Schilling 1988).

In the absence of extracellular Ca, L- and T-type channels lose their divalent cation selectivity and become highly permeable to monovalent cations (Kostyuk et al. 1983; Hess and Tsien 1984; Almers et al. 1984; McCleskey and Almers 1985). Raising the extracellular Ca concentration to micromolar levels induces divalent cation selectivity, with an apparent K_D for Ca of $0.7-2\,\mu M$. Hence, the interaction of Ca with some high-affinity binding site appears to promote a conformational change that profoundly changes the properties of the permeation pathway. Recently, Pietrobon et al. (1988) have directly demonstrated that a permeant Ca ion, by interaction with a site within the channel, allosterically destabilizes a protonated site at the channel surface. It is intriguing that within the same range of extracellular Ca concentrations we have observed distinct changes in the cooperative binding of dihydropyridine ligands to intact cardiac cells (Kokubun et al. 1986). The [^{3}H]isradipine binding curve in the presence of the channel activator S-202-791 changed from a sigmoid to a nearly hyperbolic shape when extracellular Ca was reduced from $5\,\mu M$ to about $0.01\,\mu M$ by complexation to ethylene glycol tetra-acetic acid (EGTA). Raising the extracellular Ca concentration to millimolar levels also suppressed sigmoidicity. The maximal binding capacity for dihydropyridines was somewhat larger in the presence of Ca than in its absence (Porzig and Becker 1988). Nevertheless it is clear that the fluxes of monovalent cations through Ca channels at very low extracellular Ca concentrations are still sensitive to inhibition by dihydropyridines and phenylalkylamines (Carbone and Lux 1988).

In neuronal L-type Ca channels the sea snail toxin, ω-conotoxin GVIA, (ω-CgTX), seems to detect Ca-dependent changes of the conformational state. With Ca as the main charge carrier ω-CgTX blocks the channels slowly but quasi-irreversibly. When the charge carrier is Na at low extracellular Ca concentration, this block is not maintained even though ω-CgTX remains bound to the channel protein (Carbone and Lux 1989).

High millimolar concentrations of divalent cations partially antagonize the effect of organic Ca channel blockers (Fleckenstein 1977; Lee and Tsien 1983). This is reminiscent of the inhibitory action of high Ca concentrations on the effects of local anaesthetics (Hille 1984). Indeed, a similar mechanism seems to be responsible in both cases (Kass and Krafte 1987). Due to a screening effect on negative surface charges, divalent cations shift the steady-state inactivation curve for the Ca current to the right. Stronger depolarizations are necessary to reach half maximal inactivation. By contrast, Ca channel blockers all shift the inactivation curve to the left, such that inactivation occurs already at more negative potentials (see above). These two effects tend to compensate each other when channel blockers are used in the presence of divalent cations ($2-20$ mM) (Kass and Krafte 1987).

3.5 Calcium Channels and the Modulated Receptor Hypothesis

Analysis with cDNA has shown marked amino acid sequence homologies between the dihydropyridine-binding protein and the Na channel protein (Tanabe et al. 1987; for review see Glossmann and Striessnig 1989). On the basis of the structural similarities it is not surprising that a detailed functional analysis of the interaction of Ca channel blockers with the channel protein has revealed strong analogies to the mechanism of action of local anaesthetics (Sanguinetti and Kass 1984a; Janis and Triggle 1984; Hondeghem and Katzung 1984; Kokubun et al. 1986). Most of the relevant results with Ca channel blockers can be explained within the framework of the modulated receptor hypothesis developed by Hille (1977) and Hondeghem and Katzung (1977). It uses a model where the time- and voltage-dependent state changes of the ion channel are coupled to conformational changes of the drug receptor to describe the interactions of local anaesthetics with Na channels. Thus the three principal conformations of the channel, resting (R), open (O) and inactivated (I), are thought to differ in their drug (D)-binding kinetics. R and I are two different conformations of a closed state. The frequency dependence of drug action can then be explained by a slow dissociation rate of the DI complex compared with the DR or DO complexes. Similarly, the shift of the steady-state inactivation curve towards more polarized potentials would be a consequence of the stabilization of the channel in the I state. This stabilization would be due to the formation of relatively long-lived DI complexes. Recent studies on the voltage dependence of allosteric interactions, between different Ca channel ligands in intact cells, suggest that more than one type of DI complex may exist (Porzig and Becker 1988). In fully depolarized cells, where all channels are inactivated, verapamil still decreased the binding affinity of radiolabelled isradipine. Analogous results have been obtained repeatedly in membrane homogenates (for review see Godfraind et al. 1986; Glossmann and Striessnig 1989). Conversely, *d-cis*-diltiazem is reported to enhance dihydropyridine affinity in fragmented membranes, but not in depolarized intact cells. These allosteric interactions between different ligands in depolarized membranes indicate that the functionally inactivated channel can assume drug-induced conformations which are clearly different from the voltage-induced states.

The "modulated receptor" can be considered as a special case of an allosteric protein. Its conformation is not only modulated by ligand binding, but is predominantly determined by membrane voltage. For a given membrane potential the steady-state frequency distribution of channel conformations (R-O-I) assumes a characteristic value. Allosteric modulation then means a drug-induced shift of this steady-state value (allosteric constant) in favour of either the conducting or non-conducting states (Kokubun et al. 1986; Porzig and Becker 1988). This concept is particularly suitable for describing the interactions between drugs and the L-type Ca channels because the specific drug receptors on the channel protein are accessible for direct characterization and analysis. Thus, it is clear from the available evidence that membrane potential changes indeed affect the binding properties of the drug receptor protein rather than modulating only receptor accessibility (Hondeghem and Katzung 1984). The described voltage-dependent allosteric interactions between different types of Ca channel blockers and activators as well as the strong stereospecificity of Ca channel ligands (Bellemann et al. 1983; Glossmann et al. 1985b; for review see Godfraind et al. 1986) are strong arguments in favour of this notion. By contrast, little is known about the binding site for local anaesthetics within the Na channel because their relatively low affinity has not allowed direct in situ binding studies. However, indirect evidence suggests that the drug-binding proteins in both Na and Ca channels share a number of properties.

3.6 Comparative Aspects of Sodium and Calcium Channels

3.6.1 Interaction with Toxins

The sodium channel protein interacts with various neurotoxins at five different sites (for reviews see Catterall 1987, 1988). Allosteric coupling has been shown between the toxin binding site 2 (the "batrachotoxin site"), on one hand, and the toxin binding site 3 (the "α-scorpiotoxin site") as well as the local anaesthetic site on the other hand (Catterall 1977, 1981; Postma and Catterall 1984; Sheldon et al. 1987). Analogous allosteric interactions have been discovered between several new peptide toxins (see Table 3) and the "classical" Ca channel blockers and activators. Gonioporatoxin (from a toxic coral, Qar et al. 1986), atrotoxin (from the rattlesnake *Crotalus atrox*, Hamilton et al. 1985; Lacerda and Brown 1986) and taicatoxin (from the Australian taipan snake, Brown et al. 1987) have all been shown to interact allosterically with the dihydropyridine-binding site. Gonioporatoxin and atrotoxin are both activators, and taicatoxin is a blocker of L-type Ca channels. The effect of taicatoxin is voltage dependent, whereas the effect of the two activating toxins seems not to depend on membrane voltage.

3.6.2 Stereospecificity

As mentioned above, the actions of Ca channel blockers are strongly stereospecific. Enantiomers may differ in their potency by more than two orders of magnitude (Bellemann et al. 1983; Hof et al. 1986; Godfraind et al. 1986). Significant, though less prominent, stereospecific differences in potency have also been observed with the enantiomeric local anaesthetics RAC 109 and RAC 421 (Yeh 1980; Postma and Catterall 1984; Bolger et al. 1987) or tecainid (Sheldon et al. 1988).

3.6.3 Cross-reactivity of Drugs

An increasing number of drugs and toxins have been shown to cause comparable functional changes in both channels, usually, but not always, with different potencies. The toxins veratridine, batrachotoxin, cevadin and α-dihydrograyanotoxin, all known to increase Na channel currents by inhibiting inactivation, seem to block Ca channels in a neuroblastoma cell line (Romey and Lazdunski 1982). Although the affected channel type was not defined at the time of that work (L- and T-type channels are both present in these cells), later studies confirmed that veratridine and batrachotoxin inhibit L-type currents in GH_4C_1 pituitary cells, albeit in micromolar concentrations (Enyeart et al. 1987). The two toxins also inhibit the Bay K 8644-stimulated Ca uptake in depolarized neuroblastoma cells (Kongsamut et al. 1985b).

The anticonvulsant drug phenytoin, a known blocker of myocardial and neuronal Na channels (Sanchez-Chapula and Josephson 1983; Willow et al. 1985; Willow 1986), is equally potent as a blocker of cardiac Ca channels (Scheuer and Kass 1983). This may be due to a direct interaction with the dihydropyridine-binding site (Messing et al. 1985; Yatani et al. 1986b). The cardiotonic agent S ($-$) DPI 201-106 that prolongs the open state of myocardial Na channels, in somewhat higher concentrations also acts as a blocker of sarcolemmal Ca channels. This latter effect is not stereoselective. Its binding site on the Ca channel seems to be different from, but allosterically coupled to, the dihydropyridine-, phenylalkylamine- and benzothiazepine sites (Scholtysik and Rüegg 1987; Siegl et al. 1988). Furthermore, some local anaesthetics were shown to inhibit stereospecifically [^{3}H]nitrendipine and [^{3}H]verapamil binding to brain and cardiac membranes (Harris et al. 1985; Bolger et al. 1987). The interaction appeared to be of allosteric rather than of a competitive nature. Yet, although the inhibition of binding occurred in therapeutically relevant concentrations, it did not correlate with the Na channel blocking strength of individual compounds. Moreover, some local anaesthetics had no effect at all or even stimulated dihydropyridine binding. Therefore, the binding sites involved in this interaction are probably not analogous to the local anaesthetic site in Na channels. Direct functional evidence for a block of Ca channels by a local anaesthetic (lidocaine) was obtained in experiments with smooth muscle (Hay and Wadsworth 1982).

On the other hand, dihydropyridine Ca channel ligands were found to block cardiac Na channels in a very similar way to Ca channels, except that somewhat higher concentrations were needed (Yatani and Brown 1985; Yatani et al. 1988a). The block of whole-cell and single-channel Na currents equals the Ca channel block in its stereospecificity and voltage dependence. In addition, it can be relieved by the Ca channel activator Bay K 8644. These results suggest that functionally equivalent binding sites for dihydropyridines must exist in both channels. However, the relation of these sites with the binding site for "classical" local anaesthetics is far from clear. In a functional comparison of different Ca channel ligands with lidocaine for their local anaesthetic effects in a rat phrenic nerve preparation, verapamil (V), methoxyverapamil (M) and flunarizine (F) were found to have an activity that was stronger than (V, M) or equal (F) to that of lidocaine. However, the dihydropyridine derivative nifedipine was without effect in the preparation (Hay and Wadsworth 1982). Probably these findings reflect functional differences between myocardial and neuronal Na channels. Finally, a blocking action on both, voltage-dependent Na and Ca channels, has also been observed for bepridil (Yatani et al. 1986a). This non-dihydropyridine compound with vasodilatory and antiarrhythmic properties (Fleckenstein 1977) was described originally as a fairly specific Ca channel ligand with binding properties closely related to d-cis-diltiazem (Galizzi et al. 1986a; Balwierczak et al. 1986; Janis et al. 1987; Glossmann and Striessnig 1988).

3.7 Accessibility of Drug-Binding Sites on Calcium Channels In Vivo

The modulated receptor hypothesis as proposed by Hille (1977) suggests that local anaesthetics, whether charged or neutral, block Na channels by inter- acting with a single specific receptor on the channel protein rather than by physically plugging open channels. Moreover, it is assumed that this receptor site is accessible via the open channel for charged compounds and via dif- fusion within the membrane lipid bilayer for lipophilic uncharged com- pounds.

   Is this second postulate compatible with the behaviour of Ca channel li- gands? Highly lipophilic compounds are found among the dihydropyridines, most of which do not carry a charge at physiological pH values. Membrane partition coefficients ranging between 5000 an 150000 have been determined for a number of dihydropyridines (Herbette and Katz 1987). Consequently, under steady-state conditions, a subnanomolar concentration in the aqueous phase (i.e. close to the apparent K_D value of many dihydropyridines) would correspond to a micromolar concentration in the membrane phase (Rhodes et al. 1985; Herbette and Katz 1987; Lüllmann and Mohr 1987). The high de- gree of drug accumulation in the membrane led to the suggestion of a general perturbation hypothesis to account for the action of dihydropyridines on Ca channels (Lüllmann and Mohr 1987). The same kind of argument was for- warded 17 years ago to explain the action of local anaesthetics (Seeman 1972). Stereospecificity of action, high-affinity binding and allosteric effects all ar- gue against such an unspecific perturbation mechanism. However, it is rather difficult to prove experimentally that the receptor indeed "sees" the concen- trations in the membrane phase rather than the one in the aqueous phase. One way to do this is to measure the rates of drug receptor binding. Provided this rate is diffusion limited, it should be much faster if the drug approaches the receptor via the hydrophobic pathway rather than by the hydrophilic pathway. Unfortunately, the on-rates appear to be slower than the calculated maximum rate for aqueous diffusion and, hence, are probably not diffusion limited (Rhodes et al. 1985; Herbette and Katz 1987). Yet, the membrane pathway for dihydropyridines is supported by other circumstantial evidence. Kokubun and Reuter (1984) have shown in patch clamp experiments on cultured rat myocar- dial cells that a single Ca channel within a cell-attached patch, isolated from the extracellular medium by the high-resistance contact between patch pipette and membrane, was still accessible to dihydropyridine channel activators. Furthermore, channel block by amlodipine, a dihydropyridine derivative with a pK value of 8.6 and hence, predominantly charged at pH 7.4, is very slowly reversible at physiological pH values (Burges et al. 1987; Kass et al. 1988). However, the block is rapidly reversible at pH 10 where the drug is in its highly lipophilic uncharged form and can readily leave its binding site via the mem- brane pathway.

The results with amlodipine do not necessarily imply binding to a site within the hydrophilic channel pore. In its ionized form at pH 7.4, amlodipine had little effect if administered from the inside but was an effective blocker if present in the external medium. Provided the receptor is situated close to the external face of the lipid bilayer, the charged compound could reach its target via the membrane phase by partitioning into the bilayer with the ionized group positioned close to the polar heads of the lipid molecules (Kass and Arena 1989). In earlier studies with a quaternary analogue of nifedipine, no significant channel block was observed following external administration (Uehara and Hume 1985). However, in this case partitioning into the bilayer might not have been possible because the permanently charged group was attached directly to the benzene ring rather than to the side chain as in amlodipine.

The position of drug-binding sites within the Ca channel has also been evaluated by using permanently charged derivatives of phenylalkylamines and dihydropyridines. Unlike amlodipine, the phenylalkylamine D890 containing a quarternary ammonium residue was ineffective when administered externally in isolated guinea pig myocytes. Yet, it caused quasi-irreversible channel blockade when injected intracellularly. By contrast, the parent compound gallopamil (D 600) caused a reversible channel block irrespective of the method of administration (Hescheler et al. 1982). Similar results were later obtained by Affolter and Coronado (1986) with Ca channels from skeletal muscle transverse tubules, reconstituted into membrane bilayers. Together with the strong use dependence of phenylalkylamine action (Sanguinetti and Kass 1984a; Uehara and Hume 1985), these observations suggest that opening of the channel is required at least for charged phenylalkylamines to reach their target site. Conversely, the negative inotropic potency of various Ca channel blockers, including verapamil and d-cis-diltiazem, is well correlated with the fraction of each drug that is protonized at physiological pH values (Mannhold et al. 1984). Hence, it seems that for these drugs (but not for nifedipine) the charged form is the functionally active molecule.

3.8 Direct Effects on the Calcium Channel Protein by Miscellaneous Drugs

In addition to the "classical" organic Ca channel blockers, several other agents have been found to block voltage-dependent L-type Ca channels in neuronal and muscle cells. Neuroleptics of the diphenylbutylpiperidine series like pimozide and fluspirilene have been studied most extensively. These compounds are usually categorized as dopamine receptor blocking agents. In voltage clamp experiments, fluspirilene was found to block transverse tubular Ca channels in skeletal muscles with high potency. The IC_{50} value $(0.1-0.2 \text{ n}M)$ was not significantly dependent on membrane voltage (Galizzi et al. 1986b). Diphenylbutylpiperidines seem to be much less potent in neuronal tissue and smooth muscle. IC_{50} values close to $100 \text{ n}M$ were reported for the inhibition of voltage-dependent ^{45}Ca uptake in GH_4C_1 pituitary cells (Enyeart et al. 1987) and in smooth muscle cells (Flaim et al. 1985). The claim that these effects are due to a direct interaction with the channel protein is based on binding studies with fluspirilene in transverse tubular membranes showing a 1:1 stoichiometry with devapamil (desmethoxyverapamil) binding and a good correlation between apparent K_D and IC_{50} values (Galizzi et al. 1986b). Moreover, diphenylbutylpiperidines have been shown to inhibit noncompetitively the binding of dihydropyridines and of devapamil, possibly through allosteric interactions between the respective binding sites (Gould et al. 1983; Qar et al. 1987). Another drug with neuroleptic properties, reserpine, also has Ca channel blocking properties in pituitary cells and in smooth muscle that can be distinguished from its well-known inhibitory effect on synaptic catecholamine storage (Casteels and Login 1983; Login et al. 1985). This effect, which required micromolar reserpine concentrations, has not yet been studied at the level of drug receptors. At least for diphenylbutylpiperidines, the Ca channel blocking effect occurs at therapeutic concentrations. Yet, its contribution, if any, to the overall antipsychotic effect of these drugs is still completely unknown. Other potent neuroleptics like chlorpromazine and haloperidol have only weak Ca channel blocking activities (Flaim et al. 1985; Enyeart et al. 1987).

Other drugs with inhibitory effects on voltage-dependent Ca channels include tetrandrine, an alkaloid from a Chinese medical herb (King et al. 1988), aminoglycoside antibiotics (Wright and Collier 1977; Adamo and Durrett 1978; Knaus et al. 1987) and the calmodulin antagonist calmidazolium (Enyeart et al. 1987). However, except for tetrandrine, which has been shown to compete specifically for the diltiazem-binding site in cardiac membrane preparations (King et al. 1988), no detailed information is available on the precise mechanism of the observed Ca channel block.

3.9 Organic Toxins with Specific Effects on L-Type Calcium Channels

Compared with the rich choice of natural toxins affecting the Na channel of excitable tissues, the Ca channel toxicology is much less developed. Nevertheless, the screening of animal toxins has revealed several interesting peptide compounds with preferential action on Ca channels. These toxins, including the channel activators gonioporatoxin and atrotoxin as well as the channel blockers taicatoxin and ω-conotoxin, are discussed fully in the accompanying contribution by Glossmann and Striessnig (1989) and in three recent reviews (Hamilton and Perez 1987; Wu and Narahashi 1988; Gray et al. 1988). A short overview is given in Table 3. It is usually not possible to establish a direct effect on the channel protein from the toxin effect on intact cells, like changes in contractility of cardiac and smooth muscle cells, changes in neurotransmitter release or in ^{45}Ca fluxes and Ca currents. For the toxins mentioned above, the notion of a direct effect is supported by the following additional findings. Gonioporatoxin (Qar et al. 1986), atrotoxin and taicatoxin (Hamilton et al. 1985; Lacerda and Brown 1986) all appear to interact allosterically with the dihydropyridine-binding site in myocardial cells. Moreover, the effect of taicatoxin on Ca currents in cardiac cells appears to be voltage dependent (Brown et al. 1987). Patch clamp experiments can be used to exclude a possible involvement of second messenger systems by comparing the effects of toxin administration within the patch pipette with a toxin administration exclusively to the membrane area outside the pipette (Brum et al. 1984). Except for ω-conotoxin (McCleskey et al. 1987), this test has not been performed systematically in other toxin studies.

4 Indirect Modulation of L-Type Calcium Channel Function by Drugs

In the past several years it has become increasingly clear that a number of hormones, neurotransmitters, and drugs have profound effects on Ca channel function in intact cells by interfering with endogenous regulatory mechanisms rather than with binding sites at the channel protein. Three major modulatory mechanisms have been established: (1) Changes in receptor density or affinity, (2) interaction of the channel with transmembrane signalling proteins of the G-protein family, and (3) phosphorylation of the channel protein by endogenous protein kinases. The latter two regulatory mechanisms have been evaluated predominantly by electrophysiological recording of whole-cell or single-channel currents. They are fully discussed in the chapter by Pelzer in this volume. I shall deal here in detail only with the first mechanism. The second mechanism will be discussed only as far as drug effects on intact cells are concerned. However, in Table 2 I have summarized a number of whole-cell studies dealing with drug or hormone effects which indirectly affect the function of voltage-dependent Ca channels. Most of these compounds seem to modulate protein kinase or G-protein interactions with the channel protein. However, it is important to realize that many of these effects have been observed only in one or two different tissues. In view of the heterogeneity of Ca channels, even within the L-type channel class, it is often not clear whether an effect can be generalized or is specific for one channel subtype. Therefore, in Table 2

I have always indicated the tissues in which the experiments have been performed.

4.1 Effects on Channel Density

Acute as well as long-term changes in channel densities have been observed. However, compared with the modulation of channel opening and closing kinetics, this mechanism seems to play only a minor role in the regulation of voltage-dependent Ca channels. In frog ventricular cells, β-adrenergic stimulation causes a large increase in voltage-dependent Ca currents thought to be due, at least in part, to a recruitment of "unavailable" channels (Bean et al. 1984). By contrast, Ca channels in mammalian cardiac tissue respond to catecholamines exclusively by an increase in their overall opening probability (Cachelin et al. 1983; see accompanying review by Pelzer). The mechanism by which Ca channel density could be enhanced in frog cardiac cells remains unknown. cAMP- or Ca-mediated phosphorylation reactions may play a role. In chick myotubes, short-term stimulation (30 min) of the Ca- and phospholipid-dependent protein kinase C caused a more than twofold increase in dihydropyridine-binding capacity and ^{45}Ca uptake (Navarro 1987). In the same preparation, long-term treatment with the β-adrenoceptor agonist isoproterenol or with other compounds known to raise intracellular cAMP was also found to induce a threefold increase in the density of dihydropyridine-binding sites (Schmid et al. 1985). Similarly, in bag cell neurons from the abdominal ganglion of the sea snail *Aplysia*, stimulation of protein kinase C, within a few minutes, causes the recruitment of a previously "masked" population of Ca channels. The kinetic properties of this new channel class are significantly different from those of the channel species active under control conditions (Kaczmarek 1987). The physiological relevance of these findings is difficult to assess since the change in Ca channel properties induced by protein kinase activation varies in different tissues (Kaczmarek 1987). In PC12 cells a maximal 20% increase in the density of dihydropyridine-binding sites associated with a comparable increase in ^{45}Ca fluxes has been observed during 90-min treatment with the lectin concanavalin A (Greenberg et al. 1987). Con A interacts specifically with carbohydrates on the cell surface. Hence, these findings point to a possible link between membrane carbohydrates and Ca signal transduction mechanisms

Several studies have assessed long-term regulation of Ca channel density. A decrease or increase in receptor density is a well-known adaptive phenomenon with most hormone receptor-effector systems exposed to chronic hyper- or hypostimulation. In analogy to hypostimulation by blocking receptors with hormone antagonists, the effect of chronic treatment with dihydropyridine channel blockers on myocardial channel density has been studied in rats (Nishiyama et al. 1986). Hyperstimulation was mimicked by chronic depolarization of PC12 cells in tissue culture (Delorme et al. 1988; Delorme and McGee 1986). A 2-week treatment with the channel blocker nifedipine was without effect on myocardial dihydropyridine-binding sites. No significant change of receptor density and no rebound effect, after the treatment had ended, were observed under conditions where both phenomena could be recorded after chronic β-adrenoceptor blockade (Nishiyama et al. 1986). On the other hand, depolarization for up to 6 days induced by high K concentrations in the culture medium caused a loss of dihydropyridine-binding sites and a concomitant decrease in ^{45}Ca influx. Similar changes could be produced by a 30-h treatment with the Ca inophore ionomycin (Delorme et al. 1988). These results suggest that the transient increase in intracellular Ca was the relevant signal for the reduction in channel density. The mechanism of this effect remains unclear. The long recovery period (24 h) points to a definite loss of channels that has to be replaced by resynthesis. These results have not yet been reproduced in other tissues and, therefore, cannot be generalized. In fact, our own attempts to show a similar effect of chronic depolarization in tissue-cultured rat cardiac cells were unsuccessful (H. Porzig and C. Becker, unpublished). A Ca_i-induced loss of Ca channels is also not easily compatible with recent results in cardiac sarcolemma vesicle preparations, showing a clear requirement for intracellular Ca to maintain dihydropyridine binding (Schilling 1988). More studies in intact cells are needed before the physiological role of intracellular Ca or of phosphorylation reactions on the regulation of Ca channel density in vivo can be fully evaluated.

A few reports show chronic effects of hormones and neurotransmitters on Ca channel density. In aneurally cultured human muscle a 14-day treatment with insulin alone or together with two other growth factors caused a more than twofold increase in dihydropyridine-binding sites and ^{45}Ca fluxes (Desnuelle et al. 1987). Treatment of rats with thyroxine was found to reduce the number of dihydropyridine-binding sites in cardiac tissue by 30% whereas the thyreostatic propylthiouracil caused an increase in binding site density (Hawthorn et al. 1988).

Other in vivo treatments that have been shown to enhance dihydropyridine binding include the destruction of sympathetic nerve endings in rats or chicken by injection of 6-hydroxydopamine (Renaud et al. 1984; Skattebøl and Triggle 1986), chronic treatment with reserpine (Power and Colucci 1985), and induction of morphine tolerance in mice (Ramkumar and El-Fakahany 1984). 6-Hydroxytryptamine increased [^{3}H]nitrendipine binding to myocardial tissue by 31%. Reserpine caused a 2.7-fold receptor increase in rat seminal vesicles. In morphine-tolerant mice, receptors in a whole brain homogenate went up by 60%. Finally, in brain homogenate from mice, a significant increase in the receptor affinity for dihydropyridines was observed after prolonged (up to 2 months) administration of chlorpromazine (Ramkumar and El-Fakahany 1985). However, in none of these cases has it been shown that the effect was specific for Ca channels. In fact, simultaneous effects on nicotinic Ach-receptors were observed with insulin, on α_1-adrenoceptors with reserpine, and on β-receptors with thyroxine and 6-hydroxydopamine (Desnuelle et al. 1987; Powers and Colucci 1985; Skattebøl and Triggle 1986; Hawthorn et al. 1988). Therefore, the mechanism causing the apparent changes in channel density is unclear and may well represent a secondary consequence of some hormone-induced change in protein synthesis.

4.2 G-Protein-Mediated Indirect Modulation of Calcium Channels

A detailed account of G-protein effects on L-type Ca channels is given in the chapter by Pelzer and in recent reviews by Rosenthal and Schultz (1988), Rosenthal et al. (1988a), Hosey and Lazdunski (1988) and Levitan (1988). It is important to note that in addition to indirect hormonal regulation of Ca channels via G-protein-coupled receptors, the effect of directly acting drugs like channel-activating and -blocking dihydropyridines is also sensitive to modulation by G-proteins. In intact cells, involvement of G-proteins in drug action on Ca channels can be inferred from the effects of intracellular administration of non-hydrolyzable GTP analogues, in particular GTPγS, or from the actions of the G-protein-specific pertussis and cholera toxins. Such studies have suggested that the effect of dihydropyridine Ca channel activators like Bay K 8644, at least in neuronal cells, critically depends on the presence of activated G_i or G_o-type G-protein (Scott and Dolphin 1988). Intracellular GTPγS strongly enhanced the stimulatory effect of ($\pm$)-Bay K 8644, and pertussis toxin (PTX) completely abolished the stimulatory component of Bay K 8644 action, leaving intact its channel-blocking activity (see Hadley and Hume 1988). In another study the same authors found that intracellular GTPγS converted the blocking action of D 600, nifedipine and d-cis-diltiazem on L-type currents in dorsal root ganglia into an activating activity that was inhibited by pertussis toxin (Scott and Dolphin 1987). G-protein (type G_s)-mediated amplification of the stimulatory effect of Bay K 8644 has also been observed with myocardial Ca channels (Yatani et al. 1987, 1988b). Moreover, in isolated myocardial membrane patches, GTPγS prevented the usual rapid rundown (irreversible inactivation) of Ca channels. The opening probability of Ca channels from skeletal muscle t-tubules being reconstituted into lipid bilayers was enhanced by activated G_s-protein or by its α subunit over and above the value obtained with Bay K 8644 alone (Yatani et al. 1988b). Second messengers like cAMP appeared not to be involved in these reactions. These results are all compatible with the assumption that G-proteins help to maintain the Ca channel protein in an activatable state, perhaps by stabilizing the resting or open states of the channel. If this interpretation is correct, one would expect G-proteins also to antagonize the effects of Ca channel blockers which seem to act by stabilizing the inactivated channel state. Such a loss of blocking potency has been observed for dihydropyridines, verapamil and *d-cis*-diltiazem in dorsal root ganglia (Scott and Dolphin 1987). In the 235-1 pituitary cell line Schettini et al. (1988) have seen that an inhibitory effect of the dihydropyridine derivative nicardipine on maitotoxin-induced Ca influx was abolished by pertussis toxin. However, these reports are somewhat anecdotal and a systematic evaluation of the effects of G-proteins on the voltage-dependent interactions between Ca channels and activating or blocking ligands is still lacking. It would be particularly interesting to see whether endogenous G-protein activation affects voltage-dependent binding of dihydropyridines in intact cells. In a sarcolemmal preparation from smooth muscle, the GTP analogue GppNHp was recently shown to reduce the apparent affinity of the channel activator Bay K 8644. The affinity of the blocker nimodipine remained unchanged (Higo et al. 1988). By contrast, a study with rat synaptic membranes provided indirect evidence for an increase in the affinity of Bay K 8644 induced by the non hydrolyzable GTP analog GMP-PNP (Bergamschi et al. 1988). An earlier study in t-tubule membranes from rabbit skeletal muscle suggested that GTP or GppNHp reduced significantly the affinity of part of the high-affinity verapamil-binding sites. Again, no effect on the binding of the dihydropyridine channel blocker nitrendipine was observed (Galizzi et al. 1984). Overall, the role of G-proteins for the physiological functioning of L-type Ca channels and its interplay with channel phosphorylation remains unclear. No experiments have been performed in intact cells under conditions where any endogeneous activity of the G_s protein has been excluded. Indeed, this may be impossible if a protein sequence homologous to the functional substructure of a G-protein forms part of one of the subunits of the channel protein.

Two other classes of drugs, benzodiazepines and opioids, have effects on neuronal and GH_3-pituitary L-type Ca channels that seem to be mediated by G-proteins (probably type G_o). Benzodiazepines inhibit depolarization-induced ^{45}Ca uptake in GH_3 cells (Gershengorn et al. 1988) and nerve terminals most likely via an interaction with type B (low-affinity) GABA receptors. These receptors are known to be linked to Ca channels (Dunlap 1981; Dunlap and Fischbach 1981; Taft and De Lorenzo 1984; Désarmenien et al. 1984; Deisz and Lux 1985; Tsunoo et al. 1986). The same component of Ca uptake is also inhibited by organic Ca channel blockers. An involvement of G-protein in the effect of benzodiazepines has not been demonstrated directly but can be inferred from the observation that these proteins are required to couple $GABA_B$/baclofen receptors to Ca channels (Dolphin and Scott 1987; Holz et al. 1986). On the other hand, benzodiazepines do not seem to interact with non-neuronal L-type Ca channels (Holck and Osterrieder 1985). An inhibitory action of G-proteins on Ca-channels is also the most probable explanation for the GABA- or noradrenaline-induced reduction of substance P release in chick dorsal root ganglion cells (Holz et al. 1989).

In neuroblastoma-glioma hybrid cells (NG 108-15) opioids and enkephalins inhibit Ca currents through neuronal L-type Ca channels via δ-opiate receptors (Tsunoo et al. 1986; Hescheler et al. 1987; Narahashi 1988). In addition, stimulatory effects on Ca channels via μ-opiate receptors have also been reported (Lorentz et al. 1988).

The inhibitory effect of D-Ala-D-Leu enkephalin (DADLE) could be abolished by pertussis toxin and was restored by intracellular injection of G_i or G_o. Since G_o was tenfold more effective, it was suggested that it is G_o that couples δ-opioid receptors to Ca channels (Hescheler et al. 1987). The opioid effect on Ca channels appears to be voltage dependent. The block increased with hyperpolarization and decreased with depolarization (Tsunoo et al. 1986; Narahashi 1988). Up to now such coupling of δ-opioid receptors to L-type Ca channels has only been observed in neuroblastoma cell lines. Therefore, it would be premature to attempt any generalizations. Indeed, contradictory observations have been made in primary cultures of dorsal root ganglion cells from mice. In this preparation Ca current was reduced specifically by κ-opioid receptor agonists acting on N-type Ca channels rather than by interactions of δ-receptors with L-channels (Werz and MacDonald 1985; North 1986; Gross and MacDonald 1987). Nevertheless, the latter type of interaction could perhaps explain recent observations by Contreras et al. (1988), suggesting a significant effect of organic Ca channel antagonists on morphine-induced analgesia and tolerance in mice: d-cis-diltiazem, flunarizine, nicardipine and verapamil all increased the analgesic effect of morphine. Nifedipine had an antagonistic effect. All blockers, except diltiazem, reduced tolerance development, induced by a single dose of a slow-release morphine preparation. However, more systematic studies are clearly needed to assess the functional consequences and physiological relevance of opioid-mediated effects on Ca channels.

5 Pharmacology of T-Type Calcium Channels

A low-threshold, rapidly inactivating T-type Ca current has been described in most cells that have a voltage-gated Ca conductance pathway (for review see Miller 1987 a, b; Tsien et al. 1988; Hosey and Lazdunski 1988). But there are few reports on specific modulation of this current component by drugs or neurotransmitters. Moreover, in most cases where drug effects on T-channels have been tested, their functional significance for the overall drug action remained unclear. The relevant reports all rely on electrophysiological dissection of Ca current components by a voltage clamp technique to define isolated effects on T-type currents. No high-affinity ligand is known which could be used to analyse T-type channels. Nevertheless it is quite clear that T- and L-type channels are different proteins, being the product of different genes. In myodysgenic mice, a mutant that lacks L-type Ca channels and currents in skeletal muscle cells, T-type currents are fully preserved (Beam et al. 1986). A few drugs inhibit T-type currents preferentially and have only low potencies with L-type channels. This is the case for phenytoin, which blocks transient Ca current in N1E-115 neuroblastoma cells at concentrations ($3-100\,\mu M$) not affecting the slowly inactivating current component. The block was use- and voltage-dependent, increasing at more depolarized holding potentials (Twombly et al. 1988). Other examples include retinoic acid (Bosma and Sidell 1988), tetramethrin (Tsunoo et al. 1985) and the diuretic compound amiloride (Tang et al. 1988). Retinoic acid was shown to inhibit a T-Type Ca current in the MHY 206 mouse hybridoma cell line. Fifty percent inhibition required $50\,\mu M$ retinoic acid. This concentration is sufficient to block cell prolifera- tion, but it is not known whether the two effects are causally related. In the case of amiloride it is not clear whether the inhibition of myocardial T-cur- rents results from a direct effect or from an increase in the intracellular H^+ concentration. The reported K_D value for the effect on the T-channel ($30\,\mu M$) is well within the concentration range known to inhibit the H^+-Na^+ exchange carrier (for review see Benos 1982; Lazdunski et al. 1985). Among the anorganic cations, nickel in low concentrations ($40\,\mu M$) was found to block the T-type current in rabbit sinoatrial node cells preferentially, without significant effect on the L-type current (Hagiwara et al. 1988). A few other drugs or inorganic ions block T- and L-channels at about similar concentra- tions. In guinea pig myocardial cells $5-10\,\mu M$ flunarizine or its congener cin- narizine block both channel types in a use-dependent way. In the same prepa- ration T-channels are not blocked by $10\,\mu M$ verapamil (Tytgat et al. 1988; van Skiver et al. 1988). In rabbit sinoatrial node, cadmium and cobalt were found to block both channel types with about equal potency (Hagiwara et al. 1988).

Indirect inhibition of T-type currents in intact cells, perhaps through sec- ond messenger or G-protein-mediated mechanisms, have also been reported. In clonal pituitary GH_3/B_6 cells, the antioestrogen tamoxifen ($0.1\,\mu M$) re- versibly inhibited both slow (L-type) and fast (T-type) inactivating Ca conduc- tances (Sartor et al. 1988). Full inhibition of L-type currents was associated with 60% – 80% inhibition of T-type currents. The mechanism of this effect is not clear. It was not mediated via estrogen receptors. A tamoxifen-induced inhibition of protein kinase C (Su et al. 1985) also appears unlikely because a similar inhibition of T- and L-type currents could be reached in this cell line by activating the kinase with 1-oleoyl-2-acetyl-glycerol (Marchetti and Brown 1988). In neuronal cells, a functional coupling of T- (and L-) channels to a G-protein (possibly G_o) is suggested in studies with the putative neurotrans- mitter neuropeptide Y (NPY). In rat dorsal root ganglion cells, NPY inhibited

T- and L-type currents. In both cases pertussis toxin abolished this inhibitory response which, however, reappeared after perfusion of the cell with the active α-subunit of G_o (Ewald et al. 1988a; Walker et al. 1988).

Another neurotransmitter shown to affect T-type Ca currents is dopamine. In chick dorsal root ganglia the rapidly inactivating low-threshold Ca channels were almost irreversibly inhibited by dopamine. Under the same conditions the slowly inactivating high-threshold channels only showed a reversible slowing of their activation kinetics (Marchetti et al. 1986).

The physiological relevance of these observations depends very much on the relative contribution of the various Ca current components to neuronal function, e.g. neurotransmitter release. Some aspects of functional associations of different types of Ca channels are discussed below.

6 Pharmacology of N-Type Calcium Channels

This type of high-threshold current with inactivation rates slower than T- and faster than L-type currents has been described on the basis of single-channel patch clamp analysis in neuronal cells (Fox et al. 1987a, b). In different types of neurons these channels display an unusually large range of inactivation time constants (50 – 500 ms) and single-channel conductances (11 – 20 pS) (Tsien et al. 1988). Therefore, to treat N-channels as a homogeneous class is certainly an oversimplification (see Glossmann and Striessnig 1989). The ongoing controversy of whether L- and N-type current components can indeed both be detected in macroscopic currents (Swandulla and Armstrong 1988; Tsien et al. 1988) shows that electrophysiological criteria alone are not sufficient to characterize Ca channel subtypes. L-type channels have always been defined by their sensitivity towards organic Ca channel blockers. Until recently, no pharmacological tools were available to dissect high-voltage activated Ca current components that were resistant to dihydropyridines or verapamil. The discovery that such neuronal Ca channel subtypes can be labelled by the snail venom ω-conotoxin (Kerr and Yoshikami 1984; for review see Olivera et al. 1985; Gray et al. 1988) has considerably boosted Ca channel pharmacology (compare the accompanying review by Glossmann and Striessnig). It also provides a means to test the physiological function of channel subtypes. Earlier functional studies seemed to show that ω-CgTX blocked neuronal L- and N-type channels nearly irreversibly but had little effect on T-channels (McCleskey et al. 1987; Miller 1987a, b; Tsien et al. 1988). However, it was soon discovered that this apparent block of L-type channels was strictly confined to neuronal tissue. L-type currents in myocardial tissue or in vascular smooth muscle were not affected (McCleskey et al. 1987). Evidence is now accumulating that neuronal L-type channels, too, may be insensitive to ω-

CgTX. This issue is controversial because it is apparently very difficult to record "pure" neuronal L-type currents, free from a "contaminating" N-type current component (Plummer et al. 1989). Functional studies and binding experiments strongly support the notion that L- and N-type channels correspond to two distinctly different neuronal channel populations. ω-CgTX, but not dihydropyridines, strongly inhibits neurotransmitter release in central neurons (Kerr and Yoshikami 1984; Reynolds et al. 1986 and see below). The regional distribution of ω-CgTX-binding sites in brain differs significantly from the one for dihydropyridines (Wagner et al. 1988). Moreover, the density of ω-CgTX-binding sites 10- to 80-fold higher than the one for dihydropyridines (Wagner et al. 1988).

Binding studies in brain homogenates revealed very high affinities with apparent K_D values in the picomolar range (Abe et al. 1986; Cruz and Olivera 1986; Knaus et al. 1987; Wagner et al. 1988). However, functional studies in intact cells or nerve terminals usually required concentrations higher by one to three orders of magnitude (Reynolds et al. 1986; Oyama et al. 1987; McCleskey et al. 1987). This was readily explained by different compositions of the experimental media: Binding was studied in low ionic strength solutions, but function in intact cells had to be assessed in isotonic media. Mono- and divalent cations exert a strong inhibitory effect on ω-CgTX binding (Cruz and Olivera 1986; Abe et al. 1986; Wagner et al. 1988). IC_{50} values for divalent alkaline earth cations range between 0.2 and 0.6 mM, for monovalent alkali cations between 31 and 47 mM (Wagner et al. 1988). This non-competitive interaction of cations with toxin binding has severely hampered binding studies in intact cells. Such studies have been performed recently by Martin-Moutot et al. (1989) in primary rat neuronal cells and in our own laboratory (Porzig et al. 1989). Both groups were interested in developmental aspects of Ca channel regulation. Martin-Montot et al. report a maximal binding capacity of 60 fmol/mg protein. We have used living PC12 cells to assess the regulation of Ca channel number during nerve growth factor (NGF)-induced differentiation. Undifferentiated PC12 cells bind maximally $12-15$ fmol/10^6 cells ω-CgTX (corresponding to about 30 fmol/mg protein) and about 7 fmol/10^6 cells of the dihydropyridine isradipine. Comparable results have been obtained in PC12 cell homogenates by Sher et al. (1988) for ω-CgTX and by Toll (1982), Albus et al. (1984) and Messing et al. (1985) for dihydropyridines. Cultivation of PC12 cells in the presence of NGF for 4 days caused a differentiation into neuron-like cells with extensive neurite growth and a doubling of specific ω-CgTX binding. Under the same conditions we observed only small changes in the density of dihydropyridine-binding sites. Two recent pharmacological studies support the view that a new type of dihydropyridine-insensitive Ca channel is expressed in NGF-treated cells (Takahashi et al. 1985; Kongsamut and Miller 1986). Both studies measured depolarization-induced neurotransmitter release from PC12 cells. Ca, entering through voltage-sensitive channels, triggers this release reaction. In undifferentiated cells the neurotransmitter release is completely blocked by dihydropyridines, whereas in NGF-treated cells dihydropyridines reduced the release only by about 40%. On the other hand, Co^{2+}, a universal inhibitor of Ca channels, completely blocked the reaction in both differentiated and undifferentiated cells (Takahashi et al. 1985). A preferential incorporation of N-type Ca channels in the presence of NGF is also supported by recent electrophysiological evidence (Streit and Lux 1987; Plummer et al. 1989). The latter study reported a marked increase in the fraction of Ca channel current that could be inhibited by ω-CgTX.

From our own studies it seems most likely that the induction of N-type channels is a specific effect of NGF rather than being associated with morphological differentiation (i.e. stop of cell division, growth of cell soma, formation of growth cones, development of neurites and dendrites). In a PC12 cell line which was not responsive to NGF, ω-CgTX-binding site density was not increased upon NGF-independent differentiation induced by ouabain. In summary, functional as well as binding experiments suggest the presence of a distinct dihydropyridine-insensitive Ca channel population in neuronal cells that may carry N-type currents and is specifically labelled by ω-CgTX.

Another blocker of N-type currents with a presumably direct action on the channel may be represented by the lanthanoid gadolinium. In the neuroblastoma-glioma hybrid cell line NG 108-15 gadolinium seems to block N-type Ca currents rather specifically (Docherty 1988). However, the dihydropyridine-sensitive component of total Ca current was not assessed in this report. Therefore, it is not clear whether the effects of gadolinium and of dihydropyridines would be additive. Moreover, the N-current block by gadolinium needs confirmation in other neuronal tissues.

Indirect modulation of N-channels via secondary receptor mechanisms has been reported repeatedly. However, in many studies the contributions of L- and N-current components to the overall hormone- or drug response were not clearly separated. Therefore, some of the effects on Ca currents mediated by hormone- or neurotransmitter receptors and described above in the context of neuronal L-channel modulation may well involve responses partially or entirely due to changes in N-channel currents. Specific inhibition of N-type Ca currents was observed with dynorphin A acting on κ-opioid receptors in mouse dorsal root ganglion cells (Gross and MacDonald 1987; Werz and MacDonald 1985), with NPY in myenteric plexus (Hirning et al. 1988), with adenosine in hippocampal neurons (Madison et al. 1987), with noradrenaline acting on α-adrenoceptors in frog sympathetic ganglia (Lipscombe and Tsien 1988) and with acetylcholine acting on muscarinic M_1 receptors in rat sympathetic neurons (Wanke et al. 1987). The latter effect could be mimicked by intracellular perfusion with GTPγS and was eliminated by pertussis toxin. cAMP and activators of protein kinase C were ineffective. Hence, N-type Ca current modulation by muscarinic agonists in these cells appears to be mediated by a G-protein (type G_i or G_o). On the other hand, evidence is accumulating that the inhibition of neuronal Ca currents (L- or N-type) by other neurotransmitters, in particular noradrenaline, may not be mediated by a direct interaction of G-proteins with the channel but requires the activation of PKC as an intermediary step. In the latter case, G-proteins would play an indirect role by activating the synthesis of a diffusible second messenger (Rane and Dunlap 1986; Rane et al. 1989).

In principle, it is conceivable that the same mechanisms govern the inhibitory interactions of hormones and neurotransmitters with neuronal L- and N-type channels (compare discussion on p. 236/237). However, it is probably fair to state that somewhat stronger experimental evidence supports G-protein-dependent regulation of N-type Ca channels. The contribution of L-type channels in neurotransmitter effects on neuronal Ca currents is still controversial.

7 Functional Role of Different Types of Calcium Channels

Selective modulation of Ca channels by specific drugs is the most important technique by which the contribution of individual channel types to the overall function of a cell can be assessed. Such studies have shown that in most cases biological functions like secretion, transmitter release or contraction are not exclusively associated with individual channel types. It seems that, at least in part, tissue distribution of channels determines their functional role (for review see Miller 1987 a, b; Tsien et al. 1988; Hirning et al. 1988).

T-channels are associated with typical regenerative phenomena that require current activation at negative membrane potentials and rapid inactivation. Using low concentrations of Ni^{2+} and of tetramethrin (Tsunoo et al. 1985) as relatively specific inhibitors, Hagiwara et al. (1988) have shown that T-type current contributes importantly to the generation of pacemaker potentials in the rabbit sinoatrial node. It is also possible that the bursting behaviour of various neurons in the mammalian CNS involves regenerative Ca currents through T-channels (Miller 1987 a, b; Tsien et al. 1988). Moreover, in some cells, e.g. fibroblasts (Chen et al. 1988; Peres et al. 1988), certain smooth muscle cells (Sturek and Hermsmeyer 1986), adrenal glomerulosa cells (Cohen et al. 1988), T-type Ca currents seem to constitute a major voltage-gated Ca influx pathway and, hence, are essential for contractile or secretory events in these cells. However, a detailed analysis of the contribution of T-currents to Ca-dependent cellular functions must await the advent of specific inhibitors.

N-type Ca currents appear to represent a prominent component of voltage-dependent Ca influx in neuronal and in some secretory cells. It is quite characteristic that in many neuronal preparations (tissue cultured cells, brain slices, sympathetic nerve endings in peripheral tissues) where cells contain both L- and N-type channels, the depolarization-induced secretion of neurotransmitters is not, or only partially, reduced by organic blockers of L-type channels. By contrast, secretion is, often but not always, stimulated in those preparations by L-channel activators like Bay K 8644. On the other hand, secretion can be strongly inhibited by ω-CgTX (Dooley et al. 1988; Hirning et al. 1988; Maggi et al. 1988; Barnes and Davies 1988; for reviews see Miller and Freedman 1984; Reynolds et al. 1986; Miller 1987a, b; Tsien et al. 1988). However, it should be noted that effectiveness of ω-CgTX and ineffectiveness of dihydropyridines to inhibit secretion does not automatically exclude an important role of L-channel Ca current. The effect of ω-CgTX is voltage independent. It may be easily observed using field stimulation at strongly negative resting potentials. Such conditions do not favour the voltage-dependent blocking action of dihydropyridines which may be visible only when the release of neurotransmitter is provoked by K-induced depolarization (Rane et al. 1987). In the case of rat sympathetic neurons the preferential coupling of N-channels with secretion is all the more surprising because total voltage-dependent net Ca influx measured with the fura-2 technique was inhibited by 50% in the presence of dihydropyridine channel blocking (Hirning et al. 1988). Macroscopic spatial inhomogeneities of channel distribution are not a very likely explanation for this phenomenon. Patch clamp and fura-2 studies suggested that both channel types co-existed on all parts of the sympathetic neuron, including growth cones (Thayer et al. 1987; Lipscombe et al. 1988). Possibly, a preferential interaction exists between N-channels and catecholamine-containing storage vesicles, but experimental evidence for this assumption is scarce. Nevertheless, it seems that the release of catecholamines from differentiated neuronal preparations containing such transmitters is predominantly controlled by voltage-dependent Ca entry through N-type channels

By contrast, secretory events in other neuronal and in particular endocrine tissues can be elicited also by Ca influx through L-type channels (Miller 1987a, b; Tsien et al. 1988). It is important to note in this context that the association of Ca channel types with specific cell functions is not necessarily invariant. At least two well-documented cases exist where such association is shifted in the course of cell differentiation. The release of the neurotransmitter substance P from embryonic chick dorsal root ganglia and rat sensory neurons is inhibited by nifedipine in a voltage-dependent manner (Perney et al 1986; Rane et al. 1987). Yet, dihydropyridines do not seem to affect sub-

stance P release from slices of adult rat spinal cord (Miller 1987b). Catecholamine release from undifferentiated PC12 cells can be almost completely
abolished by dihydropyridines (Garcia et al. 1984; Takahashi et al. 1985;
Kongsamut and Miller 1986). Similarly, noradrenaline release from cultured
adrenal chromaffin cells, the parent cell of the PC12 tumor cell line, is highly
sensitive to dihydropyridines (Ceña et al. 1983; Garcia et al. 1984). However,
after NGF-induced differentiation, neurotransmitter release appears to be
controlled predominantly by Ca flowing through N- rather than L-type channels (see above). It is tempting to suggest that the loss of L-channel-dominat-
ed release reactions has a more general significance for ontogenic development and cell differentiation.

As described in the first part of this review, Ca current through L-type
channels has been found to be closely linked to contraction in cardiac and
smooth muscle (for reviews see Godfraind et al. 1986; Janis et al. 1987; Tsien
1987; Hosey and Lazdunski 1988). In many cases, Ca entry via L-type channels is also required for hormone secretion in endocrine cells. This has been
studied most extensively in normal rat anterior pituitary cells, in human pituitary tumor cells, and in rat pituitary-derived permanent cell lines (e.g. GH_3
and GH_4C_1). These cells have two populations of Ca channels that seem to
fit into the T- and L-type categories (Matteson and Armstrong 1986; Cota
1986; Cohen and McCarthy 1987; Yamashita et al. 1988). Depolarization-induced Ca influx and hormone secretion is blocked with nanomolar concentrations of the dihydropyridine derivatives, nimodipine, nisoldipine or
nifedipine (Tan and Tashjian 1984a, b; Enyeart et al. 1985; Stojiković et al.
1988) and stimulated by the L-channel activator Bay K 8644 (Enyeart et al.
1986; Stoijiković et al. 1988). However, these results certainly do not prove an
exclusive role for dihydropyridine-sensitive channels in stimulus-secretion
coupling. In an interesting study by DeRiemer (1989), the contribution of Ca
flux through L- and T-channels to the secretory response of pituitary cells has
been analyzed in detail. The author compared pituitary growth hormone re-
leasing cells and prolactin secreting cells. The former cells have predominantly L-channels and few T-channels, whereas the latter have both types in about
equal amounts. Selective inhibition of T-channels with ethanol had a marked
inhibitory effect on prolactin secretion, but not on growth hormone secretion.
It seems likely that a preferential coupling of L-channels to growth hormone
secretion has nothing to do with any specific association of secretory vesicles
with channels but simply results from the quantitative predominance of Ca
entry through this pathway, in particular during prolonged depolarization. Ca
entry through T-channels may gain relatively more weight during short-term
depolarizations.

The release of parathyroid hormone (PTH) which is negatively coupled to
Ca entry is partially inhibited by L-channel activators like (+)-S-202-791 and
activated by dihydropyridine channel blockers (Fitzpatrick et al. 1986). Recent

evidence suggests that this L-type channel in parathyroid cells is not only functionally but also immunologically closely related to skeletal muscle L-channels. A specific polyclonal antibody against the Ca channel in rat muscle t-tubular membranes was shown to activate parathyroid Ca channels and, consequently, to inhibit PTH release (Fitzpatrick et al. 1988).

8 Conclusions

Our knowledge about specific pharmacological modulation of Ca channel function on the level of intact cells is growing at a remarkably fast rate. Yet, the rapid proliferation of mainly electrophysiologically defined new Ca channel types has been matched only partially by an increasing arsenal of drugs that can be used to probe the role of defined channels in cellular function. Selective modulation of one or the other Ca channel has been discovered repeatedly as an interesting side effect of compounds which are in use for some other purpose, not thought to be related to Ca channel modulation. However, a systematic search for Ca channel ligands on the basis of information on their molecular structure is not yet feasible.

Some of the newly described channel varieties are still inaccessible to differential labelling by drugs. Thus the large family of dihydropyridine-sensitive L-type channels with its tissue-specific subtypes (L_m, L_h, L_n) cannot be clearly dissected by pharmacological means. The analysis of Ca channels from different sources with the methods of molecular genetics will probably further boost the number of channel subtypes. However, the tremendous therapeutic potential of this development can be exploited only if highly subtype-selective drugs can be found. Ca plays a fundamental role as charge carrier and second messenger, in particular for neurotransmitter release and other neuronal functions. Therefore, any non-selective modulation of neuronal Ca channels has a potential for side effects that makes it rather useless for therapeutic purposes. The wide therapeutic applicability of dihydropyridines resulting from their preferential affinity for peripheral L-type channels and the marked voltage dependency of their action is a fortunate coincidence. Future research will show whether a similarly successful approach is possible in neuropharmacology.

Acknowledgements. The work of the author was supported by grants from the Swiss National Science Foundation (No. 3.059-0.84 and 3.078-0.87). I am grateful to Prof. H. Reuter for a critical reading of the manuscript.

References

Abe T, Koyano K, Saisu H, Nishiuchi Y, Sakakibara S (1986) Binding of ω-conotoxin to receptor sites associated with the voltage-sensitive calcium channel. Neurosci Lett 71: 203–208

Adams HR, Durrett LR (1978) Gentamicin blockade of slow Ca^{++} channels in atrial myocardium of guinea pigs. J Clin Invest 62:241–247

Affolter H, Coronado R (1986) Sidedness of reconstituted calcium channels from muscle tranverse tubules as determined by D600 and D890 blockade. Biophys J 49:767–772

Ahmad SN, Miljanich GP (1988) The calcium channel antagonist, ω-conotoxin and electric organ nerve terminals: binding and inhibition of transmitter release and calcium influx. Brain Res 453:247–256

Albus W, Habermann E, Ferry DR, Glossmann H (1984) Novel 1,4-dihydropyridine (Bay K8644) facilitates Ca-dependent [^{3}H]noradrenaline release from PC12 cells. J Neurochem 42:1186–1189

Almers W, McCleskey EW, Palade PT (1984) A non-selective cation conductance in frog muscle membrane blocked by micromolar external calcium ions. J Physiol (Lond) 353:565–583

Balwierczak JL, Grupp IL, Grupp G, Schwartz A (1986) Effects of bepridil and diltiazem on [^{3}H]intrendipine binding to canine cardiac sarcolemma. Potentiation of pharmacological effects of nitrendipine by bepridil. J Pharmacol Exp Ther 237:40–48

Barnes S, Davies JA (1988) The effect of calcium channel agonists and antagonists on the release of endogenous glutamate from cerebellar slices. Neurosci Lett 92:58–63

Beam KG, Knudson CM, Powell JA (1986) A lethal mutation in mice eliminates the slow calcium current in skeletal muscle cells. Nature 320:168–170

Bean BP (1984) Nitrendipine block of cardiac calcium channels: high affinity binding to the inactivated state. Proc Natl Acad Sci USA 81:6388–6392

Bean BP (1989) Classes of calcium channels in vertebrate cells. Ann Rev Physiol 51:367–384

Bean BP, Nowycky MC, Tsien RW (1984) β-Adrenergic modulation of calcium channels in frog ventricular heart cells. Nature 307:371–375

Bean BP (1985) Two kinds of calcium channels in canine atrial cells – differences in kinetics, selectivity and pharmacology. J Gen Physiol 86:1–30

Bechem M, Hebisch S, Schramm M (1988) Ca^{2+} agonists: new sensitive probes for Ca^{2+} channels. Trends Pharmacol Sci 9:257–261

Bellemann P (1984) Binding properties of a novel calcium channel activating dihydropyridine in monolayer cultures of beating myocytes. FEBS Lett 167:88–92

Bellemann P, Ferry D, Lübbecke F, Glossmann H (1981) [^{3}H]-Nitrendipine, a potent calcium antagonist, binds with high affinity to cardiac membranes. Arzneimittelforschung 31: 2064–2067

Bellemann P, Schade A, Towart R (1983) Dihydropyridine receptor in rat brain labeled with [^{3}H]nimodipine. Proc Natl Acad Sci USA 80:2356–2360

Benos DJ (1982) Amiloride: a molecular probe of sodium transport in tissues and cells. Am J Physiol 242:C131–C145

Bergamaschi S, Govoni S, Cominetti P, Parenti M, Trabucchi M (1988) Direct coupling of a G-protein to dihydropyridine binding sites. Biochem Biophys Res Comm 156:1279–1286

Bixby JL, Spitzer NC (1983) Enkephalin reduces quantal content at frog neuromuscular junction. Nature 301:431–432

Bkaily G, Sperelakis N, Renaud JF, Payet MD (1985) Apamin, a highly specific Ca^{2+} blocking agent in heart muscle. Am J Physiol 248:H961–H965

Bodewei R, Hering S, Schubert B, Winkler J, Wollenberger A (1985) Calcium channel block by phenytoin in neuroblastoma-glioma hybrid cells. Biomed Biochim Acta 44:1229–1238

Bolger GT, Gengo PJ, Luchowski EM, Siegel H, Triggle DJ, Janis RA (1982) High affinity binding of a calcium channel antagonist to smooth and cardiac muscle. Biochem Biophys Res Commun 104:1604–1609

Bolger GT, Gengo P, Klockowski R, Luchowski E, Siegel H, Janis RA, Triggle AM, Triggle DJ (1983) Characterization of binding of the Ca^{++} channel antagonist [H-3]nitrendipine to guinea-pig ileal smooth muscle. J Pharmacol Exp Ther 225:291–309

Bolger GT, Marcus KA, Daly JW, Skolnick P (1987) Local anesthetics differentiate dihydropyridine calcium antagonist binding site in rat brain and cardiac membranes. J Pharmacol Exp Ther 240:922–930

Bosma M, Sidell N (1988) Retionic acid inhibits Ca^{2+} currents and cell proliferation in a B-lymphocyte cell line. J Cell Physiol 135:317–323

Brown AM, Kunze DL, Yatani A (1984) The agonist effect of dihydropyridines on Ca channels. Nature 311:570–572

Brown AM, Yatani A, Lacerda AE, Gurrola GB, Possani LD (1987) Neurotoxins that act selectively on voltage-dependent cardiac calcium channels. Circ Res 61 [Suppl I]:6–9

Brum G, Osterrieder W, Trautwein W (1984) β-Adrenergic increase in the calcium conductance of cardiac myocytes studied with the patch clamp. Pflüger Arch Eur J Physiol 401:111–118

Burges RA, Gardiner DG, Gwilt M, Higgins AJ, Blackburn KJ, Campbell SF, Cross PE, Stubbs JK (1987) Calcium channel blocking properties of amlodipine in vascular smooth muscle and cardiac muscle in vitro: evidence for voltage modulation of vascular dihydropyridine receptors. J Cardiovasc Pharmacol 9:110–119

Cachelin AB, de Peyer JE, Kokubun S, Reuter H (1983) Ca^{2+} channel modulation by 8-bromocyclic AMP in cultured heart cells. Nature 304:462–464

Campbell KP, Leung AT, Sharp AH (1988) The biochemistry and molecular biology of the dihydropyridine-sensitive calcium channel. Trends Neurosci 11:425–430

Caparrotta L, Fassina G, Froldi G, Poja R (1987) Antagonism between (−)-N^6-phenylisopropyl-adenosine and the calcium channel facilitator Bay K 8644, on guinea-pig isolated atria. Br J Pharmacol 90:23–30

Carbone E, Lux HD (1988) Sodium currents through neuronal calcium channels: kinetics and sensitivity to calcium antagonists. In: Morad M, Nayler W, Kazda S, Schramm M (eds) The calcium channel: structure, function and implications. Springer, Berlin Heidelberg New York, pp 115–127

Carbone E, Lux HD (1989) ω-Conotoxin blockade distinguishes Ca from Na permeable states in neuronal calcium channels. Pflügers Arch Eur J Physiol 413:14–22

Carbone E, Morad M, Lux HD (1987) External Ni^{2+} selectively blocks the low-threshold Ca^{2+} currents of chick sensory neurons. Pflügers Arch Europ J Physiol 408:R60 (abstract)

Casteels R, Login IS (1983) Reserpine has a direct action as a calcium antagonist on mammalian smooth muscle cells, J Physiol (Lond) 340:403–414

Catterall WA (1977) Activation of the action potential Na$^+$ ionophore by neurotoxins. An allosteric model. J Biol Chem 252:8669–8676

Catterall WA (1981) Inhibition of voltage-sensitive sodium channels in neuroblastoma cells by antiarrhythmic drugs. Mol Pharmacol 20:356–362

Catterall WA (1987) Common modes of drug action on Na$^+$ channels: local anesthetics antiarrhythmics and anticonvulsants. Trends Pharmacol Sci 8:57–65

Catterall WA (1988) Structure and function of voltage-sensitive ion channels. Science 242:50–61

Cavalié A, Ochi R, Pelzer D, Trautwein W (1983) Elementary currents through Ca^{2+} channels in guinea pig myocytes. Pflügers Arch Eur J Physiol 398:284–297

Ceña V, Nicolas GP, Sanchez-Garcia P, Kirpekar SM, Garcia AG (1983) Pharmacological dissection of receptor-associated and voltage-sensitive ionic channels involved in catecholamine release. Neuroscience 10:1455–1462

Chen C, Corbley MJ, Roberts TM, Hess P (1988) Voltage-sensitive calcium channels in normal and transformed 3T3 fibroblasts. Science 239:1024–1026

Chin PJS, Tetzloff G, Chatterjee M, Sybertz EJ (1988) Phorbol 12, 13-dibutyrate, an activator of protein kinase C, stimulates both contraction and Ca^{2+} fluxes in dog saphenous vein. Naunyn-Schmiedeberg's Arch Pharmacol 338:114–120

Cognard C, Romey G, Galizzi JP, Fosset M, Lazdunski M (1986) Dihydropyridine-sensitive Ca^{2+} channels in mammalian skeletal muscle cells in culture: electrophysiological properties and interactions with Ca^{2+} channel activator (Bay K 8644) and inhibitor (PN200-110). Proc Natl Acad Sci USA 83:1518–1522

Cohen CJ, McCarthy RT (1987) Nimodipine block of calcium channels in rat anterior pituitary cells. J Physiol (Lond) 387:195–225

Cohen CJ, McCarthy RT, Barrett PQ, Rasmussen H (1988) Ca channels in adrenal glomerulosa cells: K^+ and angiotensin II increase T-type Ca channel current. Proc Natl Acad Sci USA 85:2412–2416

Contreras E, Tamayo L, Amigo M (1988) Calcium channel antagonists increase morphine-induced analgesia and antagonize morphine tolerance. Eur J Pharmacol 148:463–466

Cota G (1986) Calcium channel currents in pars intermedia cells of the rat pituitary gland. Kinetic properties and washout during intracellular dialysis. J Gen Physiol 88:83–105

Crosland RD, Hsiao TH, McClure WO (1984) Purification and characterization of β-leptinotarsin-n an activator of presynaptic calcium channels. Biochemistry 23:734–741

Cruz LJ, Olivera BM (1986) Calcium channel antagonist ω-conotoxin defines a new high affinity site. J Biol Chem 261:6230–6233

Cruz LJ, Johnson DS, Olivera BM (1987) Characterization of the ω-conotoxin target. Evidence for tissue-specific heterogeneity in calcium channel types. Biochemistry 26:820–824

Deisz RA, Lux HD (1985) γ-Aminobutyric acid-induced depression of calcium currents of chick sensory neurons. Neurosci Lett 56:205–210

Delorme EM, McGee R (1986) Regulation of voltage-dependent Ca^{2+} channels of neuronal cells by chronic changes in membrane potentials. Brain Res 397:189–192

Delorme EM, Rabe CS, McGee R (1988) Regulation of the number of functional voltage-sensitive Ca^{++} channels on PC 12 cells by chronic changes in membrane potential. J Pharmacol Exp Ther 244:838–843

De Pover A, Grupp I, Grupp G, Schwartz A (1983) Diltiazem potentiates the negative inotropic action of nimodipine in the heart. Biochim Biophys Res Commun 114:922–929

DeRiemer SA (1989) Functions for calcium channels in pituitary cells. Ann NY Acad Sci 560:413–414

DeRiemer SA, Strong JA, Albert KA, Greengard P, Kaczmarek LK (1985) Enhancement of calcium current in aplysia neurones by phorbol ester and protein kinase C. Nature 313:313–316

Désarmenien M, Feltz P, Occhipinti G, Santangelo F, Schlichter R (1984) Coexistence of $GABA_A$ and $GABA_B$ receptors on Aδ and C primary afferents. Br J Pharmacol 81:327–333

Desnuelle C, Askanas V, Engel WK (1987) Insulin enhances development of functional voltage-dependent Ca^{2+} channels in aneurally cultured human muscle. J Neurochem 49:1133–1139

DiVirgilio F, Pozzan T, Wollheim CB, Vicentini LM, Meldolesi J (1986) Tumor promoter phorbol myristate acetate inhibits Ca^{2+} influx through voltage-gated Ca^{2+} channels in two secretory cell lines, PC 12 and RINm5F. J Biol Chem 261:32–35

Docherty RJ (1988) Gadolinium selectively blocks a component of calcium current in rodent neuroblastoma glioma hybrid (NG 108-15) cells. J Physiol (Lond) 398:33–47

Dolphin AC, Scott RH (1986) Inhibition of calcium currents in cultured rat dorsal root ganglion neurones by (−)-baclofen. Br J Pharmacol 88:213–220

Dolphin AC, Scott RH (1987) Calcium channel currents and their inhibition by (−)-baclofen in rat sensory neurones: modulation by guanine nucleotides. J Physiol (Lond) 386:1–17

Dolphin AC, Forda SR, Scott RH (1986) Calcium-dependent currents in cultured rat dorsal root ganglion neurones are inhibited by an adenosine analogue. J Physiol (Lond) 373:47–61

Dooley DJ, Lupp A, Hertting G, Osswald H (1988) ω-Conotoxin GVIA and pharmacological modulation of hippocampal noradrenaline release. Eur J Pharmacol 148:261–268

Dubé GP, Baik YH, Schwartz A (1985a) Effects of a novel calcium channel agonist dihydropyridine analogue, Bay K 8644, on pig coronary artery: biphasic mechanical response and paradoxial potentiation of contraction by diltiazem and nimodipine. J Cardiovasc Pharmacol 7:377–389

Dubé GP, Baik YH, Vaghy PL, Schwartz A (1985b) Nitrendipine potentiates Bay K 8644-induced contraction of isolated porcine coronary artery: evidence for functionally distinct dihydropyridine receptor subtypes. Biochem Biophys Res Commun 128:1295–1302

Dunlap K (1981) Two types of γ-aminobutyric acid receptor on embryonic sensory neurons. Br J Pharmacol 74:579–585

Dunlap K, Fischbach GD (1981) Neurotransmitters decrease the calcium conductance activated by depolarization of embryonic chick sensory neurones. J Physiol (Lond) 317:519–535

Enyeart JJ, Aizawa T, Hinkle PM (1985) Dihydropyridine Ca^{2++} antagonists: potent inhibitors of secretion from normal and transformed pituitary cells. Am J Physiol 248:C510–C519

Enyeart JJ, Aizawa T, Hinkle PM (1986) Interaction of dihydropyridine Ca^{2+} agonist Bay-K 8644 with normal and transformed pituitary cells. Am J Physiol 250:C95–C103

Enyeart JJ, Sheu SS, Hinkle PM (1987) Pituitary Ca^{2+} channels: blockade by conventional and novel Ca^{2+} antagonists. Am J Physiol 253:C162–C170

Ewald DA, Matthies HJG, Perney TM, Walker MW, Miller RJ (1988a) The effect of down regulation of protein kinase C on the inhibitory modulation of dorsal root ganglion neuron Ca^{2+} currents by neuropeptide Y. J Neurosci 8:2447–2451

Ewald DA; Sternweis PC, Miller RJ (1988b) Guanine nucleotide-binding protein G_o-induced coupling of neuropeptide Y receptors to Ca channels in sensory neurons. Proc Natl Acad Sci USA 85:3633–3637

Fish RD, Sperti G, Colucci WS, Clapham DE (1988) Phorbol ester increases the dihydropyridine-sensitive calcium conductance in a vascular smooth muscle cell line. Circ Res 62:1049–1054

Fitzpatrick LA, Brandi ML, Aurbach GD (1986) Control of PTH secretion is mediated through calcium channels and is blocked by pertussis toxin treatment of parathyroid cells. Biochem Biophys Res Commun 138:960–965

Fitzpatrick LA, Chin H, Nirenberg M, Aurbach GD (1988) Antibodies to an α-subunit of skeletal muscle calcium channels regulate parathyroid cell secretion. Proc Natl Acad Sci USA 85:2115–2119

Flaim SF, Brannan MD, Swigart SC, Gleason MM, Muschek LD (1985) Neuroleptic drugs alternate calcium influx and tension development in rabbat thoracic aorta: effects of pimozide, penfluridol, chlorpromazine and haloperidol. Proc Natl Acad Sci USA 82:1237–1241

Fleckenstein A (1977) Specific pharmacology of calcium in myocardium, cardiac pacemakers and vascular smooth muscle. Annu Rev Pharmacol Toxicol 17:149–166

Forder J, Scriabine A, Rasmussen H (1985) Plasma membrane calcium flux, protein kinase C activation and smooth muscle contraction. J Pharmacol Exp Ther 235:267–273

Forscher P, Oxford GS (1985) Modulation of calcium channels by norepinephrine in internally dialyzed avian sensory neurons. J Gen Physiol 85:743–763

Fosset M, Jaimovich E, Delpont E, Lazdunski M (1982) [^{3}H]Nitrendipine labelling of the Ca^{2+}-channel in skeletal muscle. Eur J Pharmacol 86:141–142

Fox AP, Nowycky MC, Tsien RW (1987a) Kinetic and pharmacological properties distinguishing three types of Ca currents in chick sensory neurons. J Physiol (Lond) 394:149–172

Fox AP, Nowycky MC, Tsien RW (1987b) Single channel recordings of three types of calcium channels in chick sensory neurones. J Physiol (Lond) 394:173–200

Franckowiak G, Bechem M, Schramm M, Thomas G (1985) The optical isomers of the 1,4-dihydropyridine Bay K 8644 show opposite effects on Ca channels. Eur J Pharmacol 114:223–226

Freedman SB, Miller RJ (1984) Calcium channel activation: a different type of drug action. Proc Natl Acad Sci USA 81:5580–5583

Galizzi JP, Fosset M, Lazdunski M (1984) Properties of receptors for the Ca^{2+}-channel blocker verapamil in transverse-tubule membranes of skeletal muscle. Stereospecificity, effect of Ca^{2+} and other inorganic cations, evidence for two categories of sites and effect of nucleoside triphosphate. Eur J Biochem 144:211–214

Galizzi JP, Fosset M, Lazdunski M (1985) Characterization of the Ca^{2+} coordination site regulating binding of Ca^{2+} channel inhibitors d-cis-diltiazem (±) bepridil and (−) desmethoxyverapamil to their receptor site in skeletal muscle transverse tubule membranes. Biochem Biophys Res Commun 132:49−55

Galizzi JP, Borsotto M, Barhanin J, Fosset M, Lazdunski M (1986a) Characterization and photoaffinity labelling of receptor sites for the Ca^{2+} channel inhibitors d-cis-diltiazem, (±)-bepridil, desmethoxyverapamil and (+)-PN 200-110 in skeletal muscle transverse tubule membranes. J Biol Chem 261:1393−1397

Galizzi JP, Fosset M, Romey G, Laduron P, Lazdunski M (1986b) Neuroleptics of the diphenylbutylpiperidine series are potent Ca channel inhibitors. Proc Natl Acad Sci USA 83:7513−7517

Galizzi JP, Qar J, Fosset M, van Renterghem C, Lazdunski M (1987) Regulation of calcium channels in aortic muscle cells by protein kinase C activators (diacylglycerol and phorbol esters) and by peptides (vasopressin and bombesin) that stimulate phosphoinositide breakdown. J Biol Chem 262:6947−6950

Galvan M, Adams PR (1982) Control of calcium current in rat sympathetic neurons by norepinephrine. Brain Res 244:135−144

Garcia AG, Sala F, Reig JA, Viniegra S, Frias J, Fontériz R, Gandia L (1984) Dihydropyridine Bay-K-8644 activates chromaffin cell calcium channels. Nature 309:69−71

Garcia ML, King VF, Siegl PKS, Reuben JP, Kaczorowski GJ (1986) Binding of Ca^{2+} entry blockers to cardiac sarcolemmal membrane viscles. Characterization of diltiazem-binding sites and their interaction with dihydropyridine and aralkylamine receptors. J Biol Chem 261:8146−8157

Gershengorn MC, Thaw CN, Raaka EG (1988) Benzodiazepines modulate voltage-sensitive calcium channels in GH_3 pituitary cells at sites distinct from thyrotropin-releasing hormone receptors. Endocrinology 123:541−544

Gjörstrup P, Hardin H, Isaksson R, Westerlund C (1986) The enantiomers of the dihydropyridine derivative H160/51 show opposite effects of stimulation and inhibition. Eur J Pharmacol 122:357−361

Glossmann H, Striessnig J (1988) Calcium channels. Vitam Horm 44:155−328

Glossmann H, Striessnig J (1989) Molecular properties of calcium channels. Rev Physiol Biochem Pharmacol (this volume)

Glossmann H, Ferry DR, Lübbecke F, Mewes R, Hofmann F (1982) Calcium channels: direct identification with radioligand binding studies. Trends Pharmacol Sci 3:431−437

Glossmann H, Ferry DR, Goll A, Striessnig J, Zernig G (1985a) Calcium channels and calcium channel drugs − recent biochemical and biophysical findings. Arzneimittelforschung 35-2:1917−1935

Glossmann H, Ferry DR, Goll A, Striessnig J, Schober M (1985b) Calcium channels: basic properties as revealed by radioligand binding studies. J Cardiovasc Pharmacol 7 [Suppl 6]:S20−S30

Godfraind T, Miller R, Wibo M (1986) Calcium antagonists and calcium entry blockade. Pharmacol Rev 38:321−416

Goll A, Glossmann H, Mannhold R (1986) Correlation between the negative inotropic potency and binding parameters of 1,4-dihydropyridine and phenylalkylamine calcium channel blockers in cat heart. Naunyn-Schmiedeberg's Arch Pharmacol 334:303−312

Gould RJ, Murphy KMM, Reynolds IJ, Snyder SH (1983) Antischizophrenic drugs of the diphenylbutylpiperidine type act as Ca channel antagonists. Proc Natl Acad Sci USA 80:5122−5125

Gray WR, Olivera BM, Cruz LJ (1988) Peptide toxins from venomous conus snails. Annu Rev Biochem 57:665−700

Green FJ, Farmer BB, Wiseman GL, Jose MJL, Watanabe AM (1985) Effect of membrane depolarization on binding of [^{3}H]nitrendipine to rat cardiac myocytes. Circ Res 56:576−585

Greenberg DA, Carpenter CL, Messing RO (1986) Depolarization-dependent binding of the calcium channel antagonist (+)-[H-3] PN 200-110 to intact cultured PC 12 cells. J Pharmacol Exp Ther 328:1021–1027

Greenberg DA, Carpenter CL, Messing RO (1987) Lectin-induced enhancement of voltage-dependent calcium flux and calcium channel antagonist binding. J Neurochem 48: 888–894

Gross RA, Macdonald RL (1987) Dynorphin A selectively reduces a large transient (N-type) calcium current of mouse dorsal root ganglion neurons in cell culture. Proc Natl Acad Sci USA 84:5469–5473

Gurney AM, Nerbonne JM, Lester HA (1985) Photoinduced removal of nifedipine reveals mechanisms of calcium antagonist action on single heart cells. J Gen Physiol 86:353–380

Hadley RW, Hume JR (1988) Calcium channel antagonist properties of Bay K 8644 in single guinea pig ventricular cells. Circ Res 62:97–104

Hagiwara N, Irisawa H, Kameyama M (1988) Contributions of two types of calcium currents to the pacemaker potentials of rabbit sino-atrial node cells. J Physiol (Lond) 395:233–253

Hamilton SL, Perez M (1987) Toxins that affect voltage-dependent calcium channels. Biochem Pharmacol 36:3325–3330

Hamilton SL, Yatani A, Hawkes MJ, Redding K, Brown AM (1985) Atrotoxin: a specific agonist for calcium currents in the heart. Science 229:182–185

Hamilton SL, Yatani A, Brush K, Schwartz A, Brown AM (1987) A comparison between the binding and electrophysiological effects of dihydropyridines on cardiac membranes. Mol Pharmacol 31:221–230

Harris RA, Jones SB, Bruno P, Bylund DB (1985) Effects of dihydropyridine derivatives and anticonvulsant drugs on [^{3}H]nitrendipine binding and calcium and sodium fluxes in brain. Biochem Pharmacol 34:2187–2191

Hawthorn MH, Gengo P, Wei XY, Rutledge A, Moran JF, Gallant S, Triggle DJ (1988) Effect of thyroid status on β-adrenoceptors and calcium channels in rat cardiac and vascular tissues. Naunyn-Schmiedeberg's Arch Pharmacol 337:539–544

Hay DWP, Wadsworth RM (1982) Local anaesthetic activity of organic calcium antagonists: relevance to their actions on smooth muscle. Eur J Pharmacol 77:221–228

Herbette LG, Katz AM (1987) Molecular model for the binding of 1,4-dihydropyridine calcium channel antagonists to their receptor in the heart: drug "imaging" in membranes and consideration for drug design. In: Venter JC, Triggle D (eds) Structure and physiology of the slow inward calcium channel. Liss, New York, pp 89–108 (Receptor Biochemistry and Methodology, vol 9)

Hescheler J, Pelzer D, Trube G, Trautwein W (1982) Does the organic calcium channel blocker D 600 act from inside or outside on the cardiac cell membrane? Pflügers Arch Eur J Physiol 393:287–291

Hescheler J, Rosenthal W, Trautwein W, Schultz G (1987) The GTP-binding protein, G$_o$, regulates neuronal calcium channels. Nature 325:445–447

Hescheler J, Rosenthal W, Hinsch KD, Wulfern M, Trautwein W, Schultz G (1988) Angiotensin II-induced stimulation of voltage-dependent Ca^{2+} currents in an adrenal cortical cell line. EMBO J 7:619–624

Hess P, Tsien RW (1984) Mechanism of ion permeation through calcium channels. Nature 309:453–456

Hess P, Lansman JB, Tsien RW (1984) Different modes of channel gating behaviour favoured by dihydropyridine Ca agonists and antagonists. Nature 311:538–544

Higo K, Saito H, Matsuki N (1988) Characteristics of [^{3}H]nimodipine binding to sarcolemmal membranes from rat vas deferens and its regulation by guanine nucleotide. Jpn J Pharmacol 48:213–222

Hille B (1977) Local anesthetics: hydrophilic and hydrophobic pathways for the drug receptor reaction. J Gen Physiol 69:497–515

Hille B (1984) Ionic channels of excitable membranes. Sinauer, Sunderland MA, pp 316–320

Hirning LD, Fox AP, McCleskey EW, Olivera BM, Thayer SA, Miller RJ, Tsien TW (1988) Dominant role of N-type Ca^{2+} channels in evoked release of norephinephrine from sympathic neurons. Science 239:57–61

Hof RP, Rüegg UT, Hof A, Vogel A (1985) Stereoselectivity at the calcium channel: opposite action of the enantiomers of a 1,4-dihydropyridine. J Cardiovasc Pharmacol 7:689–693

Hof RP, Hof A, Rüegg UT, Cook NS, Vogel A (1986) Stereoselectivity at the calcium channel: different profiles of hemodynamic activity of the enantiomers of the dihydropyridine derivative PN 200-110. J Cardiovasc Pharmacol 8:221–226

Holck M, Osterrieder W (1985) The peripheral high affinity benzodiazepine binding site is not coupled to the cardiac Ca^{2+} channel. Eur J Pharmacol 118:293–301

Höltje HD, Marrer S (1987) A molecular graphics study on structure-action relationships of calcium-antagonistic and agonistic 1,4-dihydropyridines. J Comp aided Mol Des 1:23–30

Holz GG, Rane SG, Dunlap K (1986) GTP-binding proteins mediate transmitter inhibition of voltage-dependent calcium channels. Nature 319:670–672

Holz GG, Kream RM, Spiegel A, Dunlap K (1989) G-proteins couple α-adrenergic and GABA$_b$ receptors to inhibition of peptide secretion from peripheral sensory neurons. J Neurosci 9:657–666

Hondeghem LM, Katzung BG (1977) Time and voltage-dependent interactions of antiarrhythmic drugs with cardiac sodium channels. Biochem Biophys Acta 472:373–398

Hondeghem LM, Katzung BG (1984) Antiarrhythmic agents: the modulated receptor mechanism of action of sodium and calcium channel-blocking drugs. Annu Rev Pharmacol Toxicol 24:387–423

Hosey MM, Lazdunski M (1988) Calcium channels: molecular pharmacology, structure and regulation. J Membr Biol 104:81–106

Janis RA, Triggle DJ (1984) 1,4-Dihydropyridine Ca^{2+} channel antagonists and activators: a comparison of binding characteristics with pharmacology. Drug Dev Res 4:257–274

Janis RA, Sarmiento JG, Maurer SC, Bolger GT, Triggle DJ (1984a) Characteristics of the binding of [H-3]nitrendipine to rabbit ventricular membranes – modification by other Ca^{++} channel antagonists and by the Ca^{++} channel agonist Bay K-8644. J Pharmacol Exp Ther 231:8–15

Janis RA, Rampe D, Sarmiento JG, Triggle DJ (1984b) Specific binding of a calcium channel activator [^{3}H] Bay K 8644 to membranes from cardiac muscle and brain. Biochem Biophys Res Commun 121:317–323

Janis RA, Silver PJ, Triggle DJ (1987) Drug action and cellular calcium regulation. Adv Drug Res 16:309–589

Johansen J, Taft WC, Yang J, Kleinhans AL, de Lorenzo RJ (1985) Inhibition of Ca^{2+} conductance in identified leech neurons by benzodiazepines. Proc Natl Acad Sci USA 82:3935–3939

Kaczmarek LK (1987) The role of protein kinase C in the regulation of ion channels and neurotransmitter release. Trends Neurosci 10:30–34

Kanaya S, Arlock P, Katzung BG, Hondeghem LM (1983) Diltiazem and verapamil preferentially block inactivated cardiac calcium channels. J Mol Cell Cardiol 15:145–148

Kass RS (1987) Voltage-dependent modulation of cardiac calcium channel current by optical isomers of Bay K 8644: implications for channel gating. Circ Res 61 [Suppl 1]:1–5

Kass RS, Arena JP (1989) Influence of pH$_0$ on calcium channel block by amlodipine, a charged dihydropyridine compound. Implications for location of the dihydropyridine receptor. J Gen Physiol 93:1109–1127

Kass RS, Krafte DS (1987) Negative surface charge density near heart calcium channels. Relevance to block by dihydropyridines. J Gen Physiol 89:629–644

Kass RS, Arena JP, DiManno D (1988) Block of heart calcium channels by amlodipine: Influence of drug charge on blocking activity. J Cardiovasc Pharmacol 12 (Suppl 7):S45–S49

Kawashima Y, Ochi R (1988) Voltage-dependent decrease in the availability of single calcium channels by nitrendipine in guinea-pig ventricular cells. J Physiol (Lond) 402:219–235

Kerr LM, Yoshikami D (1984) A venom peptide with a novel presynaptic blocking action. Nature 308:282–284

King VF, Garcia ML, Himmel D, Reuben JP, Lam YKT, Pan JX, Han GQ, Kaczorowski GJ (1988) Interaction of tetrandrine with slowly inactivating calcium channels. Characterization of Ca channel modulation by an alkaloid of Chinese medical herb origin. J Biol Chem 263:2238–2244

Knaus HG, Striessnig J, Koza A, Glossmann H (1987) Neurotoxic aminoglycoside antibiotics are potent inhibitors of [^{125}I]-omega-conotoxin GVIA binding to guinea-pig cerebral cortex membranes. Naunyn-Schmiedeberg's Arch Pharmacol 336:583–586

Kobayashi M, Ohizumi Y, Yasumoto T (1985) The mechanism of action of maitotoxin in relation to Ca^{2+} movements in guinea-pig and rat cardiac muscles. Br J Pharmacol 86:385–391

Kobayashi M, Kondo S, Yasumoto T, Ohizumi Y (1986) Cardiotoxic effects of maitotoxin, a principal toxin of seafood poisoning, on guinea pig and rat cardiac muscle. J Pharmacol Exp Ther 238:1077–1083

Kojima I, Shibata H, Ogata E (1986) Pertussis toxin blocks angiotensin II-induced calcium influx but not inositol triphosphate production in adrenal glomerulosa cell. FEBS Lett 204:347–351

Kokubun S, Reuter H (1984) Dihydropyridine derivatives prolong the open state of Ca channels in cultured cardiac cells. Proc Natl Acad Sci USA 81:4824–4827

Kokubun S, Prod'hom B, Porzig H, Reuter H (1986) Studies on Ca channels in intact cardiac cells: voltage-dependent effects and cooperative interactions of dihydropyridine enantiomers. Mol Pharmacol 30:571–584

Kongsamut S, Miller RJ (1986) Nerve growth factor modulates the drug sensitivity of neurotransmitter release from PC 12 cells. Proc Natl Acad Sci USA 83:2243–2247

Kongsamut S, Kamp TJ, Miller RJ, Sanguinetti MC (1985a) Calcium channel agonist and antagonist effects of the steroisomers of the dihydropyridine 202-791. Biochem Biophys Res Commun 130:141–148

Kongsamut S, Freedman SB, Simon BE, Miller RJ (1985b) Interaction of steroidal alkaloid toxins with Ca channels in neuronal cell lines. Life Sci 36:1493–1501

Kostyuk PG, Mironov SL, Shuba YM (1983) Two ion-selecting filters in the calcium channel of the somatic membrane of mollusc neurons. J Membr Biol 76:83–93

Kunze DL, Hamilton SL, Hawkes MJ, Brown AM (1987) Dihydropyridine binding and calcium channel function in clonal rat adrenal medullary tumor cells. Mol Pharmacol 31:401–409

Lacerda AE, Brown AM (1986) Atrotoxin increases probability of opening of single Ca channels in cultured rat ventricular cells. Biophys J 49:174a

Lamb GD, Walsh T (1987) Calcium currents, charge movement and dihydropyridine binding in fast- and slow-twitch muscles of rat and rabbit. J Physiol (Lond) 393:595–617

Laurent S, Kim D, Smith TW, Marsh JD (1985) Inotropic effect, binding properties and calcium flux effects of the calcium agonist CGP 28392 in intact cultured embryonic chick ventricular cells. Circ Res 56:676–682

Lazdunski M, Frelin C, Vigne P (1985) The sodium/hydrogen exchange system in cardiac cells: its biochemical and pharmacological properties and its role in regulating internal concentrations of sodium and internal pH. J Mol Cell Cardiol 17:1029–1042

Lee KS, Tsien RW (1983) Mechanism of calcium channel blockade by verapamil, D 600, diltiazem and nitrendipine in single dialysed heart cells. Nature 302:790–794

Lee RT, Smith TW, Marsh JD (1987) Evidence for distinct calcium channel agonist and antagonist binding sites in intact cultured embryonic chick ventricular cells. Circ Res 60: 683–691

Levitan IB (1988) Modulation of ion channels in neurons and other cells. Annu Rev Neurosci 11:119–136

Lewis DL, Weight FF, Luini A (1986) A guanine nucleotide-binding protein mediates the inhibition of voltage-dependent calcium current by somatostatin in a pituitary cell line. Proc Natl Acad Sci USA 83:9035–9039

Lipscombe D, Tsien RW (1987) Noradenaline inhibits N-type Ca channels in isolated frog sympathetic neurons. J Physiol (Lond) 390:84P

Lipscombe D, Madison DV, Poenie M, Reuter H, Tsien RY, Tsien RW (1988) Spatial distribution of calcium channels and cytosolic calcium transients in growth cones and cell bodies of sympathetic neurons. Proc Natl Acad Sci USA 85:2398–2402

Ljung B, Kjellstedt A, Orebäck B (1987) Vascular versus myocardial selectivity of calcium antagonists studied by concentration-time-effect relations. J Cardiovasc Pharmacol 10[Suppl 1]:S34–S39

Login IS, Judd AM, Cronin MJ, Yasumoto T, Macleod RM (1985) Reserpine is a calcium channel antagonist in normal and GH_3 rat pituitary cells. Am J Physiol 248:E15–E19

Lorentz M, Hedlund B, Århem P (1988) Morphine activates calcium channels in cloned mouse neuroblastoma cell lines. Brain Res 445:157–159

Loutzenhiser R, Rüegg U, Hof RP (1984) Studies on the mechanism of action of the vasoconstrictive dihydropyridine CGP 28392. Eur J Pharmacol 105:229–237

Lüllmann H, Mohr K (1987) High and concentration-proportional accumulation of [^{3}H]-nitrendipine by intact cardiac tissue. Br J Pharmacol 90:567–574

Maan AC, Hosey MM (1987) Analysis of the properties of binding of calcium-channel activators and inhibitors to dihydropyridine receptors in chick heart membranes. Circ Res 61:379–388

Madison DV, Fox AP, Tsien RW (1987) Adenosine reduces an inactivating component of calcium current in hippocampal CA3 neurons. Biophys J 51:30a (abstract)

Maggi CA, Patacchini R, Santicioli P, Lippe IT, Guiliani S, Geppetti P, Del Bianco E, Selleri S, Meli A (1988) The effect on omega conotoxin GVIA, a peptide modulator of the N-type voltage sensitive calcium channels, on motor responses produced by activation of efferent and sensory nerves in mammalian smooth muscle. Naunyn-Schmiedeberg's Arch Pharmacol 338:107–113

Mannhold R, Rodenkirchen R, Bayer R, Haas W (1984) The importance of drug ionization for the action of calcium antagonists and related compounds. Arzneimittelforschung 34:407–410

Marchetti C, Brown AM (1988) Protein kinase activator 1-oleoyl-2-acetyl-sn-glycerol inhibits two types of calcium current in GH_3 cells. Am J Physiol 254:C206–C210

Marchetti C, Carbone E, Lux HD (1986) Effects of dopamine and noradrenaline on Ca channels of cultured sensory and sympathetic neurons of chick. Pflügers Arch Eur J Physiol 406:104–111

Martin-Moutot N, Marqueze B, Azais F, Seagar M, Couraud F (1989) Properties of the calcium channel associated ω-conotoxin receptor in rat brain. Ann NY Acad Sci 560:53–55

Masuda MO, de Magalhães-Engel G, Barbose Moreira AP (1987) Characterization of isolated ventricular myocytes: two levels of resting potential. J Mol Cell Cardiol 19:831–840

Matteson DR, Armstrong CM (1986) Properties of two types of calcium channels in clonal pituitary cells. J Gen Physiol 87:161–182

McCleskey EW, Almers W (1985) The Ca channel is a large pore. Proc Natl Acad Sci USA 82:7149–7153

McCleskey EW, Fox AP, Feldman DH, Cruz LJ, Olivera BM, Tsien RW, Yoshikami D (1987) ω-Conotoxin: direct and persistent blockade of specific types of calcium channels in neurons but not muscle. Proc Natl Acad Sci USA 84:4327–4331

Messing RO, Carpenter CL, Greenberg DA (1985) Mechanism of calcium channel inhibition by phenytoin: comparison with classical calcium channel antagonists. J Pharmacol Exp Ther 235:407–411

Messing RO, Carpenter CL, Greenberg DA (1986a) Inhibition of calcium flux and calcium channel antagonist binding in the PC 12 cell line by phorbol esters and protein kinase C. Biochem Biophys Res Commun 136:1049–1056

Messing RO, Carpenter CL, Diamond I, Greenberg DA (1986b) Ethanol regulates calcium channels in clonal neural cells. Proc Natl Acad Sci USA 83:6213–6215

Mestre M, Carriot T, Néliat G, Uzan A, Renault C, Dubroeucq MC, Guérémy C, Doble A, LeFur G (1986a) PK 11195 an antagonist of peripheral type benzodiazepine receptors modulates Bay K 8644 sensitive but not β- or H_2-receptor sensitive voltage-operated calcium channels in the guinea pig heart. Life Sci 39:329−340

Mestre M, Belin C, Uzan A, Renault C, Dubroeucq MC, Gueremy C, LeFur G (1986b) Modulation of voltage-operated, but not receptor-operated, calcium channels in the rabbit aorta by PK 11195, an antagonist of peripheral type benzodiazepine receptors. J Cardiovasc Pharmacol 8:729−734

Miller RJ (1987a) Multiple calcium channels and neuronal function. Science 235:46−52

Miller RJ (1987b) Calcium channels in neurons. In: Venter JC, Triggle D (eds) Structure and physiology of the slow inward calcium channel. Liss, New York, pp 161−246 (Receptor biochemistry and methodology, vol 9)

Miller RJ, Freedman SB (1984) Are dihydropyridine binding sites voltage-sensitive calcium channels? Life Sci 34:1205−1222

Mir AK, Spedding M (1987) Calcium antagonist properties of diclofurime isomers. II. Molecular aspects: allosteric interactions with dihydropyridine recognition sites. J Cardiovasc Pharmacol 9:469−477

Monod J, Wyman J Changeux JP (1965) On the nature of allosteric transitions: a plausible model. J Mol Biol 12:88−118

Morel N, Godfraind T (1987) Prolonged depolarization increases the pharmacological effect of dihydropyridines and their binding affinity for calcium channels of vascular smooth muscle. J Pharmacol Exp Ther 243:711−715

Motomura S, Hashimoto K, Hashimoto K (1987) Effects of Bay K 8644 on the coronary vascular selectivity of the dihydropyridine Ca antagonists in the canine isolated blood-perfused papillary muscle preparation. J Cardiovasc Pharmacol 10:627−635

Mudge AW, Leeman SE, Fischbach GD (1979) Enkephalin inhibits release of substance P from sensory neurons in culture and decreases action potential duration. Proc Natl Acad Sci USA 76:526−530

Murphy KMM, Snyder SH (1982) Calcium antagonist receptor binding sites labeled with [^{3}H]nitrendipine. Eur J Pharmacol 77:201−202

Narahashi T (1988) Drugs acting on calcium channels. In: Baker PF (ed) Calcium in drug actions. Springer, Berlin Heidelberg New York, pp 255−274 (Handbuch der experimentellen Pharmakologie, vol 83)

Navarro J (1987) Modulation of [^{3}H]dihydropyridine receptors by activation of protein kinase C in chick muscle cells. J Biol Chem 262:4649−4652

Nelson MT, Standen NB, Brayden JE, Worley JF (1988) Noradrenaline contracts arteries by activating voltage-dependent calcium channels. Nature 336:382−385

Nishiyama T, Kobayashi A, Haga T, Yamazaki N (1986) Chronic treatment with nifedipine does not change the number of [^{3}H]dihydroalprenolol binding sites. Eur J Pharmacol 121:167−172

North RA (1986) Opioid receptor types and membrane ion channels. Trends Neurosci 9:114−117

Olivera BM, Gray WR, Zeikus R, McIntosh JM, Varga J, Rivier J, de Santos V, Cruz LJ (1985) Peptide neurotoxins from fish-hunting conesnails. Science 230:1338−1343

Osugi T, Imaizumi T, Mizushima A, Uchida S, Yoshida H (1986) 1-Oleoyl-2-acetyl-glycerol and phorbol diester stimulate Ca^{2+} influx through Ca^{2+} channels in neuroblastoma x glioma hybrid NG 108-15 cells. Eur J Pharmacol 126:47−52

Oyama Y, Tsuda Y, Sakakibara S, Akaike N (1987) Synthetic ω-conotoxin: a potent calcium channel blocking neurotoxin. Brain Res 424:58−64

Peres A, Sturani E, Zippel R (1988) Properties of the voltage-dependent calcium channel of mouse Swiss 3T3 fibroblasts. J Physiol (Lond) 401:639−656

Perney TM, Hirning LD, Leeman SE, Miller RJ (1986) Multiple calcium channels mediate neurotransmitter release from peripheral neurons. Proc Natl Acad Sci USA 83:6656−6659

Pietrobon D, Prod'hom B, Hess P (1988) Conformational changes associated with ion permeation in L-type calcium channels. Nature 333:373−376

Plummer MR, Logothetis DE, Hess P (1989) Elementary properties and pharmacological sensitivities of calcium channels in mammalian peripheral neurons. Neuron 2:1453–1463

Porzig H, Becker C (1988) Potential-dependent allosteric modulation of 1,4-dihydropyridine
binding by d-(cis)-diltiazem and (±)-verapmil in living cardiac cells. Mol Pharmacol
34:172–179

Porzig H, Becker C, Reuter H (1989) Effects of NGF-induced differentiation on two classes of
Ca channels in living PC 12 cells. Experientia 45:A28 (abstract)

Postma SW, Catterall WA (1984) Inhibition of binding of [^{3}H]batrachotoxin Δ20-α-benzoate
to sodium channels by local anaesthetics. Mol Pharmacol 25:219–227

Powers RE, Colucci WS (1985) An increase in putative voltage-dependent calcium channel number following reserpine treatment. Biochem Biophys Res Commun 132:844–849

Qar J, Schweitz H, Schmid A, Lazdunski M (1986) A polypeptide toxin from the coral
Goniopora. Purification and action on Ca^{2+} channels. FEBS Lett 202:331–336

Qar J, Galizzi JP, Fosset M, Lazdunski M (1987) Receptors for diphenylbutylpiperidine
neuroleptics in brain, cardiac and smooth muscle membranes. Relationship with receptors for
1,4-dihydropyridines and phenylalkylamines and with Ca^{2+} channel blockade. Eur J Pharmacol 141:261–268

Ramkumar V, El-Fakahany EE (1984) Increase in [^{3}H]nitrendipine binding sites in the brain in
morphine-tolerant mice. Eur J Pharmacol 102:371–373

Ramkumar V, El-Fakahany EE (1985) Changes in the affinity of [^{3}H]nimodipine binding sites
in the brain upon chlorpromazine treatment and subsequent withdrawal. Res Commun Chem
Pathol Pharmacol 48:463–466

Rampe D, Triggle DJ (1986) Benzodiazepines and calcium channel function. Trends Pharmacol
Sci 7:461–464

Rane SG, Dunlap K (1986) Kinase C activator 1,2-oleoyl acetylglycerol attenuates voltage-dependent calcium current in sensory neurones. Proc Natl Acad Sci USA 83:184–188

Rane SG, Holz GG, Dunlap K (1987) Dihydropyridine inhibition of neuronal calcium current
and substance P release. Pflügers Arch Eur J Physiol 409:361–366

Rane SG, Holz GG, Anderson CS, Dunlap K (1989) Calcium channel modulation via G-proteins and protein kinase C. 4th Int. Symp. Calcium Antagonists, Florence May 1989 (ab-
stracts p 5–7)

Renaud JF, Kazazoglou T, Schmid A, Romey G, Lazdunski M (1984) Differentiation of receptor
sites for [^{3}H]nitrendipine in chick hearts and physiological relation to slow Ca^{2+} channel
and to excitation contraction coupling. Eur J Biochem 139:673–681

Reuter H (1983) Calcium channel modulation by neurotransmitters, enzymes and drugs. Nature
301:569–574

Reuter H (1984) Ion channels in cardiac cell membranes. Annu Rev Physiol 46:473–484

Reuter H (1987) Modulation of ion channels by phosphorylation and second messengers. News
Physiol Sci 2:168–171

Reuter H, Stevens CF, Tsien RW, Yellen G (1982) Properties of single calcium channels in cardiac
cell culture. Nature 297:501–504

Reuter H, Porzig H, Kokubun S, Prod'hom B (1985) 1,4-Dihydropyridines as tools in the study
of Ca^{2+} channels. Trends Neurosci 8:396–400

Reuter H, Kokubun S, Prod'hom B (1986) Properties and modulation of cardiac calcium channels. J Exp Biol 124:191–202

Reynolds IJ, Gould RJ, Snyder SH (1983) [^{3}H]Verapamil binding sites in brain and skeletal
muscle: regulation by calcium. Eur J Pharmacol 95:319–321

Reynolds IJ, Wagner JA, Snyder SH, Thayer SA, Olivera BM, Miller RJ (1986) Brain voltage-
sensitive calcium channel subtypes differentiated by ω-conotoxin fraction G VI A. Proc Natl
Acad Sci USA 83:8804–8807

Rhodes DG, Sarmiento JG, Herbette LG (1985) Kinetics of binding of membrane-active drugs
to receptor sites: diffusion limited rates for a membrane bilayer approach of 1,4-dihydropyri-
dine calcium channel antagonists to their active site. Mol Pharmacol 27:612–623

Rios E, Brum G (1987) Involvement of dihydropyridine receptors in excitation-contraction coupling in skeletal muscle. Nature 325:717−720

Rogart RB, de Bruyn Kops A, Dzau VJ (1986) Identification of two calcium channel receptor sites for [^{3}H]nitrendipine in mammalian cardiac and smooth muscle membrane. Proc Natl Acad Sci USA 83:7452−7456

Romey G, Lazdunski M (1982) Lipid-soluble toxins thought to be specific for Na$^+$ channels block Ca^{2+} channels in neuronal cells. Nature 297:79−80

Rosenthal W, Schultz G (1988) Funktionen guaninnucleotid-bindender Proteine bei der rezeptorvermittelten Modulation spannungsabhängiger Ionenkanäle. Klin Wochenschr 66:557−564

Rosenthal W, Hescheler J, Trautwein W, Schultz G (1988a) Control of voltage-dependent Ca^{2+} channels by G protein-coupled receptors. FASEB J 2:2784−2790

Rosenthal W, Hescheler J, Hinsch KD, Spicher K, Trautwein W, Schultz G (1988b) Cyclic AMP-independent dual regulation of voltage-dependent Ca^{2+} currents by LHRH and somatostatin in a pituitary cell line. EMBO J 7:1627−1634

Sanchez-Chapula J, Josephson IR (1983) Effect of phenytoin on the sodium current in isolated rat ventricular cells. J Mol Cell Cardiol 15:515−522

Sanguinetti MC, Kass RS (1984a) Voltage-dependent block of calcium channel current in the calf cardiac Purkinje fiber by dihydropyridine calcium channel antagonists. Circ Res 55:336−348

Sanguinetti MC, Kass RS (1984b) Regulation of cardiac calcium channel current and contractile activity by the dihdropyridine Bay K 8644 is voltage-dependent. J Mol Cell Cardiol 16:667−670

Sanguinetti MC, Krafte DS, Kass RS (1986) Voltage-dependent modulation of Ca channel current in heart cells by Bay K 8644. J Gen Physiol 88:369−392

Sarmiento JG, Janis RA, Katz AM, Triggle DJ (1984) Comparison of high affinity binding of calcium channel blocking drugs to vascular smooth muscle and cardiac sarcolemmal membranes. Biochem Pharmacol 33:3119−3123

Sartor P, Vacher P, Mollard P, Dufy B (1988) Tamoxifen reduces calcium currents in clonal pituitary cell line. Endocrinology 123:534−540

Schettini G, Meucci O, Florio T, Grimaldi M, Landolfi E, Magri G, Yosumoto T (1988) Pertussis toxin pretreatment abolishes dihydropyridine inhibition of calcium flux in the 235-1 pituitary cell line. Biochem Biophys Res Commun 151:361−369

Scheuer T, Kass RS (1983) Phenytoin reduces calcium current in the cardiac Purkinje fiber. Circ Res 53:16−23

Schilling WP (1988) Effect of divalent cation chelation on dihydropyridine binding in isolated cardiac sarcolemmal vesicles. Biochem Biophys Acta 943:220−230

Schilling WP, Drewe JA (1986) Voltage-sensitive nitrendipine binding in an isolated cardiac sarcolemma preparation. J Biol Chem 261:2750−2758

Schmid A, Renaud JF, Lazdunski M (1985) Short term and long term effects of adrenergic effectors and cyclic AMP on nitrendipine-sensitive voltage-dependent Ca^{2+} channels of skeletal muscle. J Biol Chem 260:13041−13046

Scholtysik G, Rüegg P (1987) DPI 201-106. In: Scriabine A (ed) New Cardiovascular Drugs 1987. Raven Press, New York pp 173−188

Schramm M, Towart R (1988) Calcium channels as drug receptors. In: Baker PF (ed) Calcium in drug actions. Springer, Berlin Heidelberg New York, pp 89−114 (Handbook of experimental Pharmacology, vol 83)

Schramm M, Thomas G, Towart R, Franckowiak (1983) Novel dihydropyridine with positive inotropic action through activation of Ca channels. Nature 303:535−537

Schwartz A, Grupp IL, Grupp G, Williams J, Vaghy PL (1984) Effects of dihydropyridine calcium channel modulators in the heart: pharmacological and radioligand binding correlations. Biochem Biophys Res Commun 125:387−394

Schwartz LM, McCleskey EW, Almers W (1985) Dihydropyridine receptors in muscle are voltage-dependent but most are not functional calcium channels. Nature 314:747−751

Scott RH, Dolphin AC (1987) Activation of a G protein promotes agonist responses to calcium channel ligands. Nature 330:760–762

Scott RH, Dolphin AC (1988) The agonist effect of Bay K 8644 on neuronal calcium channel currents is promoted by G-protein. Neurosci Lett 89:170–175

Seeman P (1972) The membrane actions of anesthetics and tranquilizers. Pharmacol Rev 24:583–655

Sheldon RS, Cannon NJ, Duff HJ (1987) A receptor for type I antiarrhythmic drugs associated with rat cardiac sodium channels. Circ Res 61:492–497

Sheldon RS, Cannon NJ, Nies AS, Duff HJ (1988) Sterospecific interaction of tocainide with the cardiac sodium channel. Mol Pharmacol 33:327–331

Sher E, Pandiella A, Clementi F (1988) ω-Conotoxin binding and effects on calcium channel function in human neuroblastoma and rat pheochromocytoma cell lines. FEBS Lett 235:178–182

Siegl PKS, Garcia ML, King VF, Scott AL, Morgan G, Kaczorowski GJ (1988) Interactions of DPI 201-106, a novel cardiotonic agent, with cardiac calcium channels. Naunyn-Schmiedeberg's Arch Pharmacol 338:684–691

Skattebøl A, Triggle DJ (1986) 6-Hydroxydopamine treatment increases β-adrenoceptors and Ca^{2+} channels in rat heart. Eur J Pharmacol 127:287–289

Sladeczek F, Schmidt BH, Alonso R, Vian L, Tep A, Yasumoto T, Cory RN, Bockaert J (1988) New insights into maitotoxin action. Eur J Biochem 174:663–670

Spedding M, Gittos M, Mir AK (1987) Calcium antagonist properties of diclofurime isomers. I. Functional aspects. J Cardiovasc Pharmacol 9:461–468

Stojikovic SS, Izumi SI, Catt KJ (1988) Participation of voltage-sensitive calcium channels in pituitary hormone release. J Biol Chem 263:13054–13061

Streit J, Lux HD (1987) Voltage-dependent calcium currents in PC 12 growth cones and cells during NGF-induced cell growth. Pflügers Arch Eur J Physiol 408:634–641

Strong JA, Fox AP, Tsien RW, Kaczmarek LK (1987) Stimulation of protein kinase C recruits covert calcium channels in aplysia bag cell neurons. Nature 325:714–717

Sturek M, Hermsmeyer K (1986) Calcium and sodium channels in spontaneously contracting vascular muscle cells. Science 233:475–478

Su HD, Mazzei GJ, Vogler WR, Kuo JF (1985) Effect of tamoxifen, a nonsteroidal antioestrogen, on phospholipid calcium-dependent protein kinase and phosphorylation of its endogenous substrate proteins from the rat brain and ovary. Biochem Pharmacol 34:3649–3653

Sumimoto K, Hirata M, Kuriyama H (1988) Characterization of [^{3}H]nifedipine binding to intact vascular smooth muscle cells. Am J Physiol 254:C45–C52

Swandulla D, Armstrong CM (1988) Fast-deactivating calcium channels in chick sensory neurons. J Gen Physiol 92:197–218

Taft WC, DeLorenzo RJ (1984) Micromolar-affinity benzodiazepine receptors regulate voltage-sensitive calcium channels in nerve terminal preparations. Proc Natl Acad Sci USA 81:3118–3122

Takahashi M, Tsukui H, Hatanaka H (1985) Neuronal differentiation of Ca^{2+} channel by nerve growth factor. Brain Res 341:381–384

Tan KN, Tashijan AH Jr (1984a) Voltage-dependent calcium channels in pituitary cells in culture. I. Characterization by $^{45}Ca^{2+}$ fluxes. J Biol Chem 259:418–426

Tan KN, Tashijan AH Jr (1984b) Voltage-dependent calcium channels in pituitary cells in culture. II. Participation in thyrotropin-releasing hormone action. J Biol Chem 259:427–434

Tanabe T, Takeshima H, Mikami A, Flockerzi V, Takahashi H, Kangawa K, Kojima M, Matsuo H, Hirose T, Numa S (1987) Primary structure of the receptor for calcium channel blockers from skeletal muscle. Nature 328:313–318

Tang CM, Presser F, Morad M (1988) Amiloride selectively blocks the low threshold calcium channel. Science 240:213–215

Thayer SA, Hirning LD, Miller RJ (1987) The distribution of multiple types of Ca^{2+} channels in rat sympathetic neurons in vitro. Mol Pharmacol 32:579–586

Thomas G, Gross R, Schramm M (1984) Calcium channel modulation: ability to inhibit or promote calcium influx resides in the same dihydropyridine molecule. J Cardiovasc Pharmacol
6:1170−1176

Toll L (1982) Calcium antagonists. High-affinity binding and inhibition of calcium transport
in a clonal cell line. J Biol Chem 257:13189−13192

Triggle DJ, Janis RA (1987) Calcium channel ligands. Annu Rev Pharmacol Toxicol 27:347−370

Tsien RW (1987) Calcium currents in heart cells and neurons. In: Kaczmarek LK, Levitan IB
(eds) Neuromodulation. The biochemical control of neuronal excitability. Oxford University
Press, Oxford, pp 206−242

Tsien RW, Hess P, McCleskey EW, Rosenberg RL (1987) Calcium channels: mechanisms of selectivity, permeation and block. Annu Rev Biophys Biophys Chem 16:265−290

Tsien RW, Lipscombe D, Madison DV, Bley RK, Fox AP (1988) Multiple types of neuronal calcium channels and their selective modulation. Trends Neurosci 11:431−438

Tsunoo A, Yoshii M, Narahashi T (1985) Differential block of two calcium channels in
neuroblastoma cells. Biophys J 47:433a

Tsunoo A, Yoshii M, Narahashi T (1986) Block of calcium channels by enkephalin and
somatostatin in neuroblastoma-glioma hybrid NG-108-15 cells. Proc Natl Acad Sci USA
83:9832−9836

Twombly DA, Yoshii M, Narahashi T (1988) Mechanism of calcium channel block by phenytoin.
J Pharmacol Exp Ther 246:189−195

Tytgat J, Vereeke J, Carmeliet E (1988) Differential effects of verapamil and flunarizine on
cardial L-type and T-type Ca channels. Naunyn Schmiedeberg's Arch Pharmacol 337:
690−692

Uehara A, Hume JR (1985) Interactions of organic calcium channel antagonists with calcium
channels in single frog atrial cells. J Gen Physiol 85:621−648

Vaghy PL, Grupp IL, Grupp G, Schwartz A (1984a) Effects of Bay K 8644, a dihydropyridine
analog on [^{3}H]nitrendipine binding to canine cardiac sarcolemma and the relationship to a
positive inotropic effect. Circ Res 55:549−553

Vaghy PL, Grupp IL, Grupp G, Balwierczak JL, Williams JS, Schwartz A (1984b) Correlation
of nitrendipine and Bay K 8644 binding to isolated canine heart sarcolemma with their pharmacological effects on the canine heart. Eur J Pharmacol 102:373−374

Van Skiver, Spires S, Cohen CJ (1988) Block of T-type Ca channels in guinea pig atrial cells
by cinnarizine. Biophys J 53:233a

Vilven J, Leung AT, Imagawa T, Sharp AH, Campbell KP, Coronado R (1988) Interaction of
calcium channels of skeletal muscle with monoclonal antibodies specific for its
dihydropyridine receptor. Biophys J 53:556a (abstract)

Wagner JA, Snowman AM, Biswas A, Olivera BM, Snyder SH (1988) ω-Conotoxin G VI A
binding to a high-affinity receptor in brain: characterization, calcium sensitivity and
solubilization. J Neurosci 8:3354−3359

Walker MW, Ewald DA, Perney TM, Müller RJ (1988) Neuropeptide Y modulates neurotransmitter release and Ca^{2+} currents in rat sensory neurones. J Neurosci 8:2438−2446

Wanke E, Ferroni A, Malgaroli A, Ambrosini A, Pozzan T, Meldolesi J (1987) Activation of
a muscarinic receptor selectively inhibits a rapidly inactivated Ca^{2+} current in rat sympathetic neurons. Proc Natl Acad Sci USA 84:4313−4317

Werz MA, MacDonald RL (1985) Dynorphin and neoendorphin peptides decrease dorsal root
ganglion neuron calcium-dependent action potential duration. J Pharmacol Exp Ther
234:49−56

Williams JS, Grupp IL, Grupp G, Vaghy PL, Dumont L, Schwartz A, Yatani A, Hamilton S,
Brown AM (1985) Profile of the oppositely acting enantiomers of the dihydropyridine
202-791 in cardiac preparations: receptor binding, electrophysiological and pharmacological
studies. Biochem Biophys Res Commun 131:13−21

Willow M (1986) Pharmacology of diphenylhydantoin and carbamazepine action on voltage-
sensitive sodium channels. Trends Neurosci 9:147−149

Willow M, Gonoi T, Catterall WA (1985) Voltage clamp analysis of the inhibitory actions of diphenylhydantoin and carbamazepine on voltage-sensitive sodium channels in neuroblastoma cells. Mol Pharmacol 27:549–558

Wright JM, Collier B (1977) The effects of neomycin upon transmitter release and action. J Pharmacol Exp Ther 200:576–587

Wu CH, Narahashi T (1988) Mechanism of action of novel marine neurotoxins on ion channels. Annu Rev Pharmacol Toxicol 28:141–162

Yamashita N, Matsunaga H, Shibuya N, Teramoto A, Takakura K, Ogata E (1988) Two types of calcium channels and hormone release in human pituitary tumor cells. Am J Physiol 255:E137–E145

Yatani A, Brown AM (1985) The calcium channel blocker nitrendipine blocks sodium channels in neonatal rat cardiac myocytes. Circ Res 56:868–875

Yatani A, Brown AM, Schwartz A (1986a) Bepridil block of cardiac calcium and sodium channels. J Pharmacol Exp Ther 237:9–17

Yatani A, Hamilton SL, Brown AM (1986b) Diphenylhydantoin blocks cardiac calcium channels and binds to the dihydropyridine receptor. Circ Res 59:356–361

Yatani A, Codina J, Imoto Y, Reeves JP, Birnbaumer L, Brown AM (1987) A G protein directly regulates mammalian cardiac calcium channels. Science 238:1288–1292

Yatani A, Kunze DL, Brown AM (1988a) Effects of dihydropyridine calcium channel modulators on cardiac sodium channels. Am J Physiol 254:H140–H147

Yatani A, Imoto Y, Codina J, Hamilton SL, Brown AM, Birnbaumer L (1988b) The stimulatory G protein of adenylyl cyclase G_s also stimulates dihydropyridine-sensitive Ca^{2+} channels. Evidence for direct regulation independent of phosphorylation by cAMP-dependent protein kinase or stimulation by a dihydropyridine agonist. J Biol Chem 263:9887–9895

Yeh JZ (1980) Blockage of sodium channels by stereoisomers of local anesthetics. In: Fink BR (ed) Molecular mechanisms of anaesthesia. Raven, New York, pp 35–44 (Progress in anesthesiology, vol 2)

Subject Index